W9-AXI-883

NONDESTRUCTIVE EVALUATION: APPLICATION TO MATERIALS PROCESSING

Proceedings of a symposium sponsored by the Energy and Resources Activity of the American Society for Metals at the TMS Fall Meeting, Philadelphia, Pennsylvania, October 3-4, 1983

Edited by
OTTO BUCK
Ames Laboratory
Iowa State University
Ames, Iowa

and

STANLEY M. WOLF
Division of Materials Sciences
Office of Basic Energy Sciences
Department of Energy
Washington, D.C.

A Publication of

The American Society for Metals

Library of Congress Catalog Card Number: 84-071516

ISBN: 0-87170-190-1

SAN: 204-7586

Manufactured by Publishers Choice Book Mfg. Co.
Mars, Pennsylvania 16046

Table of Contents

FOREWORD . . . v

I SOLIDIFICATION

Ultrasonic Measurement of Internal Temperature Distribution
H. N. G. Wadley, S. J. Norton, F. S. Biancaniello, and R. Mehrabian . . . 3

Real-Time Ultrasonic Sensing of Arc Welding Processes
L. A. Lott, J. A. Johnson, and H. B. Smartt . . . 13

Ultrasonic Measurement of Solid/Liquid Interface Position During Solidification and Melting of Iron and Steel
R. L. Parker . . . 23

Acoustic Emission Studies of Electron Beam Melting and Rapid Solidification
R. B. Clough, H. N. G. Wadley, F. S. Biancaniello . . . 27

An Acoustic Emission Technique for Characterization of Solidification Cracking in Laser-Melted Stainless Steels
R. J. Coyle, Jr., D. G. Cruickshank, M. C. Jon, and V. Palazzo . . . 41

Ultrasonic NDE of Rapidly Solidified Products and its Application to Process Control
C. L. Friant, M. Rosen, R. E. Green, Jr., and R. Mehrabian . . . 51

Nondestructive Characterization of Rapidly Solidified Al-Mn Alloys by Ultrasonic and Electrical Methods
J. J. Smith, M. Rosen, and H. N. G. Wadley . . . 61

Ultrasonic Characterization of Oxygen Contamination in Titatium 6211 Plates and Weldments
S. R. Buxbaum and R. E. Green, Jr. . . . 71

Ultrasonic Wave Interaction in Anisotropic Systems
L. Adler, K. Bolland, A. Csakany, A. Jungman, and B. Oliver . . . 81

II POWDER METALLURGY AND CERAMICS

Application of NDE Methods to Green Ceramics : Initial Results
D. S. Kupperman, H. B. Karplus, R. B. Poeppel, W. A. Ellingson, H. Berger, C. Robbins, and E. Fuller . . . 89

Porosity Study of Sintered and Green Compacts of $YCrO_3$ Using Small Angle Neutron Scattering Techniques
K. Hardman-Rhyne, N. F. Berk, and E. D. Case . . . 103

Nondestructive Evaluation of Powder Metallurgy Materials
B. R. Patterson, K. L. Miljus, and W. V. Knopp . . . 109

Ultrasonic Measurements of Graphite Properties
S.-W. Wang and L. Adler . . . 121

III RESIDUAL STRESS

Residual Strain Measurements in Microelectronic Materials
W. E. Mayo, J. Chaudhuri, and S. Weissmann . . . 129

Effects of Microstructure on the Acoustoelastic Measurements of Stress
R. B. Thompson, J. F. Smith, and S. S. Lee . . . 137

Determination of Stress Generated by Shrink Fit
N. Chandrasekaran, Y. H. Wu, and K. Salama 147

Evaluation of Interfacial Stresses from the Amplitude of Reflected Ultrasonic Signals
D. K. Rehbein, J. F. Smith, and D. O. Thompson 155

Solving Residual Stress Measurement Problems by a New Magnetic Method
K. Tiitto 161

IV Secondary Processing and Deformation

Microstructural Evaluation of a Ferritic Stainless Steel by Small Angle Neutron Scattering
S. Kim, J. R. Weertman, S. Spooner, G. J. Glinka, V. Sikka, and W. B. Jones 169

Effects of Carbon Content on Stress and Temperature Dependences of Ultrasonic Velocity in Steels
J. S. Heyman, S. G. Allison, K. Salama, and S. L. Chu 177

Determination of Strain Distributions and Failure Prediction by Novel X-ray Methods
S. Weissmann and W. E. Mayo 185

The Use of Ultrasonic Signal Analysis Evaluation of Carbon Steel for Deformation Induced Microstructural Damage
G. H. Thomas, S. H. Goods, and A. F. Emery 197

Evaluation of Surface Machining Damage in Structural Ceramics
B. T. Khuri-Yakub, Y. Shui, and D. B. Marshall 203

A Nondestructive Near Surface X-ray Diffraction Probe
R. A. Neiser, K. S. Grabowski, and C. R. Houska 207

SUBJECT INDEX 215

Foreword

Conventional nondestructive testing of materials probes only for the existence of a defect. About ten years ago, nondestructive evaluation (NDE) began to evolve from testing with improved instrumentation along with a better understanding of materials behavior. NDE aims to detect and characterize flaws and microstructural changes in materials, and based on consideration of physical mechanisms controlling materials behavior in a specific application, to predict future performance and reliability of the component. The cost-effectiveness of NDE has been recognized in assessing the continued operational reliability of industrial plants, electricity generating stations, and military systems. Two NDE approaches are used--"off-line" periodic inspection during scheduled outages, and more recently, "on-line" monitoring.

The development of on-line monitoring will enable NDE to play an important role in materials processing by providing quantitative materials characterization in a time frame compatible with feedback loops designed for process control. The information needed from NDE includes (1) identification of desired or undesired microstructural features during processing and (2) detection and characterization of major defects. With the first item, process parameters can be controlled in the proper range of values such that defects are avoided and uniformity of properties is enhanced. With the second, the production process can be reversed (or modified during a later step) to remove the flaw.

This Proceedings consists of twenty four papers presented at a symposium held for the purpose of discussing new research and development relevant to the application of NDE to materials processing. The papers principally address the first numbered item in the previous paragraph. Emphasis has been placed on materials in four major areas: solidification (casting and welding), powder metallurgy and ceramics, residual stress, and microstructural stability during secondary processing such as heat treatment and deformation. Most of the work reported reflects an interdisciplinary approach. For example, ultrasonic detection of temperature distributions in ingots and of the liquid-solid interface in welds (the topics of the first two papers) required expertise on NDE, on materials, such as size of the molten weld pool, as well as on heat flow modelling in multiphase media to predict temperature gradients and thus to determine effective ultrasonic velocities.

Results reported herein are pertinent both to initial manufacture as well as to some aspects of repair or refurbishment of components. With respect to solidification, for example, ultrasonic detection of solid-liquid interfaces has been extended to casting and welding of steel; these pioneering efforts are still limited to simple geometries, however. Acoustic emission has been successful in real-time monitoring of cracking during rapid solidification of alloys heated with laser and electron beams. In the section on powder processing of materials, new approaches are described applying neutron, x-ray, and acoustic probes to characterize the porosity in structural ceramics in the green as well as sintered states to an extent not possible heretofore. Correlations of defects found by NDE with mechanical behavior are reported for other powder products, namely steel and graphite. Advanced methods play a key role in residual stress determination; improved quantitative measurements, based on magnetoelastic, ultrasonic, and x-ray techniques, are reported for thin-film semiconductors, cold-rolled metal plate, work rolls in rolling mills, among others. The final section on secondary processing covers a wide range of topics - heat treating, machining, deformation, and ion implantation - and shows the effectiveness of several methods in this area.

Though substantial progress has been made in research on NDE in materials processing, utilization of the results will depend upon several factors. Improvements in instrumentation and signal processing/image enhancement are foreseen so that NDE can become more routine and compatible with commercial processing. It is particularly important that the instrumentation be able to interface with automated (robotic) processing. In addition, close collaboration among the instrumentation, control systems/information processing, and materials communities is needed in both scientific and engineering efforts. We hope this Proceedings contributes towards this result.

We appreciate the contributions of the authors in presenting these papers at the symposium and the efforts of the session chairmen in leading the discussions. We acknowledge Ms. Diane Burt for typing the many letters and editorial changes in the papers. Also, we are grateful to the American Society for Metals for its support in publishing this Proceedings and note that the symposium was sponsored by the Energy and Natural Resources Activity of the ASM Materials Sciences Division.

Otto Buck
Metallurgy and Ceramics Program
Ames Laboratory
Iowa State University
Ames, Iowa

Stanley M. Wolf
Division of Materials Sciences
Office of Basic Energy Sciences
U. S. Department of Energy
Washington, DC

April 1984

I. Solidification

ULTRASONIC MEASUREMENT OF INTERNAL TEMPERATURE DISTRIBUTION

H. N. G. Wadley
Center for Materials Science
National Bureau of Standards
Washington, DC 20234

S. J. Norton
Center for Materials Science
National Bureau of Standards
Washington, DC 20234

F. S. Biancaniello
Center for Materials Science
National Bureau of Standards
Washington, DC 20234

R. Mehrabian
Center for Materials Science
National Bureau of Standards
Washington, DC 20234

ABSTRACT

The development of process control sensors is an area of research that would substantially impact improvement of productivity and quality in basic metals producing industries. The advent of powerful microcomputers now makes it possible to combine in-process measurements with process models for both feedback and feedforward control of processing. We report here initial studies directed toward the development of a sensor system for the measurement of internal temperature distribution in hot metals. The technique we are developing is based upon the observation that in metals the velocity of elastic waves (ultrasound) decreases substantially with increase in temperature. The measurement of ultrasonic velocity along a ray propagating through a metal body gives the average temperature along the ray. Using several different reconstruction methods, computer simulations are used to show that the temperature profile can be deduced from a suitably measured set of velocities propagating along different ray paths. Experimental results are then presented and used for the determination of temperature profiles in 304 stainless steel.

INTRODUCTION

Basic metals producing industries could be revolutionized by the incorporation of feedback/feedforward control systems during metals processing. The three components in such systems are: 1) sensors to make appropriate in-process measurements, 2) a process model that uses sensor data to assess the state of processing and 3) control systems that adjust process variables for optimal processing. Recent advances in the understanding of metals processing, rapid developments in artificial intelligence and the advent of inexpensive powerful computer systems have opened the way for implementation of this emerging technology. Research, however, is needed to develop measurement sensors. The American Iron and Steel Institute (AISI) has surveyed the sensor needs of the American steel industry and identified four sensors for immediate development [1]:

- Automatic detection of pipe and gross porosity in hot steel billets, blooms and slabs.
- On-line inspection for surface defects on hot and cold rolled strip.
- Rapid in-process chemical analysis of molten metal.
- Rapid measurement of temperature distribution within a solid or solidifying body of hot steel.

The National Bureau of Standards has initiated, jointly with AISI, research projects to develop two of these sensors: one for the measurement of the porosity/primary pipe and a second for the measurement of internal temperature distribution. In this paper, we concentrate only upon the temperature sensor. We review the role of a temperature sensor during metals processing, possible physical effects that might be measured to deduce temperature distributions, and recent research at NBS to explore a method based upon ultrasonic tomography.

Our objective is the development of a sensor for rapid, direct measurement of temperature distribution within a solid or solidifying body of hot metal. The sensor should ultimately be capable of measurements in the hostile environment associated with production facilities, such as continuous casters, reheat furnaces, ingot soaking pits and annealing furnaces.

One potential application is continuous casting where liquid metal breakouts can arise if the partially solidified strand is withdrawn too rapidly. Normal variations in process variables require conservative casting speeds considerably reducing productivity to avoid the catastrophic consequences of break-outs. If sensors were available to measure temperature distribution within the solidifying strand, it would enable the shell thickness and heat removal rate to be continuously monitored and used for feedback control. For this application the sensor would have to withstand an extremely hostile environment consisting of hot steam, high temperature, dust and occasional sprays of molten steel.

In a reheat furnace application, steel slabs with unknown initial temperature states are heated to an optimum temperature for rolling. Extended heating to insure a uniform temperature is wasteful of time, material and energy. A sensor to measure temperature distributions before, during and after heating would provide the required information for feedforward/feedback control of temperature. This sensor may, however, have to survive the hostile environment of the furnace and be insensitive to combustion products and flames within a reheat furnace.

The "ideal" sensor would measure temperature within ± 10 °C over a temperature range of 500 °C to 1350 °C without the need to physically contact the sample. It should also take only a few seconds to provide the temperature distribution and must be insensitive to the presence of combustion gas products or of nearby sources of heat greater than those of the steel body. Sensors for "specific" applications might not have to match all the specifications of the "ideal" sensor. The spatial resolution needed will depend upon the particular application; for example, a resolution of ± 2 mm and ± 10 °C would be appropriate for continuous casting, whereas for a reheat furnace, a reading of the minimum or average internal temperature could be sufficient.

TEMPERATURE DEPENDENT PHYSICAL PROPERTIES

While there are many physical properties that are temperature dependent, we consider here only those for which nondestructive techniques are available that probe the interior. With these restrictions, three properties show promise for temperature distribution measurement:

- Elastic constants.
- Lattice parameter.
- Phonon distribution.

Elastic constants can be measured from ultrasonic velocity experiments since:

$$\rho \upsilon_{\ell}^2 = \lambda + 2\mu \qquad (1)$$

and

$$\rho \upsilon_s^2 = \mu \qquad (2)$$

where

υ_{ℓ} = longitudinal wave velocity (ms^{-1})

υ_s = shear wave velocity (ms^{-1})

λ, μ = Lame constants (Nm^{-2})

ρ = density (kgm^{-3})

Measurements on austenitic steels have indicated that the longitudinal wave velocity decreases almost monotonically with temperature from ~ 5,800 ms^{-1} at room temperature down to ~ 4,700 ms^{-1} at the melting point (Fig. 1) [2]. The velocity in ferritic steels also decreases with temperature although the relation is complicated by the change of crystal structure from bcc to fcc. Steels, even at high temperatures, remain semitransparent to longitudinal elastic waves. They do exhibit a microstructure-sensitive attenuation that rapidly increases as the wavelength of the ultrasonic waves approaches the grain size and as the temperature increases.

The lattice parameter of iron and iron alloys increases with temperature. X-ray techniques are currently able to measure these changes accurately near the surface of metals and high energy synchrotron techniques could

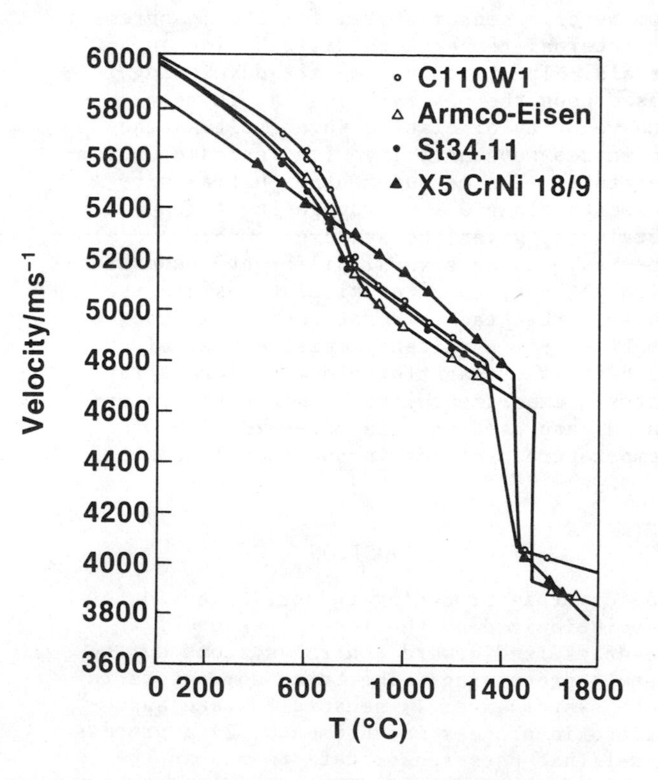

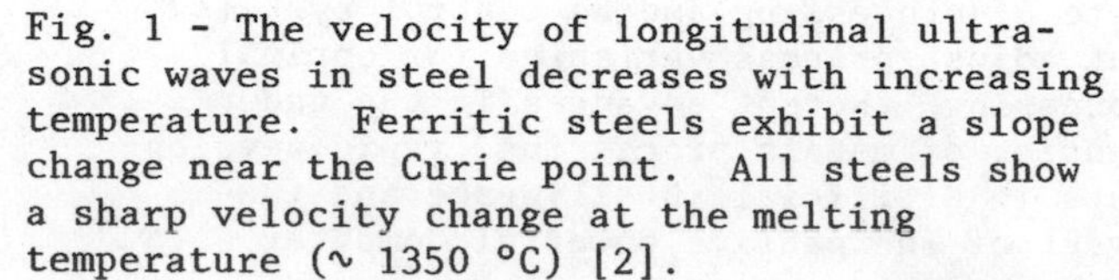

Fig. 1 - The velocity of longitudinal ultrasonic waves in steel decreases with increasing temperature. Ferritic steels exhibit a slope change near the Curie point. All steels show a sharp velocity change at the melting temperature (~ 1350 °C) [2].

conceivably probe further beneath the surface. Gamma radiation penetrates further than synchrotron generated X-rays, but the wavelengths of the gamma radiation are much smaller than the lattice parameter, so radiographic measurements of density would have to be used to deduce temperature distribution.

The phonon distribution controls the electrical resistivity and the temperature of thermal energy of neutrons injected into the steel. For resistivity, temperature-dependent dissolved impurity atom content also is important. Frequency modulation of eddy current measurements provides a potential method for deducing near surface electrical resistivity variations, while neutron thermalization appears to provide a potential method of deducing the average phonon temperature. In steel, however, the capture cross section of thermal neutrons is high. Thus, once thermalized, the neutrons may not reach the metal surface in sufficient quantities for accurate measurements.

At NBS expertise exists for ultrasonic measurements. In the past this has been developed during studies of microstructure in materials science and ultrasonic imaging in biomedicine. In addition, work had involved non-contact laser generation/detection techniques that at first sight appeared good candidates for the remote sensing ultimately required of sensors. Thus, our research has initially been directed toward the ultrasonic approach, utilizing tomographic reconstruction as a means of deducing the internal temperature distribution.

ULTRASONIC TOMOGRAPHY

Suppose on ultrasonic pulse propagates through a material taking a time τ to propagate a distance λ. Then the average velocity (temperature) of the material traversed is λ/τ. A single measurement thus enables deduction of a single quantity (average

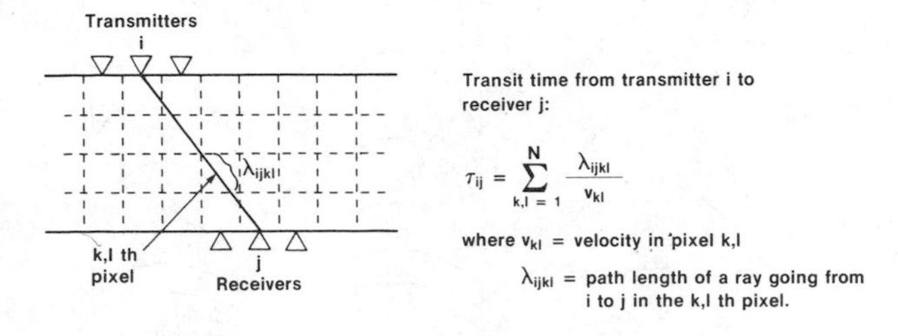

Fig. 2 - The time-of-flight of an ultrasonic pulse through one pixel is the path length divided by the average velocity in the pixel. The sum of this quantity over all N pixels intercepted by the path corresponds to the time-of-flight through the sample.

temperature). To deduce the distribution of velocity, we must decompose the material into pixels (Fig. 2). Then, if we propagate rays from transmitters i to receivers j and measure their transit times τ_{ij}, we recognize that:

$$\tau_{ij} = \sum_{k,l=1}^{N} \frac{\lambda_{ijkl}}{V_{kl}} \qquad (3)$$

where

λ_{ijkl} = path length of ray ij in the kl pixel

V_{kl} = average velocity in the pixel

In principle, since the λ_{ijkl} can be computed and τ_{ij} measured, and provided at least as many measurements as pixels are made, the average velocity in each pixel can be deduced. In practice, experimental error and microstructure effects introduce noise into the reconstruction process and over-determination (redundancy) becomes mandatory.

COMPUTER SIMULATIONS

The time constraint placed on the temperature sensors requires optimal utilization of the smallest number of time-of-flight

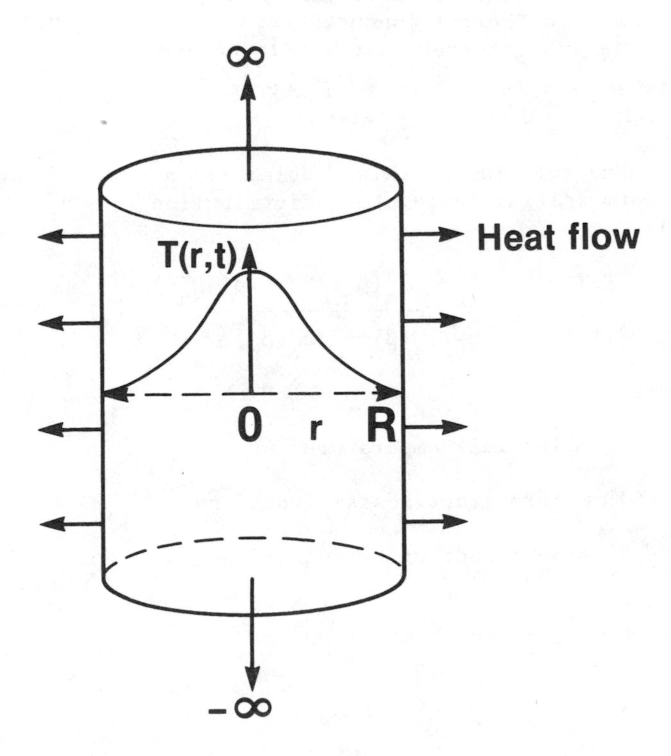

Fig. 3 - The radial flow of heat from an infinite cylinder of radius R results in a temperative profile T(r,t) that has circular symmetry.

measurements. To aid the design of a first experiment, computer simulations were used to test various methodologies. The procedure adopted consisted of simulating the cooling of an infinitely long cylinder (Fig. 3). Using heat flow theory, temperature distributions were calculated at various times. Existing ultrasonic velocity-temperature data were then used to numerically calculate transit times of rays. These transit time data were used to reconstruct a temperature distribution which was compared with that calculated from heat flow theory.

Initially, we have considered the infinite cylinder as a model of an ingot (Fig. 3). The equation governing heat flow is:

$$\frac{\partial T}{\partial t} = \alpha \left[\frac{\partial^2 T}{\partial r^2} + \frac{1}{r}\frac{\partial T}{\partial r}\right] \qquad (4)$$

subject to the conditions:

$$T = f(r),\ 0 \leq r \leq R,\ t < 0$$

$$\left.\frac{\partial T}{\partial r}\right|_{r=R} = \frac{-h}{k}\,[T - T_f], \qquad t \geq 0$$

$$\left.\frac{\partial T}{\partial r}\right|_{r=0} = 0$$

where

$f(r)$ = Initial temperature distribution
h = Heat transfer coefficient
k = Thermal conductivity
T_f = Temperature of surroundings
α = Thermal diffusivity
R = Radius of cylinder

The solution to this problem for a uniform initial temperature distribution is (4):

$$\frac{T - T_f}{T_i - T_f} = 2B \sum_{n=1}^{\infty} \frac{J_o(\beta_n r/R)\ \exp(-\beta_n^2 F)}{(\beta_n^2 + B^2) J_o(\beta_n)} \qquad (5)$$

where

T_i = Initial temperature

J_o = Zero-order Bessel function

β_n = n-th root of the equation $\beta J_1(\beta) = B J_o(\beta)$

$F = \frac{\alpha t}{R^2}$ is the Fourier number

$B = \frac{hR}{k}$ is the Biot number

Examples of temperature profiles along three rays are shown in figure 4 at various dimensionless times after the initiation of cooling.

Cylindrical symmetry precludes the necessity to use square or rectangular pixels. Radial temperature only is required from the reconstruction. The mathematical basis of the algorithm used here for reconstructing the radial temperature distribution of a steel cylinder is outlined below.

1. The velocity of sound $v(r)$ is taken to decrease with temperature in a known way and is assumed single valued.

2. The transit time τ_m is measured along N rays of length L_m by evaluating the line integral:

$$\tau_m = \int_{L_m} \frac{dl}{v(r)}, \quad m = 1,2\ldots N \qquad (6)$$

3. Expand $1/v(r)$ in an orthogonal basis set, $\phi_n(r)$:

$$\frac{1}{v(r)} = \sum_{n=1}^{N} a_n\, \phi_n(r) \qquad (7)$$

If $v(r)$ is radially symmetric, only the r dependence of temperature is required. This can be achieved in several ways. We have explored the use of basis sets consisting of either ring functions or Bessel functions. The advantage of the latter is that, for the model problem, heat flow calculations give temperature distributions in terms of linear combinations of Bessel functions.

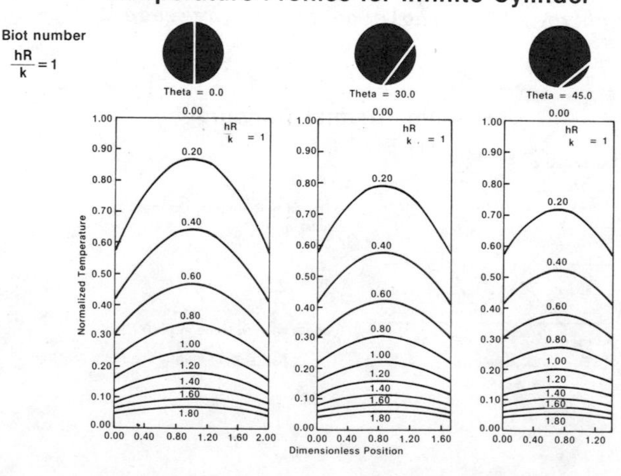

Fig. 4 - Temperature profile along three rays at various dimensionless times ($\alpha t/R^2$) after the initiation of cooling. Calculations assume the heat transfer coefficient, h, to be equal to k/R.

4. Substituting Eq. (7) into Eq. (6) gives

$$\tau_m = \sum_{n=1}^{N} a_n \int_{L_m} \phi_n(r)dl$$

$$= \sum_{n=1}^{N} a_n \Phi_{mn}, \quad m = 1,2...N \qquad (8)$$

where the Φ matrix elements are given by the line integrals of the basis functions:

$$\Phi_{mn} = \int_{L_m} \phi_n(r)dl \qquad (9)$$

The matrix elements are computed from Eq. (9) and stored. Equation (8) is a linear system of N equations which can be solved for the N unknown coefficients a_n, once the transit times τ_m are measured along N rays. The a_n are then substituted back in Eq. (7) to yield the reciprocal velocity. This is then converted to temperature using the experimentally-determined velocity versus temperature relation.

Figure 5 shows an example of a temperature reconstruction based upon four transit-time measurements for the calculation of four unknown coefficients, using a ring-function basis set. In this case, the accuracy of the reconstruction could be increased by increasing the number of rays which control the width of the ring function. Even a few rings, however, gives quite a good estimate of the temperature. In the absence of noise, there would be no need for redundancy in the inversion. In practice, the large condition-numbers of the inverse matrix result in large inaccuracy, so redundancy is advantageous.

Figure 6 shows two examples of temperature reconstructions based upon three transit-time measurements, using a Bessel function basis set. In (a), the surface temperature is assumed unknown and a small error results, because more than the first two Bessel functions of the series are required to determine the temperature field. In (b), the known surface temperature provides additional information that considerably improves the reconstruction accuracy.

Heat flow solutions for long cooling periods are dominated by the first Bessel function term of the series expansion. For these cases, as few as two rays can give remarkably good estimates of the temperature distribution.

Example of Reconstruction Using Ring Functions

For N=4

$$0.5 \begin{bmatrix} \tau_1 \\ \tau_2 \\ \tau_3 \\ \tau_4 \end{bmatrix} = \begin{bmatrix} a_1 \\ a_2 \\ a_3 \\ a_4 \end{bmatrix} \begin{bmatrix} S_{11} & S_{12} & S_{13} & S_{14} \\ O & S_{22} & S_{23} & S_{24} \\ O & O & S_{33} & S_{34} \\ O & O & O & S_{44} \end{bmatrix}$$

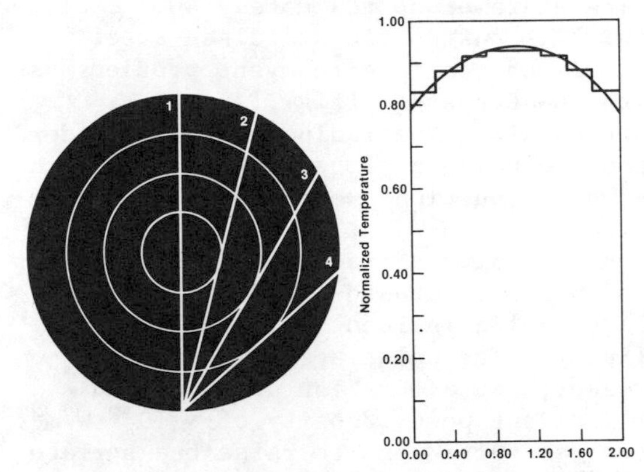

Fig. 5 - Using four time-of-flight values for the rays shown, the internal temperature distribution was reconstructed using a ring function basis. In the figure, the actual temperature (smooth curve) and reconstructed temperature (stepped curve) are shown to be within 5% of each other.

Example of Reconstruction Using Bessel Functions

N=3

(a) Unknown surface temperature (b) Independently deduced surface temperature

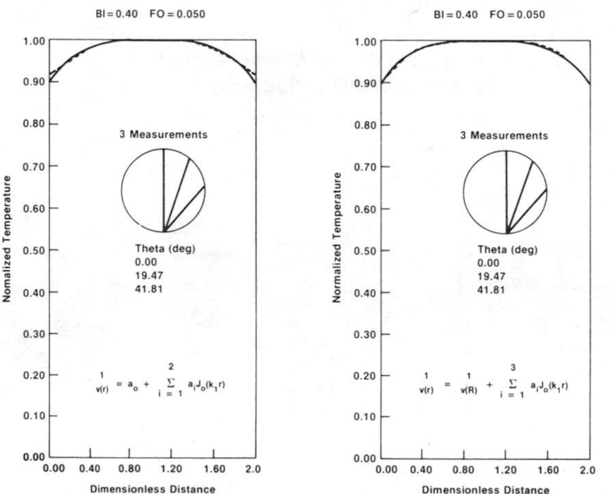

Fig. 6 - Reconstructed temperature profile (dashed curve) using three time-of-flight values and in (b), the surface temperature. A Bessel function basis set was used. Agreement with the actual temperature (solid curve) is within 2% in (a) and better than 1% in (b).

INITIAL EXPERIMENTS

From the above, it is evident that relatively few time-of-flight measurements through the circular cross section of a slowly cooling semi-infinite cylinder should be sufficient to deduce the radial temperature distribution. To verify this proposition, a type 304 (18/8) stainless steel cylinder, 60 cm in length and 15.2 cm in diameter, was heated in an induction heating coil to a temperature of approximately 400° C throughout (Fig. 7). Austenitic stainless steel was deliberately chosen to circumvent problems associated with phase changes (Fig. 1). Temperatures along the radius of the cylinder were recorded using a probe containing thermocouples whose junctions were spaced at 18 mm intervals.

High intensity ultrasonic pulses were generated using a focused (1.2 mm diameter) 30 mJ laser pulse incident upon one side of the cylinder. The pulse, generated by a Nd:YAG laser, had a duration of 25 ns and average incident power density of $\sim 10^{12}$ W/m^2. The pulse was sufficient to raise the surface temperatures of 304 stainless steel well beyond its vaporization temperature causing local material ablation, resulting in the generation of an intense ultrasonic pulse.

The laser-generated ultrasonic pulse propagated across the cylinder and was detected by a 450 kHz narrow-bandwidth transducer attached to a waveguide, designed to thermally insulate the transducer from the hot surface. The waveguide (or buffer rod) was constructed from fused quartz, a material with extremely low thermal expansion coefficient. It was ground to a conical shape to:

- Reduce dispersive wave propagation effects.
- Have as small an aperture as possible.
- Reduce thermal conduction into the rod causing its temperature to rise.

The interface between the buffer rod and the cylinder is a source of substantial signal attenuation. (For lower temperature experiments reported here a silicon based high vacuum grease was used as couplant.)

For these experiments the signal from a 450 kHz transducer, after amplification with a low-noise wideband preamplifier, was high-pass filtered at 10 kHz and digitally recorded (Fig. 7). We used an 8-bit digitizer at a 10 ns sampling rate. Digitization was initiated by an external trigger signal supplied by the Q-switch circuit of the laser, which was experimentally determined to initiate recording of waveforms 0.5 μs before the ultrasonic pulse was launched into the steel. The digitally recorded signal was stored temporarily on

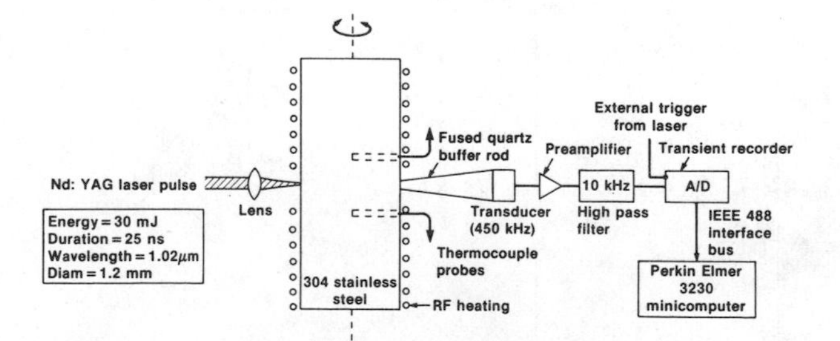

Fig. 7 - The experimental arrangement used for measuring time-of-flight values through a 304 stainless steel cylinder. Ultrasonic pulses were transmitted by a non-contact laser method and received by a piezo-electric transducer thermally insulated from the sample by a fuzed quartz buffer rod.

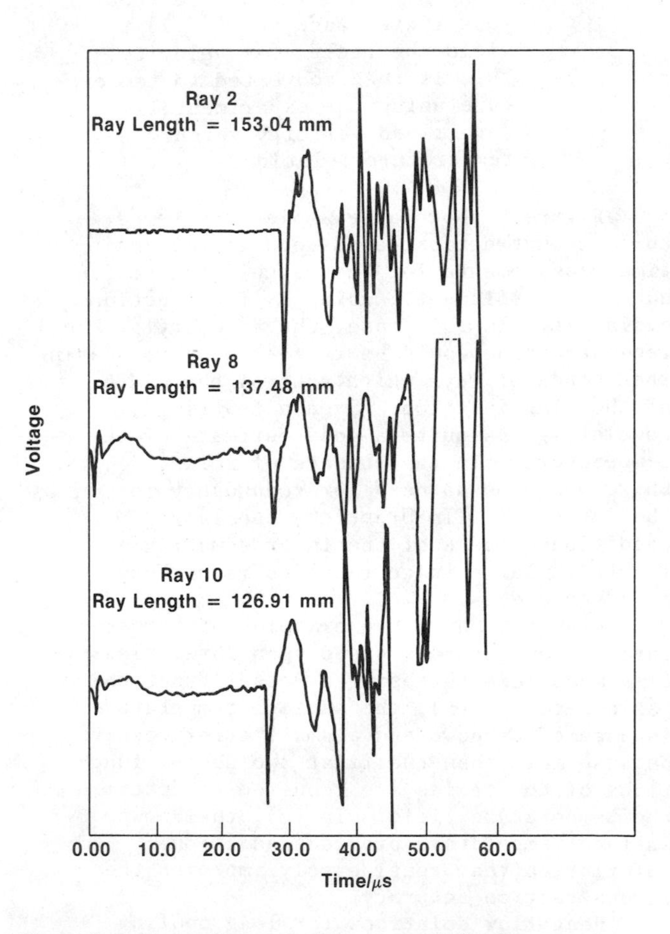

Fig. 8 - Examples of the recorded ultrasonic waveforms for three rays. Recording begins at the instant the laser pulse transmitted the ultrasonic pulse. The time-of-flight was taken to be the peak of the first arrival. The complicated later arrivals are due to echos within the sample, buffer rod and tranducer.

floppy disk, then read off-line into a minicomputer for subsequent analysis.

To deduce ray transit times in the steel, it was necessary to correct the longitudinal-wave arrival time measurements by the transit time in the buffer rod. This transit time was inferred, from pulse-echo measurements, to be 22.5 μs at room temperature and to increase with increasing buffer rod temperature. Because only comparatively low temperature measurements were made, it was sufficient to use room temperature pulse-echo data for the transit time correction.

A set of eleven ray transit times were measured on the steel cylinder by keeping the buffer rod position fixed and varying the position where the laser pulse impinged in approximately 10 mm intervals around the circumference of the cylinder. Table 1 gives the transit time data used for the reconstructions. The temperature profile determined by thermocouple probes is given in Table 2.

Table 1 - Ray Transit Data Used for Thermal Tomography

Ray No.	Path Length/mm	Transit Time/μs
1	153.40	27.95
2	153.04	27.70
3	152.03	27.50
4	150.38	27.25
5	147.95	26.75
6	145.16	26.10
7	141.78	25.60
8	137.48	24.80
9	132.76	23.95
10	126.91	22.90
11	121.64	21.85

Table 2.

Thermocouple position/mm	Indicated Temperature/°C
0 (Center)	315
16	315
33	313
49	308
76 (Surface)	229

The thermocouple is not in agreement with heat flow calculations because of the Biot number temperature variation which was not included in calculations.

DISCUSSION

If the surface temperature is known, we can evaluate a single coefficient of the basis set representation of the velocity distribution, since each ray is an independent sample of the velocity profile:

$$\frac{1}{v(r)} = \frac{1}{v(R)} + a_2 \phi_2(r) \qquad (10)$$

where

$v(R)$ = known velocity at the surface
a_2 = unknown coefficient to be determined
$\phi_2(r)$ = either a Bessel or ring function

We have assumed a linear relation between velocity and temperature,

$$T(r) = T_0 + B(v(r)-v_0) \qquad (11)$$

where T_0, B and v_0 are known constants [T_0 = 25 °C, v_0 = 5701 ms^{-1}, B = (-1/0.7) °C/ms^{-1}]. The reconstructions for all eleven rays are shown in figure 9. Considerable scatter is observed with two rays (1 and 8), which are as much as 80 °C in error. The remainder of the rays, however, were within 30 °C of the measured center temperature (Table 2).

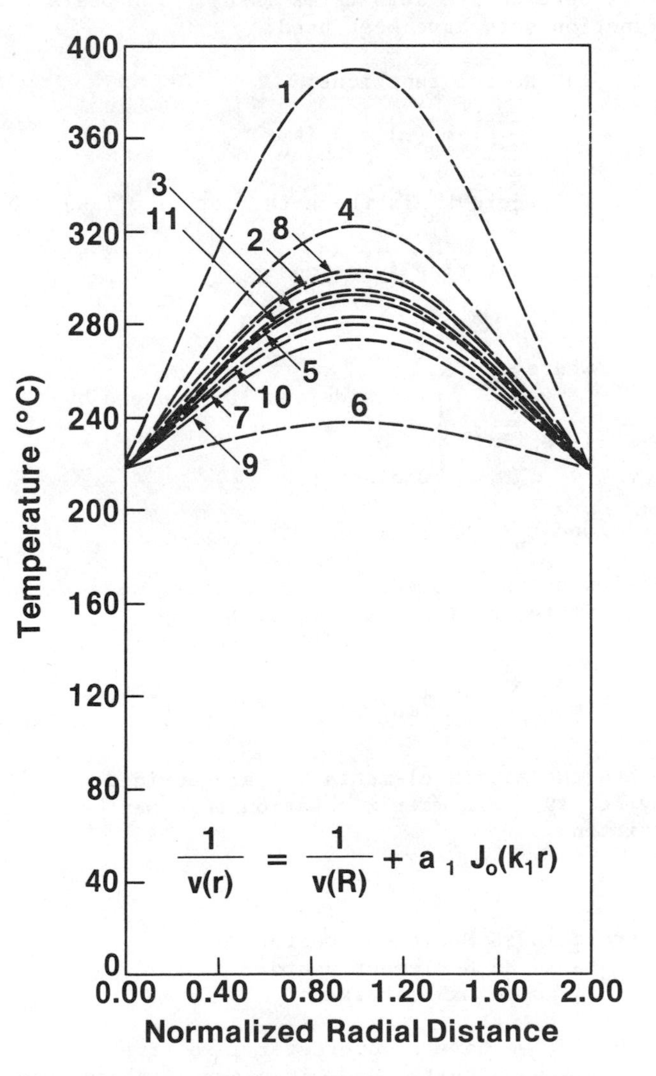

Fig. 9 - Reconstructed temperature profiles using each time of flight value to calculate on Bessel function coefficient. The scatter arises from measurement errors and grain effects.

The origin of this scatter is error in the measurement of both transit time and distance, and microstructure effects. To minimize the effect of this, we have applied a least-squares technique to the evaluation of the unknown coefficients, where now the number of measurements (M) is assumed larger than the number of coefficients (N). Suppose we measured M time-of-flight values:

$$\tau_m = \int_{L_m} \frac{dl}{v(r)}, \qquad m = 1,2,...M \qquad (12)$$

where L_m is the m-th ray path. Then, assuming the velocity to be radially symmetric, we expand $1/v(r)$ in a basis set orthogonal on a circle of radius R:

$$\frac{1}{v(r)} = \sum_{n=1}^{N} a_n \, \phi_n(r) \qquad (13)$$

This we truncate at N terms (N<M). Two basis function sets have been used:

1) Bessel functions:

$$\phi_n(r) = J_o(k_n r) \qquad (14)$$

where k_n is the n-th root of $J_o(kR) = 0$

2) Flat ring functions:

$$\phi_n(r) = \text{ring}_n(r) \qquad (15)$$

where

$$\text{ring}_n(r) = \begin{cases} 1 \text{ inside annulus bounded by radii } r_n \text{ and } r_{n+1} \\ 0 \text{ elsewhere} \end{cases}$$

and $r_n = Rn/N$

Following the procedure used earlier, we substitute Eq. (13) into (12), giving

$$\tau_m = \sum_{n=1}^{N} a_n \, \Phi_{mn}, \qquad m = 1,2...M \qquad (16)$$

where the matrix elements Φ_{mn} are defined by Eq. (9). In matrix notation this may be written:

$$\underline{\tau} = \Phi \, \underline{a}$$

where $\underline{\tau}$ is an M-element vector
$\underline{a}$ is an N-element vector
Φ is an MxN matrix

If M > N, we have an overdetermined (i.e., inconsistent) system and, in general, there is no $\underline{a}$ that solves the system (16) exactly. To find the $\underline{a}$ that is the minimum mean-square-error solution, let

$$\underline{e} = \Phi \, \underline{a} - \underline{\tau} \qquad (17)$$

Mean-square error is:

$$E = \underline{e}' \, \underline{e} = (\Phi \, \underline{a} - \underline{\tau})' \, (\Phi \, \underline{a} - \underline{\tau}) \qquad (18)$$

$$= \underline{a}' \, \Phi'\Phi \, \underline{a} - \underline{a}' \, \Phi' \, \underline{\tau} - \underline{\tau}' \, \Phi \, \underline{a} + \underline{\tau}' \, \underline{\tau}$$

where ' stands for transpose. Setting the derivative of E with respect to $\underline{a}$ equal to zero gives the minimum mean-square-error solution, $\hat{\underline{a}}$.

This results in the system of equations

$$\Phi' \, \Phi \, \hat{\underline{a}} - \Phi' \, \underline{\tau} = 0 \qquad (19)$$

or

$$\hat{\underline{a}} = (\Phi'\Phi)^{-1} \, \Phi' \, \underline{\tau} \qquad (20)$$

where $\hat{\underline{a}}$ is the desired minimum mean-square-error solution. $\Phi'\Phi$ is an NxN matrix, which in general is invertible; the matrix

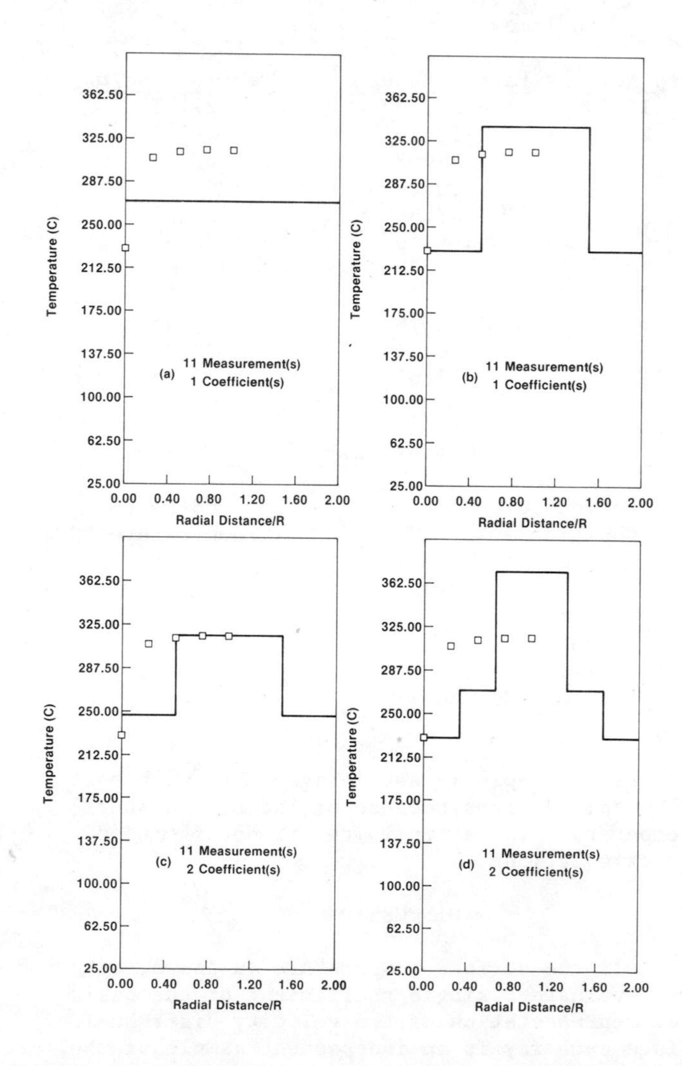

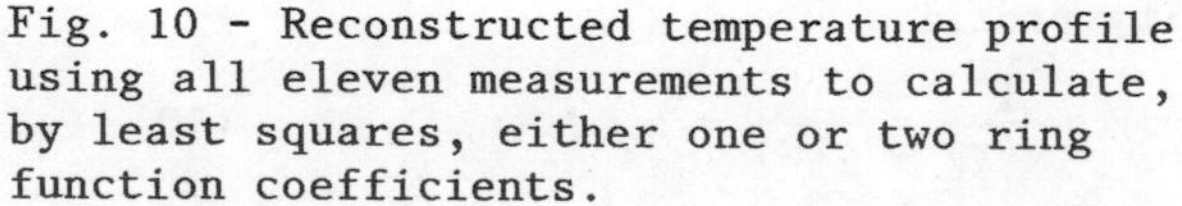
Fig. 10 - Reconstructed temperature profile using all eleven measurements to calculate, by least squares, either one or two ring function coefficients.

$(\phi'\phi)^{-1}\phi'$ is called the pseudoinverse of Eq. (16). Then, the least squares estimate for the velocity distribution v(r) is:

$$\frac{1}{\hat{v}(r)} = \sum_{n=1}^{N} \hat{a}_n \phi_n(r) \qquad (21)$$

Figure 10 shows reconstructed temperature profiles for the case of one and two ring function coefficient determinations. A single coefficient with no independent surface temperature information (Fig. 10 (a)) allows estimation of only the average temperature. The use of independent surface temperature data (Fig. 10 (b)), or the evaluation of two coefficients (Fig. 10 (c)), allows estimation of the temperature profile. Increasing the number of coefficients, which, in principle, increases the precision of the reconstructed temperature profile, actually results in worse reconstruction. This is because the condition number of the inverse matrix increases with matrix order and amplifies experimental errors.

Figure 11 shows reconstructed temperature profiles using a Bessel Function basis set. The fit is not as good as for the model problem. This is because the heat transfer coefficient varied during cooling. Then the internal temperature distribution (measured with inserted thermocouples) is not well represented by a few Bessel function terms.

The two basis sets give similar degrees of reconstruction accuracy. The advantages of the Bessel function approach that were apparent in computer simulations are mitigated in practice due in part to the more complicated heat flow.

CONCLUSIONS

1. A scheme has been developed in which a small number of time-of-flight measurements in principle could determine internal temperature profile to an accuracy of a few percent.

2. Ultrasonic time-of-flight measurements using a noncontact laser generation source have been used to reconstruct internal temperature distributions in steel to ±30 °C using as few as two basis functions.

3. Measurement error and ultrasonic grain scattering introduce noise into reconstructions. This can be reduced using least squares techniques. Further work incorporating a priori information into reconstruction algorithms may further improve reconstruction accuracy while maintaining the number of measurements constant.

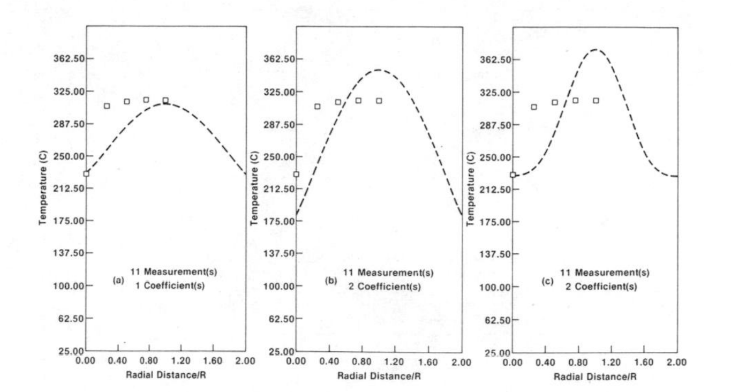

Fig. 11 - Reconstructed temperature profile using all eleven measurements to calculate, by least squares, either one or two Bessel function coefficients.

ACKNOWLEGMENTS

We wish to acknowledge valuable discussion of this work with Drs. M. Linzer, B. Droney, D. Rogers and J. Cook.

REFERENCES

1. Mehrabian, R., R. L. Whitely, E. C. van Reuth, and H. N. G. Wadley, Report of Workshop on Process Control Sensors for the Steel Industry, National Bureau of Standards, NBSIR 82-2618 (1982).
2. Kurz, W. and B. Lux, Berghuttenmann Monatsh. 114 (5), 123 (May 1969).
3. Papadakis, E. P., L. C. Lynnworth, A. K. Fowler, and E. H. Carnevale, J. Acoust. Soc. of Am. 52 (3), 850 (1972).
4. Carslaw, H. S., and J. C. Jaeger, Conduction of Heat in Solids, Clarendon Press (1959).

8311-002

REAL-TIME ULTRASONIC SENSING OF ARC WELDING PROCESSES

L. A. Lott
EG&G Idaho, Inc.
Idaho Falls, Idaho 83415

J. A. JOHNSON
EG&G Idaho, Inc.
Idaho Falls, Idaho 83415

H. B. Smartt
EG&G Idaho, Inc.
Idaho Falls, Idaho 83415

ABSTRACT

NDE techniques are being investigated for fusion zone sensing of arc welding processes for closed-loop process control. An experimental study of pulse-echo ultrasonics for sensing the depth of penetration of molten weld pools in structural metals during welding indicates that real-time ultrasonic sensing is feasible. Results on the detection of liquid/solid weld pool interfaces, the determination of interface location, and effects of high temperature gradients near the molten zones on ultrasonic wave propagation are presented. Additional work required and problems associated with practical application of the techniques are discussed.

INTRODUCTION

Modern automatic arc welding equipment is very sophisticated, controlling the major machine functions such as welding current, arc voltage, travel speed, filler wire speed, mechanical motions of the torch, etc. Some machines are even microprocessor controlled. However, with few exceptions, information about the physical state of the weld fusion zone is not used in real-time process control. (Exceptions to this are systems which use optical or IR sensors to measure the size or temperature of the fusion zone surface on the opposite side from the torch of thin materials being welded.[1]) In almost all cases, the human operator "closes the loop" manually, adjusting the welding parameters during the process based on his observations. Thus weld quality is still strongly dependent on operator skill and experience for many automated arc welding applications.

Weld bead penetration is often critical. Unfortunately, the weld operator rarely is able to determine visually the extent of penetration. A sensing technique which can directly measure the extent of weld bead penetration (preferably from the same side of the part as the welding torch) and provide real-time information for closed-loop control of the machine parameters to maintain the desired weld penetration configuration is a real need.

In addition to the sensing technology, the implementation of such a system would require

- o microprocessor control hardware (this is already available)
- o a model of process parameter/bead geometry relationships[2] (how must the welder parameters be changed to achieve a desired change in bead geometry)
- o a process sensing algorithm (what must be done with the sensor signal to extract the desired penetration information)
- o a process control algorithm (selection of the optimum parameters taking into account the process model, sensing, welding hardware and inherent process limitations).

Such an automatic welding machine is diagrammed in Fig. 1; the basic feedback loop consists of the molten weld pool, the process control sensor, the sensor data acquisition and signal processing system, the process controller, the welder power supply, the welding torch and arc, and back to the molten weld pool.

Ultrasonics shows promise for weld penetration sensing, as described in this paper. Ultrasonics has the potential to locate the edge of a molten weld zone by virtue of the unique ability of ultrasonic waves to propagate through thick metal structures and be reflected at

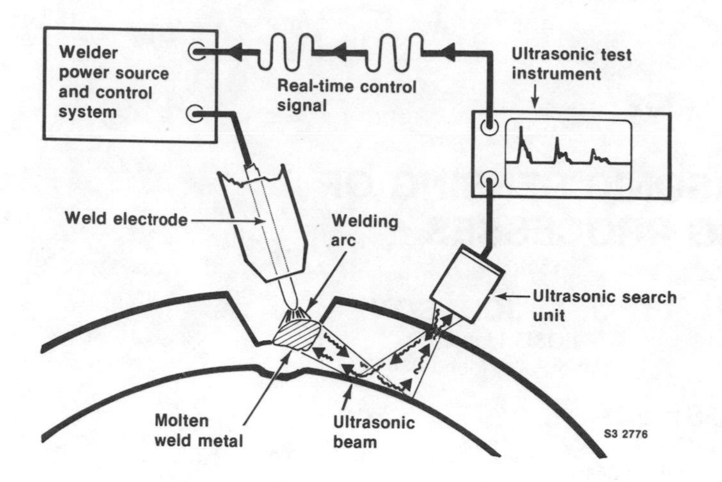

Fig. 1 - Ultrasonic controlled welding process concept

discontinuities in ultrasonic impedance such as occur at liquid/solid interfaces. By the detection and accurate timing of ultrasonic pulses reflected from the liquid/solid interface of a molten weld pool, the position of the interface could be determined and a process control signal generated to maintain the desired weld penetration. At liquid/solid interfaces of typical structural steels, the differences in density and longitudinal ultrasonic wave velocity between the liquid and solid phases are such that the amplitude reflection coefficients are approximately 0.1, or 10%.[3-7] For ultrasonic shear waves, the theoretical reflection coefficient is 1.0, or 100%, since the liquid state will not support shear wave propagation.

BACKGROUND

Ultrasonic techniques have been used to detect liquid/solid interfaces in a variety of different materials for applications including basic solidification studies, monitoring of crystal growth, and monitoring of metal casting processes. Bailey and Dula[8] followed the motion of the liquid/solid interface during the freezing of water using 1.0 MHz longitudinal ultrasonic waves incident from the liquid side. Bailey and Davila[9,10] also observed this interface during the freezing of mercury and paraffin under unidirectional heat flow conditions using 2.25 MHz longitudinal waves. Kurz and Lux[5-7] detected longitudinal wave reflection from liquid/solid interfaces in Wood's Metal, a Pb-Sn alloy, and steel. They found that reflected ultrasonic signals were reduced by the geometry of the interface and by high attenuation in the solid material at high temperatures. At low solidification rates during the casting of steel, they found that rough surfaces at the interface due to dendritic grain growth and the presence of a "mushy" transition zone between the solid and liquid reduced the reflected ultrasonic signal. They also observed a decrease in sound velocity in the solid of up to 25% caused by high temperatures which must be accounted for in thickness or interface location measurements.

Lynnworth and Carnevale[4] observed both longitudinal and shear wave reflections at liquid/solid interfaces of steel in a study of continuous casting processes. They also observed that interface roughness due to dendritic grain growth reduced the reflected ultrasonic signal. They saw little advantage of shear waves despite the theoretical 100% reflectivity at the interface. Jeskey et al.[11] considered ultrasonic pulse-echo reflectivity measurements for monitoring steel ingot solidification. However, because of ultrasound propagation problems associated with coupling, high absorption, grain scattering, and ill-defined and irregular surfaces, a method based on through-transmission transit time was used for solidification rate measurement. Parker[3] studied the feasibility of pulse-echo ultrasonics for process control in continuous casting of metals. He succeeded in determining the position of the liquid/solid interface in 99.9 Sn during both freezing and melting to ±1 mm using 5 MHz longitudinal ultrasound.

Katz[12] considered the possibility of ultrasonic measurement and control of weld penetration in a study similar to this work. He gave a good analysis and discussion of the concept and reported some experimental results on molten zones at the ends of cylindrical rods. He failed to detect molten weld pools in flat plate specimens but, nonetheless, concluded that ultrasonic weld pool dimensional measurement is potentially feasible.

The scope of the work described here is to study the problem of ultrasonically detecting, locating, and determining the dimensions of molten weld zones in structural steel. The main objectives are to determine the feasibility of the method and to obtain preliminary information needed as a first step toward developing an automatic closed-loop welding system controlled by ultrasonic sensors. The work is an experimental laboratory study in which stationary spot-on-plate welds are formed on the surface of steel specimens simultaneously with ultrasonic monitoring of the weld zone.

EXPERIMENTAL METHODS

The experimental measurements are performed with the test specimens suspended in a conventional ultrasonic immersion testing tank as shown schematically in Fig. 2. This is done in such a way that the bottom surface is immersed in water to facilitate coupling of ultrasonic waves into the sample, while leaving the top surface dry for welding. The welder and torch are common gas tungsten arc welding (GTAW) equipment and are operated manually during experimental runs. The torch and ultrasonic transducer are mechanically connected so that after initial setup and alignment, the assembly can be easily moved from point to point on the test specimen between test runs while maintaining alignment.

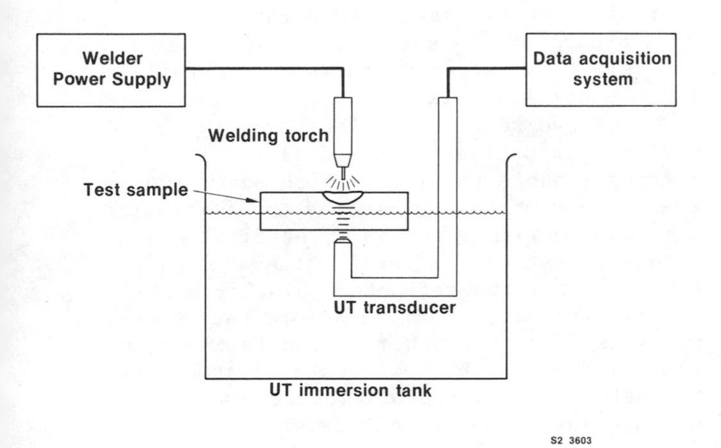

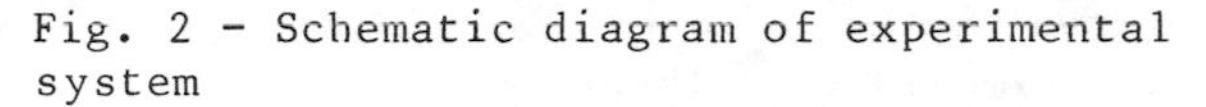
Fig. 2 - Schematic diagram of experimental system

The test specimens are 25 mm (1 in.) thick by 100 mm (4 in.) wide by 360 mm (14 in.) long plates of carbon steel or 304 stainless steel. One specimen was prepared with a series of four evenly spaced holes 1.6 mm (0.063 in.) in diameter and approximately 25 mm (1 in.) deep drilled into the side of the specimen as shown in Fig. 3. When welding on the specimen directly above the holes and monitoring with ultrasonics from directly below as shown in Fig. 3, reflected ultrasonic signals from the weld pool region are received at the same time as reflections from the side drilled holes. In this way the side drilled hole reflections serve as internal markers of known position with which the position of the weld pool reflections can be calibrated.

The data acquisition system is shown schematically in Fig. 4. Standard 0.5-in. diameter ultrasonic immersion transducers of nominal frequencies of 1, 2.25, 5, and 10 MHz are used with a standard ultrasonic testing pulser-receiver unit. The ultrasonic data acquisition is completely automatic and computer controlled. The pulser is triggered by a signal from the computer and specified portions of the received A-scans are digitized to 8-bit accuracy at a maximum rate of 20 MHz by a

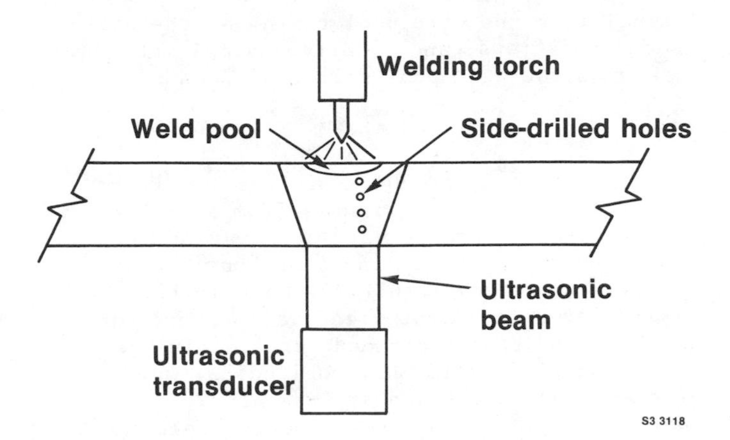

Fig. 3 - Test sample configuration

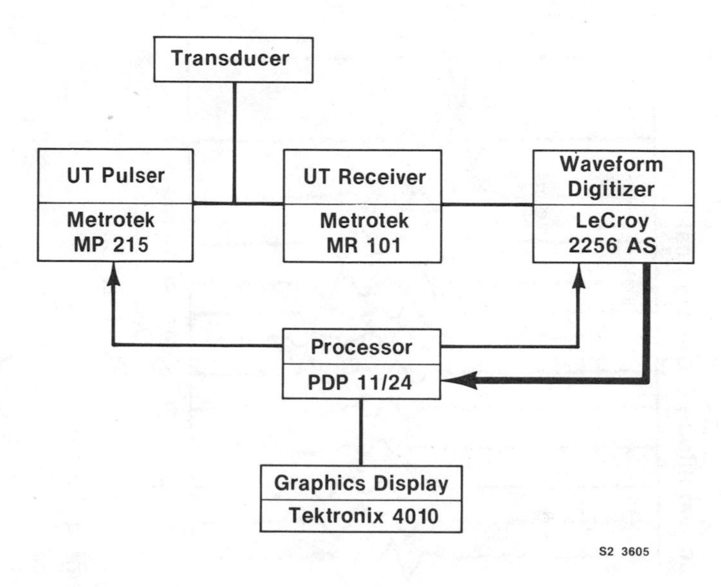

Fig. 4 - Data acquisition system

waveform digitizer whose output is then stored in the computer. The data are stored on disk or magnetic tape and recalled at a later time for display and posttest processing. A purchased interactive data processing software package (ILS, Signal Technology, Inc.) is used for data acquisition, file manipulation and management, processing, and display.

After initial setup and alignment of the welding torch, transducer, and sample, the computer is instructed to trigger the pulser at regular intervals and digitize and store windowed portions of the radio frequency A-scans. Typically, 20 or 30 A-scans are taken at 1-s intervals. Once the data acquisition cycle is initiated, the welder is turned on for a period of time, typically 10 s, and then turned off. Thus, a data record of 20 s, for example, consists of one or two A-scans showing the reflected signal from the top sample surface (no weld pool), then ten scans showing the reflected signal from the weld pool region as the pool forms and grows, and finally several more scans as the weld pool solidifies and cools.

EXPERIMENTAL RESULTS

In carbon steel and 304 stainless steel, reflected ultrasonic signals from at least two interfaces in the weld pool region are received for normally incident longitudinal waves. A typical result is shown in Fig. 5, a display of a computerized data record for a 16-s weld spot experiment on a stainless steel sample for 10 MHz longitudinal waves. The A-scans were taken at 1-s intervals before, during, and after a molten weld spot was formed in the sample.

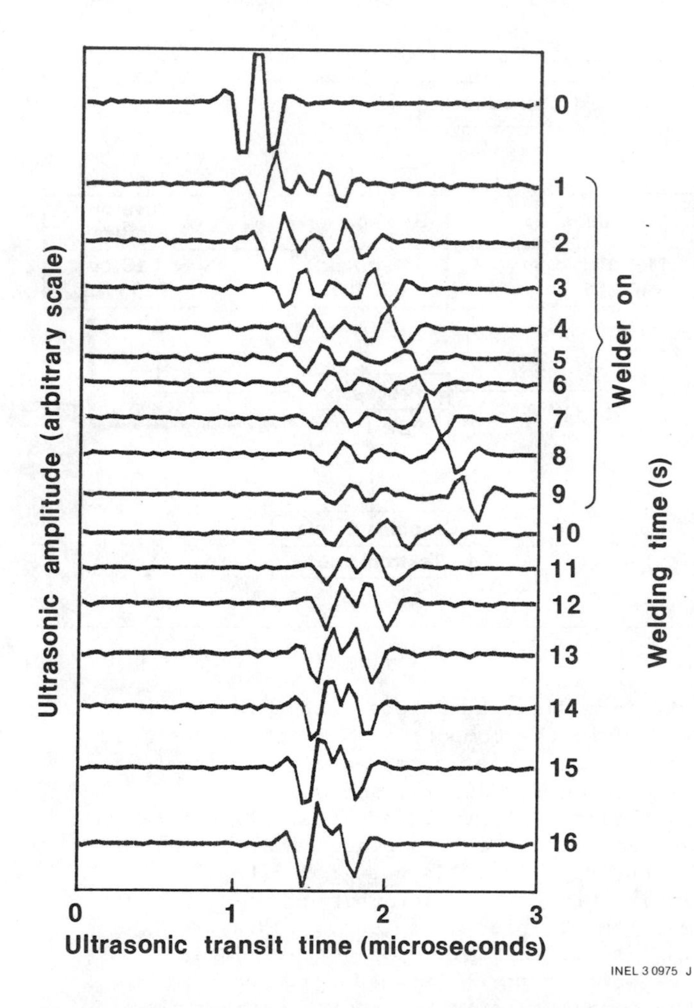

Fig. 5 - Ultrasonic A-scans of weld spot in 304 stainless steel

The first scan (time zero) was obtained before the welder was turned on and shows only the reflection from the upper solid/air surface of the sample. The next nine scans were taken while the welder was on and a molten weld pool formed and grew in size. As the pool started to form, the single reflected signal split into two reflected pulses which separated further in time as the pool grew in depth. After the welder was turned off, the two reflected pulses coalesced into a single signal as the pool solidified and cooled. The final reflected pulses are distorted because of the nonplanar surface of the weld spot and disturbance of the beam by the weld zone microstructure. Note also that in addition to separating in time during welding, both reflected pulses moved to the right (Fig. 5), corresponding to longer time delays. This effect is caused by the high temperatures present in the sample; ultrasonic wave velocities are known to decrease with increasing temperature,[5] thus producing the shift to longer time delays. This is discussed in more detail below under Analysis and Interpretation.

The two reflected pulses from the molten weld pool are more easily distinguishable in Fig. 6 which shows two of the A-scans of Fig. 5 plotted in different time and amplitude scales. The first A-scan of Fig. 6, which is the first A-scan of Fig. 5, shows the reflected pulse from the top surface of the sample before the welder was turned on; the other A-scan is the tenth scan of Fig. 5, which was taken just before the welder was turned off when the weld pool was deepest. The two pulses are easily distinguishable in the later scan and the time interval between them (approximately 1 μs) can be measured. There is a noticeable difference in shape between the two pulses: the first being similar (however attenuated) to the original solid/air reflected pulse, but the second distinctively different in both damping

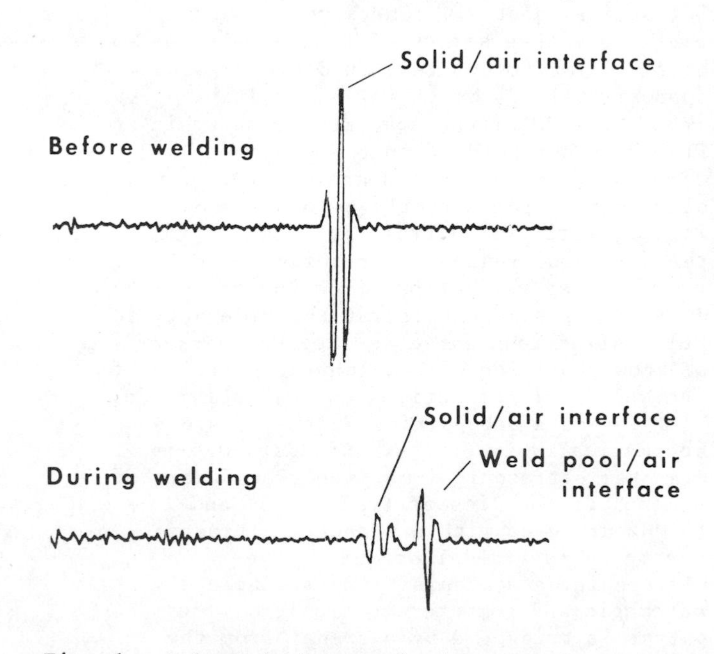

Fig. 6 - Selected ultrasonic A-scans of Fig. 2 at different amplitude and time scales (scans #0 and #9)

(number of peaks in the pulse) and phase (largest peak negative instead of positive). This implies that a pattern recognition technique based on these features might be successful in discriminating between the various signals received in an actual weld monitoring application, if necessary.

In the later scan of Fig. 6, an area of low level signal can be seen ahead in time of the two main reflecting pulses, indicating the presence of a third interface in the weld pool region. The interpretation of the ultrasonic data and the identification of the three interfaces are discussed in the next section.

ANALYSIS AND INTERPRETATION

An understanding and interpretation of the experimental data require calculation of the temperature gradients around the weld pool and of the effects of the gradients on the sound field. The gradients are calculated by numerically solving the heat flow equations. A numerical ray-tracing program is then used to determine the effects of the gradient.

TEMPERATURE GRADIENT CALCULATIONS - The thermal-hydraulics code STEALTH[13] is used to calculate the temperature gradients as a function of time. The problem is modeled by a fixed weld-pool boundary which is at the melting temperature of the steel. The heat flow from this boundary into the solid steel is calculated as a function of time, and from this the temperature distribution in the solid steel is calculated.

The mesh used in this calculation is shown in Fig. 7. (Note: Figs. 7 and 8 are oriented upside down relative to all other figures.) The horizontal axis is the radial distance from the centerline of the weld pool and the vertical axis is the distance from the top of the weld pool. The left boundary is the axis of symmetry for this axially symmetric model. The curved portion of the mesh boundary represents the liquid/solid interface which is at the melting temperature (1400°C) of 304 stainless steel. The bottom horizontal boundary, which is actually adjacent to room temperature air in the experiment, is modeled as an adiabatic boundary. The remaining boundaries are in water at 20°C and are isothermal.

The STEALTH code requires that the material properties and the initial conditions of the mesh be input. The properties are those of 304 stainless steel as given in References 14 and 15 and Table 1. The entire solid steel volume is assumed to be at 20°C at the start of the problem.

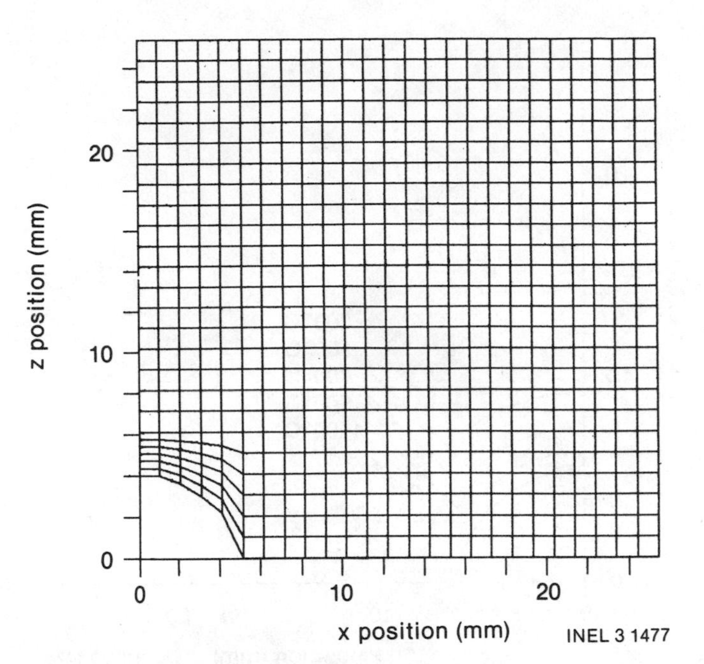

Fig. 7 - Mesh used in numerical calculation of temperature distribution

Table 1 - 304 Stainless Steel Properties

Property	Value
Density	
Solid	8.0 g/cm^3
Liquid	6.9 g/cm^3
Bulk Modulus	1.63×10^{12} dyne/cm^2
Shear Modulus	7.31×10^{11} dyne/cm^2
Heat Capacity	4.0×10^7 erg/cm^3-°C
Thermal Conductivity	1.62×10^5 erg/cm-°C

Given the initial conditions, the code calculates the temperature at the center of each mesh quadrilateral and the heat flux through each edge of each quadrilateral. This information is recalculated at time intervals of 0.01 s using the information from the previous cycle.

The results of the calculation can be plotted or printed. Fig. 8 shows the isotherms around the weld pool at 5.0 s after the start of the problem. The five contour levels correspond to 1100, 900, 700, 500, and 300°C moving away from the weld pool. The calculated temperatures at the center of each mesh cell are then used as input to the ray-tracing program described next.

RAY TRACING - The ray-tracing program calculates the effects of the temperature distribution on the rays of the sound field. In general, two effects occur: Changes in the sound speed with temperature and the consequent refraction of the sound

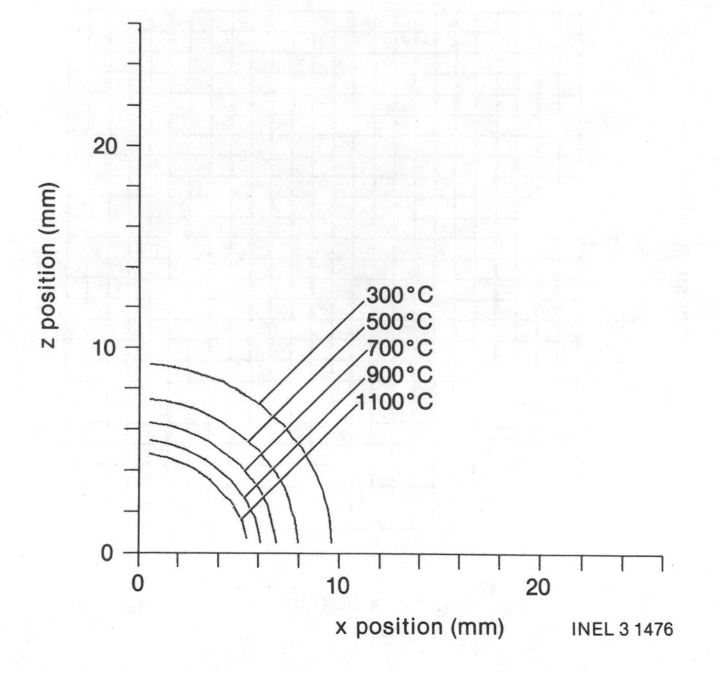

Fig. 8 - Isotherms around a 4 mm deep weld at 5.0 s

beam due to the sound-speed changes. The first affects the timing of the returning echoes and the second affects the apparent lateral position of the reflectors.

The sound field is modeled by a point source of sound at the bottom surface of the part on the centerline of the beam of the actual transducer. Rays are traced from this point into the material at any angle, accounting for the variation of the sound speed due to the temperature distribution in the part.

The material is divided up into a large number of thin layers. Within each layer the sound is assumed to travel in a straight line at a speed determined by the temperature at the center of the layer on the line of the ray. The time for the ray to traverse the layer is then calculated from the sound speed, thickness, and ray direction.

The refraction of the rays at the boundary between layers is shown in Fig. 9. A linear least-squares fit in two dimensions is made to the temperature distribution using the four temperature values nearest the intersection of the ray and the layer boundary. These values are taken from the output of the STEALTH program as described above. The expression for the temperature distribution in the neighborhood of the ray/boundary intersection is then

$$T = ax + by + c$$

where a, b, and c are the coefficients derived in the fitting process and x and y are the Cartesian coordinates. From this an

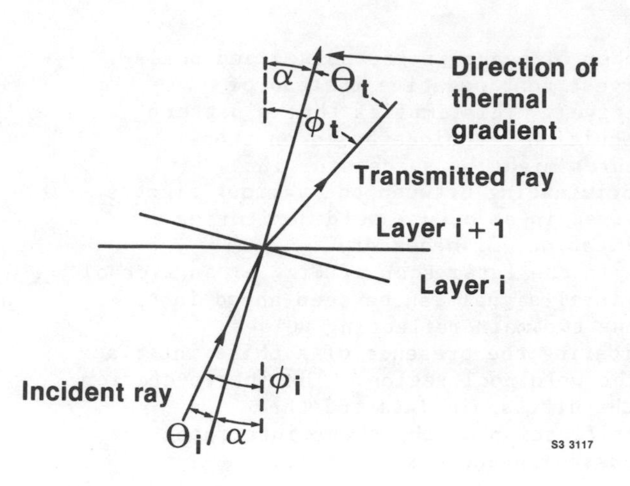

Fig. 9 - Refraction of a ray at a boundary between layers

expression for the gradient is found to be

$$\vec{\nabla}T = a\hat{i} + b\hat{j}$$

The angle of the temperature gradient relative to the axis is

$$\alpha = \mathrm{Tan}^{-1}\frac{a}{b}$$

The direction of the gradient is in the same direction as the normal to a plane surface, shown in Fig. 9. This plane surface is the boundary between the two regions of different temperature in this numerical model. It is also then the dividing line between two regions with different sound speeds. Thus the incident and refracted angles used in Snell's law must be calculated relative to the normal to this plane or relative to the direction of the temperature gradient.

The incident ray is assumed to be at angle ϕ relative to the y axis as shown in Fig. 9. The angle relative to the surface normal as required by Snell's law to calculate the refracted angle in the next layer is:

$$\Theta_i = \phi_i - \alpha$$

Using Snell's law the refracted angle relative to the surface normal can be found

$$\sin\Theta_t = \frac{V_t}{V_i}\sin\Theta_i$$

where V_i is the sound speed in the incident layer and V_t is the sound speed in the next layer. Finally the angle relative to the y axis in the next layer is calculated.

$$\phi_t = \Theta_t + \alpha$$

The sound speeds are calculated from the temperature at the center of the layer along the line of the incident ray using a straight-line fit to the measurements of Kurz and Lux for X5 CrNi 18/9 steel.[5] These data are assumed to approximate that for 304 stainless steel.

For rays entering the liquid phase, the same procedure is followed except that the temperature is assumed to be constant in the weld pool. This temperature is chosen to be approximately the average temperature of the pool or 1900°C. By extrapolating the data of Kurz and Lux[5] to 1900°C, the velocity in the liquid is found to be 3.6 mm/s. The path of any ray entering the liquid is extended 0.7 mm above the level surface of the part to account for the meniscus formed above the pool due to the expansion of the steel upon melting.

The calculation starts by assuming a given position of the point source and of the initial ray angle. The ray is then traced through successive layers until it reaches the top surface of the steel block. The time for the ray to reach any point in the block may be determined and from that the round-trip time for an echo returning from that point calculated. The relative time from different reflectors can then be determined and compared to experimental values.

RESULTS - Fig. 10 shows three rays traced through the steel. The center ray is directed normally to the surface of the part, travels to the weld pool (represented by the dotted line), into the pool, and finally to the top surface. The actual sound field would be partially reflected at the positions of the side-drilled holes, at the solid/liquid interface, and at the top surface. The other rays are at the smallest angle such that the rays will just miss the weld pool and intersect the top surface after traveling only through the solid.

The interpretation of the observed echoes is based on the times for the rays to reach each interface as calculated by the ray tracing program and are shown in Table 2. Fig. 11 shows the interpretation of the observed echoes. The times of the echoes are calculated relative to the time for the echo from the side-drilled hole that is second from the top. These times are then compared to those observed in the experiment and the correlation that is shown in Table 2 is obtained.

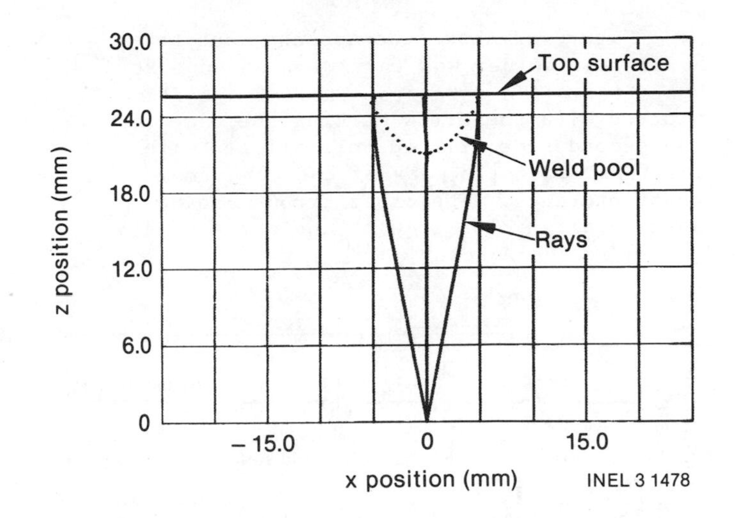

Fig. 10 - Three rays traced from the bottom of the part through and around the weld pool

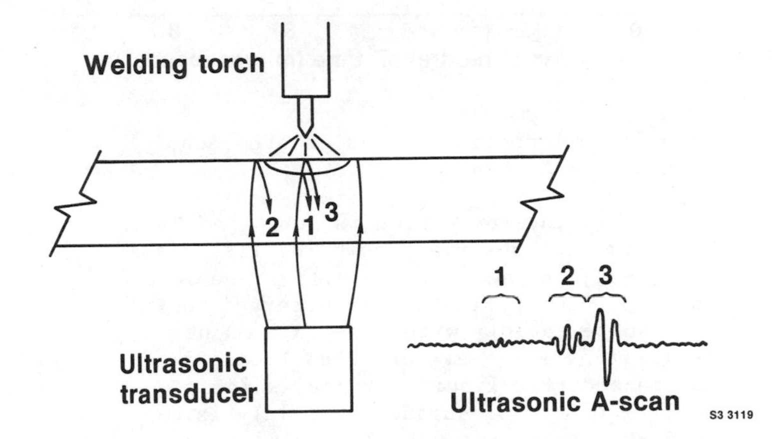

Fig. 11 - Interpretation of the three observed echoes

Table 2 - Calculated and observed echo times (μs)

Interface	Relative Time (calculated)	Relative Time (observed)
2nd side-drilled hole	0.0	0.0
1st side-drilled hole	1.83	1.85
Liquid/solid interface (1)	2.42	2.60
Solid/air interface (2)	4.16	4.10
Liquid/air interface (3)	5.04	4.75

Figure 12 shows two A-scans, one taken before the welder was turned on and one at 5 s after. The data from Table 2 are also plotted with the arrow marking the time for the second side-drilled hole aligned with the first positive peak of the echo from that interface. The other arrows can be seen to align closely with the corresponding echoes in the A scan.

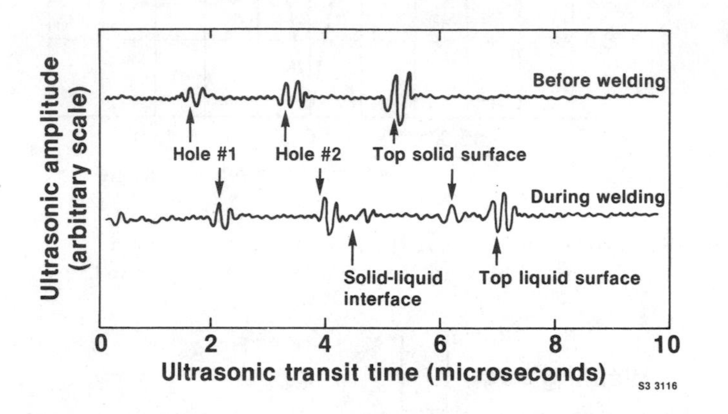

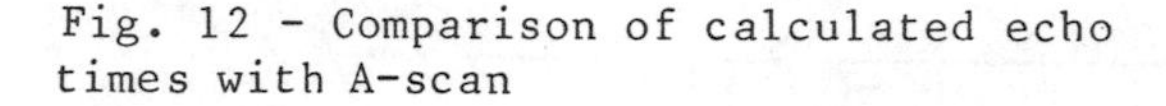

Fig. 12 - Comparison of calculated echo times with A-scan

This interpretation is supported by two experimental observations. The last signal received, as seen on an oscilloscope screen during the experiment, was observed to fluctuate rapidly with time in a manner suggesting a signal reflected from a disturbed free liquid surface, which a molten weld pool surface certainly is. Unfortunately, the data acquisition system can take data only about every second and this dynamic action does not appear in the digitized A scans. The small echo due to the liquid/solid interface was observed to move gradually earlier in time relative to the side-drilled hole, corresponding to the expected gradual increase in the depth of the pool.

These signals can be used to measure the weld-pool depth. Unfortunately, the signal from the liquid/solid interface is small and is not always present. However, the time difference between the other two echoes is approximately proportional to the pool depth. The time difference can be calculated relative to the time for the echo from the solid/liquid interface

$$t_3 = t_1 + 2D/V_\ell$$
$$t_2 = t_1 + 2D/V_s$$

where the subscripts refer to the rays shown in Fig. 11 and D is the pool depth. However, the time through the pool must be modified to account for the meniscus. For a pool depth of 4 mm, the meniscus has been calculated to be 0.7 mm high. The path length through the liquid is then increased in proportion to the pool depth and the time t_3 is recalculated:

$$\Delta D = \frac{0.7}{4} D$$

$$t_3 = t_1 + \frac{2(D + \Delta D)}{V}$$

The time difference and the pool depth can be calculated.

$$t_3 - t_2 = 2D \, \frac{1.175 V_s - V_\ell}{V_s V_\ell}$$

$$D = \frac{V_s V_\ell}{2(1.175 V_s - V_\ell)} \Delta t$$

$$= 3.95 \, \Delta t$$

where the average value for the speed V_s = 5.0 mm/μs around the pool is determined from the ray tracing program and the average value through the pool is the same as estimated above (V_ℓ = 3.6 mm/μs).

The results from three weld pools of different depths, made by using three different welder power inputs, are summarized in Fig. 13. The A-scans recorded just before the welder was turned off are shown along with a photomicrograph of the solidified fusion zone for each weld pool. (Note that it is much easier to identify the position of the reflected pulses when the complete series of A-scans can be seen and compared.) A plot of ultrasonic transit times versus weld pool depths for the data of Fig. 13 is shown in Fig. 14. The result is a linear plot with a slope of 4.0 compared to 3.95 calculated above.

SUMMARY AND CONCLUSIONS

We have demonstrated that molten weld pools in structural steels can be detected ultrasonically and the data understood, both qualitatively and quantitatively, in terms of the weld pool configuration and the surrounding temperature gradients. Using longitudinal ultrasonic waves, information about weld pool depth, or penetration, for a simple weld configuration can be obtained during welding which might be used for closed-loop process control.

Before practical application of the techniques on realistic welding processes, however, additional work needs to be done in several areas. The high temperature environment near a weld zone presents both hardware and data interpretation problems. The hardware problems are associated with the process of coupling ultrasonic energy into a hot test piece. These problems should be solvable by the use of cooled ultrasonic search heads, momentary contact

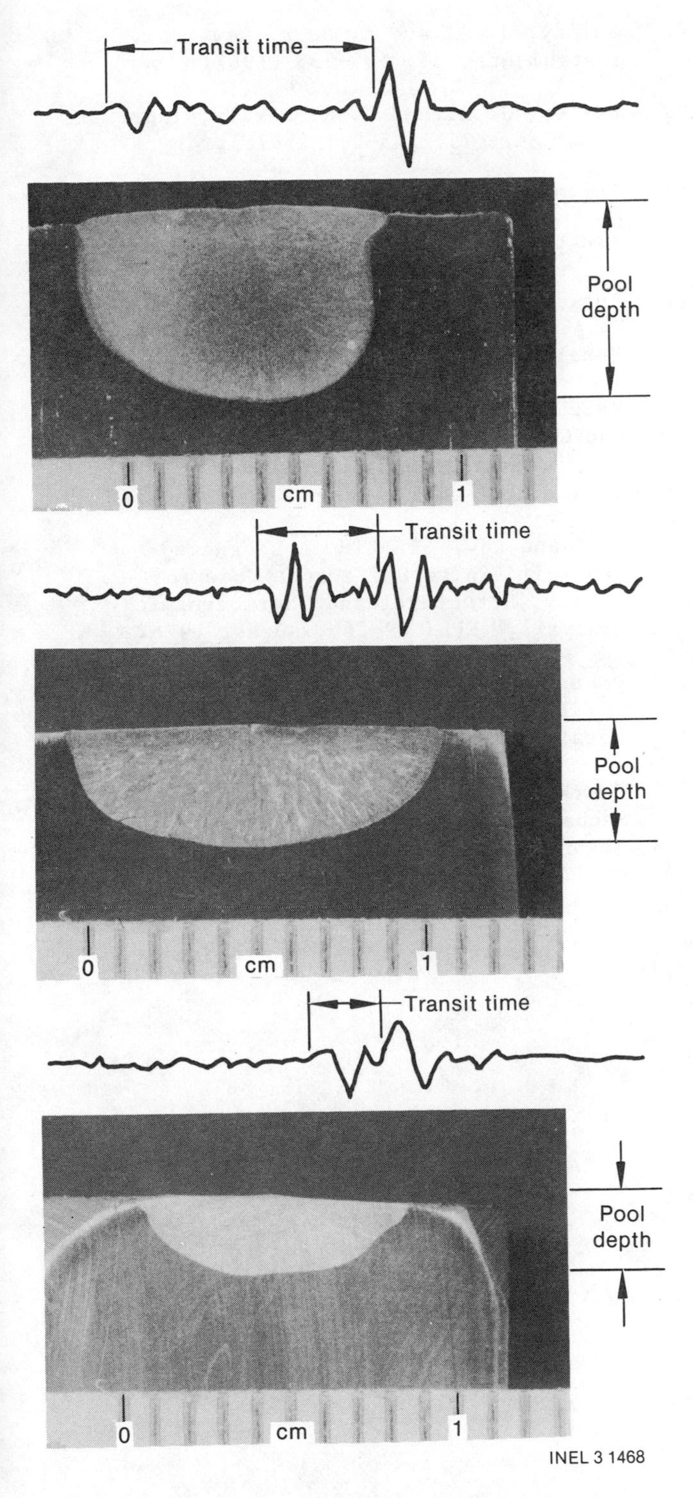

Fig. 13 - Comparison of three weld pools of different depths

techniques, or noncontact techniques such as electromagnetic acoustic transducers or laser techniques for generation and detection of ultrasonic waves. For welding geometries more complex than the stationary flat plate weld spots considered in this study, a more detailed knowledge of the effects of temperature gradients on wave

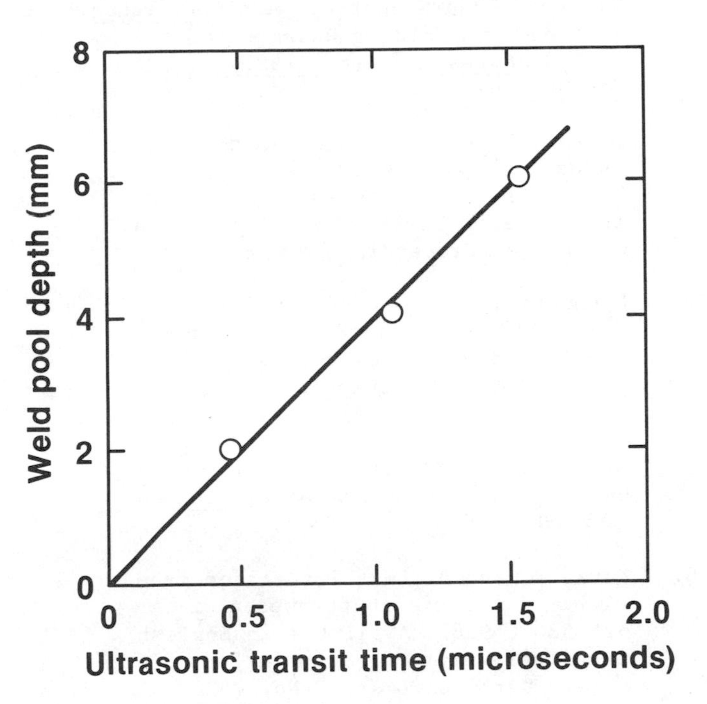

Fig. 14 - Correlation of weld-pool depth and time difference between echoes

propagation will be necessary. Realistic weld configurations will also likely require the use of ultrasonic shear waves entering from the top surface of the structure being welded, and an investigation of weld zone detection by shear waves will be necessary.

The fact that the ultrasonic signals reflected from liquid/solid interfaces of molten weld pools are quite small will likely require the use of sophisticated signal processing techniques for detection and analysis of the data. Development of such techniques remains to be done. Finally, much work must be done to develop algorithms for specific process control applications, i.e., exactly how must the ultrasonic data be acquired, interpreted in terms of weld zone configuration, and converted to an optimum process control signal.

ACKNOWLEDGMENT

This work was supported by the U.S. Department of Energy under DOE Contract No. DE-AC06-76ID01570.

REFERENCES

1. Vilkas, E. P., Welding Journal, 45, 410-16 (1966).

2. Smartt, H. B. and J. F. Key, "An Investigation of Factors Controlling GTA Weld Bead Geometry," ASM Conference on Trends in Welding Research in the United States, November 1981, New Orleans, LA.

3. Parker, R. L. , "Physics in the Steel Industry," pp. 254-271, F. C. Schwerer (ed.), (1982) published by American Institute of Physics, New York, as AIP Conference Proceedings No. 84.

4. Lynnworth, L. C. and E. H. Carnevale, "Proceedings of the Fifth International Conference on Nondestructive Testing," pp. 300-307, The Queen's Printer, Ottawa, Canada (1969).

5. Kurz, W. and B. Lux, Berg-und Huttenmannische Monatshefte, 114, 123-130 (1969) (in German).

6. Kurz, W. and B. Lux, Translated from Archiv f. d. Eisenhuttenwesen, 39, 521-530 (1968), Available in English from Henry Brutcher Translations, ASM, Metals Park, Ohio as HB No. 7528.

7. Kurz, W. and B. Lux, Zeits. Metallkunde, 57, 70-73 (1968), (in German).

8. Bailey, J. A. and A. Dula, Rev. Sci. Instruments, 38, 535-538 (1967).

9. Bailey, J. A. and J. R. Davila, Appl. Sci. Res., 25, 245-261 (1971).

10. Bailey, J. A. and J. R. Davila, Journal of the Institute of Metals, 97, 248-251 (1969).

11. Jeskey, G. V., L. C. Lynnworth, and K. A. Fowler, AFS International Cast Metals Journal, 2, 26-30 (1977).

12. Katz, J. M., "Ultrasonic Measurement and Control of Weld Penetration," M.S. thesis, Massachusetts Institute of Technology, 1982.

13. Hoffmann, R.,"STEALTH, A Language Explicit Finite-Difference Code for Solids, Structural and Thermodynamic Analysis," EPRI NP-260, August 1976.

14. "Metals Handbook," 9th Edition, Vol. 3, pp. 34-35, American Society of Metals, Metals Park, Ohio (1980).

15. "Marks' Standard Handbook for Mechanical Engineers," p. 5-5, Theodore Baumeister (ed.), McGraw-Hill, New York (1978).

ULTRASONIC MEASUREMENT OF SOLID/LIQUID INTERFACE POSITION DURING SOLIDIFICATION AND MELTING OF IRON AND STEEL

R. L. Parker
Metallurgy Division
Center for Materials Science
National Bureau of Standards
Washington, D.C. 20234

ABSTRACT

The solidification and melting of iron and stainless steel have been studied using a pulse-echo ultrasonic flaw detector, with longitudinal waves between 1 and 10 MHz. The change in acoustic impedance at the solid/liquid interface causes a portion of the beam energy to be reflected. A transducer contacts the cold end of a 3/4" by 5" cylindrical specimen. A Bridgman gradient furnace employing RF heating of a graphite susceptor provides unidirectional melting/solidification. It has been possible to follow both melting and freezing by varying the position of the RF heating coil and photographing the corresponding motion of the S/L echo on the oscilloscope. Some of the causes of the relatively weak echo observed, and of the high beam attenuation, are described, along with some experiments to improve signal/noise in this system. The aim, in part, is to provide a reliable, real-time, in-situ and nondestructive measurement of skin thickness in the continuous casting of steel, to help prevent breakouts.

BASIS OF EFFECT

Figure 1 illustrates the principle of ultrasonic flaw detection. Such flaw detectors generate a short burst of ultrasound, using an electrical pulse fed into a piezoelectric transducer. The ultrasound, generally in the frequency range 1-10 MHz, is then directed into the material to be tested. Echoes resulting from reflection of the sound by defects return to the sending transducer, which then acts as a receiver, converting them to electrical pulses. Appropriate circuitry then displays both the original pulse and the echoes on an oscilloscope screen, where the time, and consequently the distance, between the original pulse and echo can be determined.

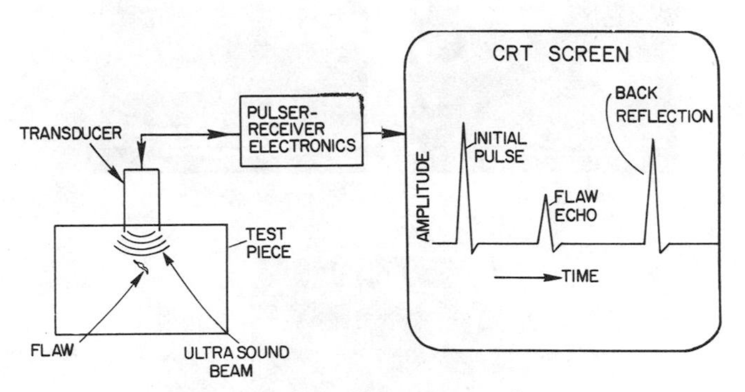

Fig. 1. Ultrasonic flaw detection

The solid-liquid interface in a melting or freezing metal should also produce a reflection, since for most metals both longitudinal sound velocity and density are lower in the liquid than in the solid. Their product, called the acoustic impedance, is about 13-15% less in the liquid, in a typical case. It can then be calculated that for normal incidence of longitudinal waves, about 10% of the pressure amplitude, or 1% of the energy, would be reflected from a planar solid-liquid interface in a pure metal. Previous work by Bailey [1] on mercury, Kurz et al [2] on steel, and Parker [3] on tin, has confirmed this prediction.

RESULTS FOR TIN

Figure 2 shows an example of the results [3] obtained for tin, using a vertical Bridgman gradient furnace, with a specimen 3/8" by 8" long in a graphite crucible. The transducer is coupled to the cool end. The frequency is 5 MHz. In figure 2c are shown multiple exposures with pictures of the oscilloscope screen taken at one minute intervals with one second exposures, after quickly lowering the crucible 1/2" with respect to the furnace. The initial solid-liquid interface position is given by the left-most peak. The distance traveled by the interface in the first minute is greatest, with successive distance intervals reduced, as the interface moves 1/2 inch. Due to limitations on heat flow rates in the furnace, the interface cannot respond instantaneously to the crucible motion, and the interface position vs. time is shown to reflect approximately the exponential type return to a steady state expected from heat flow considerations. We have also made some tests from the hot end (at 2.25 MHz) by dipping the transducer - with a one inch protective heat-shield buffer rod - directly into the melt, with similar results. Bailey's [1] tests on Hg were also made from the liquid side.

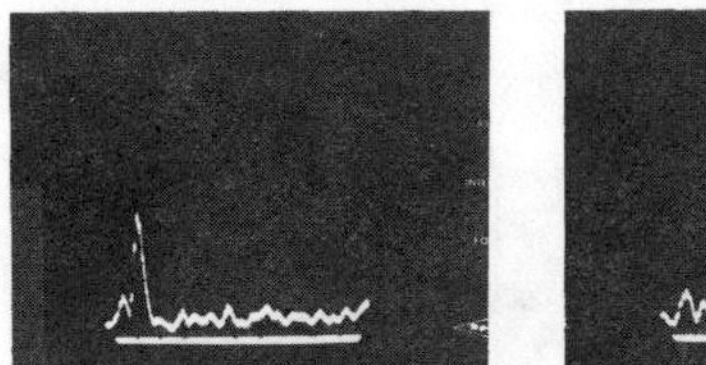

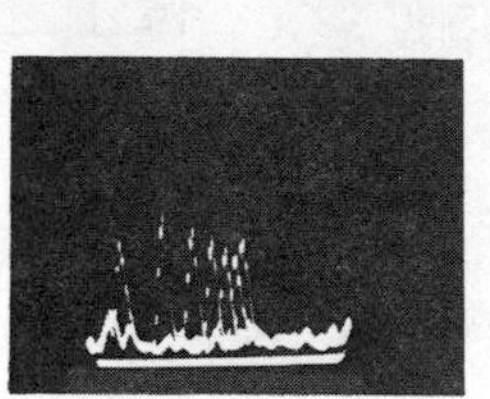
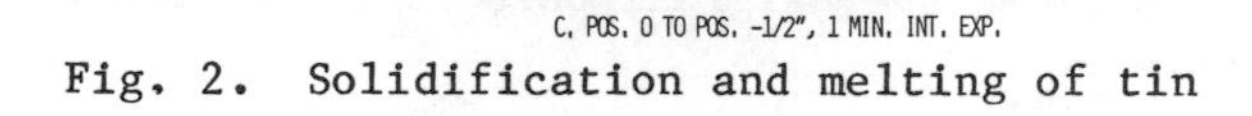

Fig. 2. Solidification and melting of tin

PRIOR TESTS ON STEEL

Examination of the oscilloscope photos given in Kurz et al [2] for low carbon steel at 1 MHz certainly shows echoes that indicate interface position and motion. They are not, however, very sharp. The beam was strongly attenuated, both by scattering and absorption in the bulk solid before and after reaching the interface, and also by the relatively low reflectivity of the interface as noted above.

SOME NBS TESTS ON SIGNAL ENHANCEMENT

Prior to our beginning work on molten steel, it seemed important to be able to improve the low signal/noise ratios in steel that Kurz et al [2] had observed. We have used computer techniques that were not available when Kurz's tests were made (in 1968). Figure 3 (performed at room temperature in 304-stainless steel at 10 MHz to simulate the high temperature scattering problem) shows that this can be done using the technique of signal averaging, as first shown by Kraus and Goebbels [4]. The "grass," or noise largely due to backward Rayleigh scattering from the steel grains, can obscure the identification of the desired backwall or interface echo (fig. 3A). This grass, while not time dependent, is however space dependent, and scanning the transducer, while storing the signals in a computer, can convert the grass to incoherent signals which tend to average out, while the desired backwall signal remains (arrows in figs. 3B, 3C for 5 and 32 scans, respectively). The equipment required to do this is not expensive - it includes a digital waveform recorder and a small Z-80 type computer.

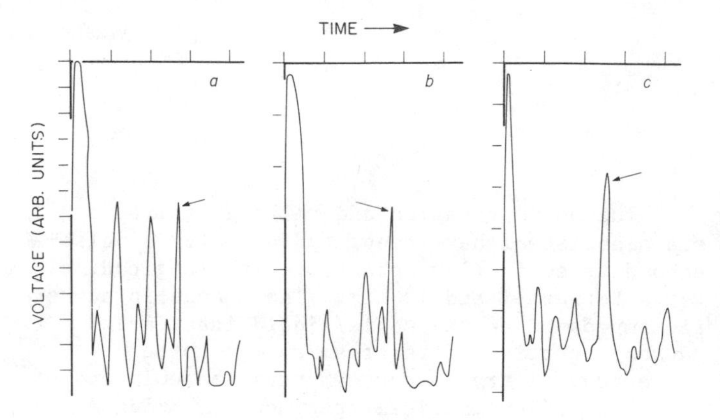

Fig. 3. Results of computer-aided digital signal averaging with strong Rayleigh backscattering in 304 stainless steel at 10 MHz, 1" thick specimen, grain size ASTM 4 (100 μm).
a. 1 scan b. 5 scans c. 32 scans

NBS TESTS ON HOT STEEL

In order to make ultrasonic measurements of the skin thickness in a steel continuous caster, four things are needed: (1) couple an ultrasonic signal to a hot, scaly, moving steel surface, (2) propagate it to the solid-liquid interface (mushy zone in the case of alloys), (3) obtain an echo from that interface and (4) identify the echo return in the presence of substantial noise from Rayleigh grain scattering. As noted above, computer averaging is effective on # (4). The work of Cole [5] and of Kawashima et al [6] on EMAT transducers using non-contact Lorentz-force generation of ultrasonic pulses in hot steel billets appears to have solved the first (coupling) problem.

To examine the ultrasonic propagation through very hot steel and reflection from the solid-liquid interface, an RF induction furnace, capable of melting steel in a unidirectional (Bridgman) geometry in an argon atmosphere, while coupled to a conventional transducer, has been set up and put into operation. Using 1 MHz longitudinal waves, we have made measurements of the solid-liquid interface location and motion in pure iron rods 3/4" OD by 5" long, and in 304 stainless rods 1 1/4" OD by 5" long, in the presence of substantial grain growth. Figure 4 shows the backwall reflection and S/L interface position as the RF coil is raised and lowered, for pure iron.

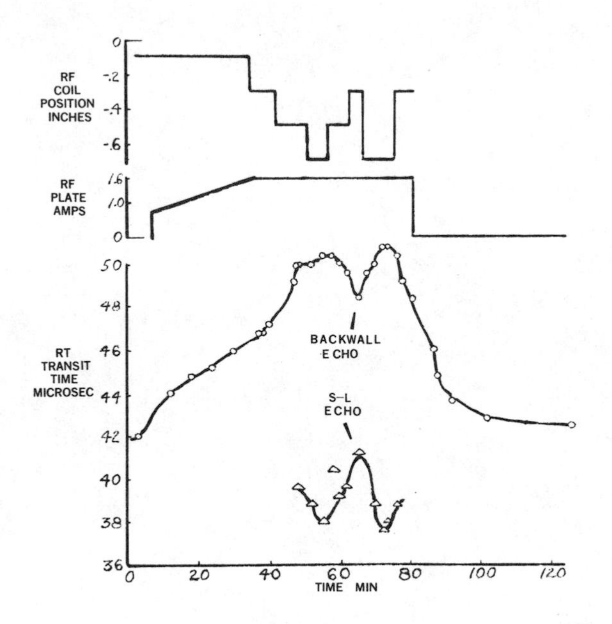

Fig. 4. Solidification and melting of iron.

CONCLUSION

In order to develop a reliable pulse-echo system for determing skin-thickness in continuous casting of steel, it is essential to determine the degree of transmission through bulk, hot steel and the degree of reflection from the solid-liquid interface. These quantities are frequency dependent and can be time-dependent in the case of grain growth and interface motion. An apparatus has been constructed to do this, and measurements are underway on iron and steel.

REFERENCES

1. Bailey, J. A., and J. R. Davila, "An Investigation of the Solidification of Mercury by an Acoustic Technique," J. Inst. Metals, Vol. 97, p. 248 (1969).

2. Kurz, W., and B. Lux, "Ortung der Erstarrungsfront in Stahl durch Ultraschall," Archiv. fur Eisenhuttenwesen, V. 39, p. 521 (1968). (Available in English from Henry Brutcher Translations, ASM, Metals Park, Ohio as H. B. #7528).

3. Parker, R. L., "Ultrasonic Measurement of Solid/Liquid Interface Position during Solidification and Melting of Metals," AIP Conference Proceedings No. 84, (Physics in the Steel Industry), New York, 1982.

4. Kraus, S., and K. Goebbels, in "Ultrasonic Materials Characterization," NBS Spec. Pub. 596, ed. H. Berger and M. Linzer, 1978, p. 551 (1980).

5. Cole, P. T., Ultrasonics 16, 151 (1978). See also, G. J. Parkinson and D. M. Wilson, British J. of NDT, 19, 178 (1977).

6. Kawashima, K. et al, Nippon Steel Technical Report No. 15, June, 1980.

ACOUSTIC EMISSION STUDIES OF ELECTRON BEAM MELTING AND RAPID SOLIDIFICATION

R. B. Clough
Center for Materials Science
National Bureau of Standards
Washington, D.C. 20234

H. N. G. Wadley
Center for Materials Science
National Bureau of Standards
Washington, D.C. 20234

F. S. Biancaniello
Center for Materials Science
National Bureau of Standards
Washington, D.C. 20234

ABSTRACT

Acoustic emission is well suited for monitoring electron beam induced melting and rapid solidification, providing a real time. volumetric survey of the processes. The present study examines the origin of acoustic emission during melting and rapid resolidification of Al and Al-4.5% Cu using heat flow theory and microstructure characterization. Specimens were heated to a steady state using a pulsed electron beam and acoustic emission measured during heating and subsequent cooling. The heat flux level was systematically increased so that both solid state heating and melting could be independently studied.

The electron beam, because of its relatively slow rise time, was found not to produce acoustic emission directly. The motion of the liquid-solid interface, for the same reason, also fails to generate detectable acoustic emission. Acoustic emissions were found to be generated by plastic deformation and sometimes crack growth. The emission was strongly influenced by the heat flux level (ultimate temperature), prior cold work, alloying, and heat treatment. A thermal stress model was used to develop an understanding of plastic deformation/fracture initiation and propagation during the heating and cooling cycle.

INTRODUCTION

Directed high energy sources (lasers and electron beams) are currently being evaluated for rapid surface melting and resolidification of a range of engineering alloys (1). In these processes, the bulk substrate, in intimate contact with an electron beam melted surface layer acts as the quenching medium. This results in high liquid-solid interface velocities and short solidification times during cooling after electron beam heating has terminated (2,3). These rapidly solidified layers exhibit enhanced wear and corrosion resistance arising, it is believed, from improved chemical homogeneity and the fine scale of the microstructure (1). These materials offer promise for applications such as bearing surfaces, cutting tool faces and corrosion-resistant parts.

At present, critical process variables such as liquid-solid interface velocity or the cooling rate can only be deduced from the one- and two-dimensional heat flow models of melting and resolidification. There are no _in situ_ methods of monitoring the important process variables. If measured, they could provide feedback/feedforward information for in-process control of microstructure and the elimination of flaws: a real-time monitoring and diagnostic technique is needed.

Acoustic emission (AE) is an attractive potential candidate for such an application. However, the origin of acoustic emission during electron beam melting/solidification is not well established. When AE is correlated with heat flow theory and microstructural examination, considerable light can be shed on operative defect mechanisms, such as plastic deformation or crack initiation and propagation. This paper illustrates the application of this method to pure Al and Al-4.5 wt.% Cu.

Heat Flow Theory. Heat flow for finite diameter electron beams can be modeled using two-dimensional heat flow theory. Results of calculations for two-dimensional heat flow for rapid surface solidification in pure Al are given in Ref.(3). Similar calculations for Al-4.5% Cu are given in Ref.(8,9). The results indicate a two-dimensional steady-state heat flow is ultimately attained when the heat absorbed over the circular region exposed to the beam is exactly offset by heat conduction into the substrate interior. At steady state, the maximum temperature is located at the surface at the center of the beam spot (T(0,0)); the thermal profile then remains fixed (with only static stresses) from which solidification proceeds.

Two dimensionless variables are used

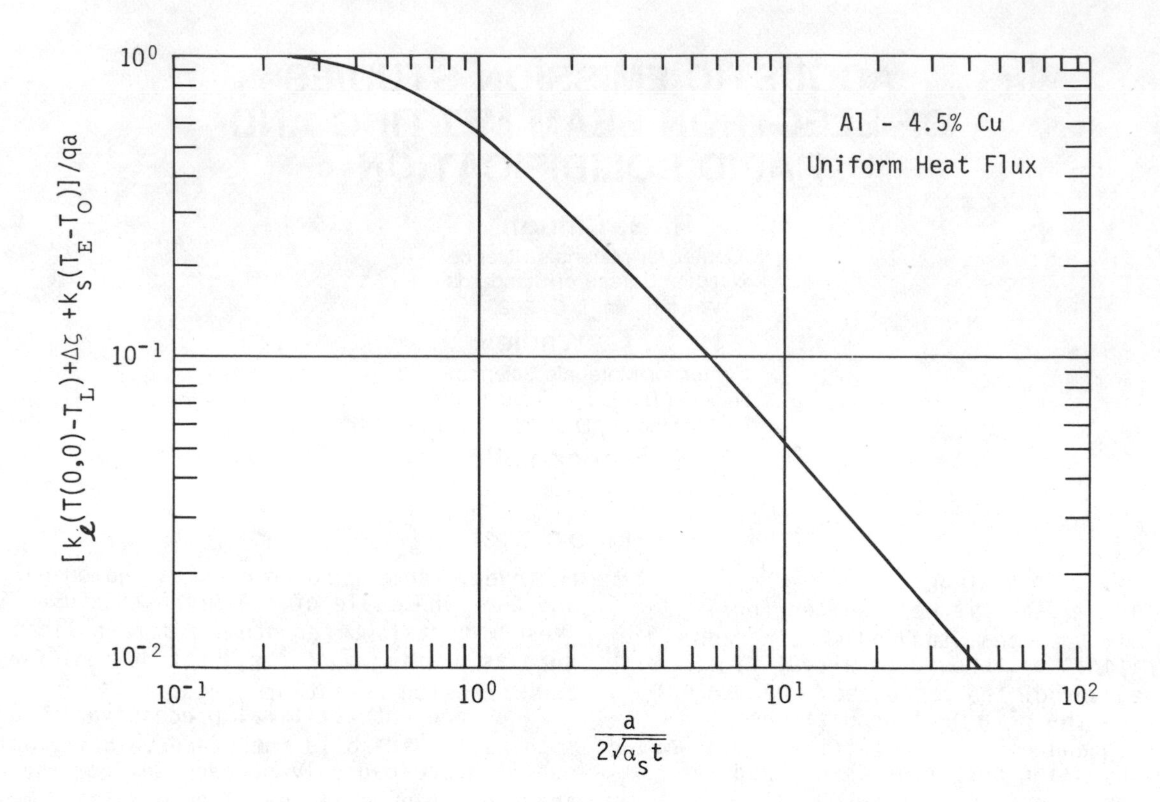

FIG.1. Temperature at the center of the circular region during melting of a semi-infinite Al-4.5 wt % Cu alloy substrate as a function of uniform absorbed heat flux (q), radius of the circular region (a) and time (t) (8).

Table 1

Absorbed Uniform Heat Flux to Reach Critical Temperatures (2,3,8,9)

Phenomenon	Temperature °C		Absorbed heat flux*, qa (W/m)	
	Al	Al-4.5% Cu	Al	Al-4.5% Cu
Melting initiated	660	536	1.4×10^5	9.6×10^4
Melting completed	660	647	2.3×10^5	1.5×10^5
Vaporization	2447	2368	4.7×10^5	3.4×10^5

*Assumes initial temperature to be 24°C.

to describe heat flow in the solid state:

$$\text{Dimensionless temperature} = k_s(T(0,0)-T_o)/qa \qquad (1)$$
$$\text{Dimensionless spot size} = a/\sqrt{4\alpha_s t}$$

where k_s is thermal conductivity of the solid, T_o is the initial substrate temperature, q and a are the absorbed heat flux and electron beam radius, α_s is the thermal diffusion coefficient of the solid, and t is the time measured from the instant the beam was switched on. When the dimensionless temperature equals 1, steady state is established. If T(0,0) is equal to the melting point or solidus temperature (for an alloy), the product qa corresponds to that above which melting will initiate at the beam center. As qa is increased, melting will spread outward from the center of the spot to cover the entire circular area. This is reached when

$$(k_\ell(T(0,0)-T_\ell) + \Delta\zeta + k_s(T_s-T_o))/qa = 1$$

where subscripts ℓ and s stand for liquid and solid, respectively, and T_ℓ and T_s are the liquidus and solidus temperatures. $\Delta\zeta$ is a term accounting for heat conduction in the "mushy" region and is calculated to be 6.2×10^4 W/m in Al-4.5% Cu (8).

Figure 1 shows the relationship between dimensionless temperature, beam radius and time during heating in the alloy (8). (The corresponding curve for pure Al is given in Ref.(3)). Using these figures one can calculate the qa level at which melting initiates, that at which it fills the beam spot and that for vaporization to occur. (Table 1).

In designing the acoustic emission experiment, our first concern has been to allow sufficient time during heating for steady state temperature to be obtained so that a constant stress is reached before solidification commences. For a given beam radius a, the time to reach steady state can be taken as that at which $a/\sqrt{4\alpha_s t} < 0.5$, as can be seen in Fig. 1, or $t_{ss} > a^2/\alpha_s$. For both materials this is approximately 13 ms. Our use of a 77 ms pulse meets the requirement for attainment of steady state conditions.

Our second concern is that, during cooling, the measurement period should extend over the cooling time of the melted region. Using two-dimensional heat flow equations (3), we examined the thermal effects of a 77 ms pulse in pure Al heated to 650°C. Following the heat pulse, it cooled to 114°C within 20 ms, and to 45°C within 100 ms. We used a 77 ms measurement period which covers the majority of the cooling period.

Microstructure. Microstructure, temperature and thermal stress determine the defect mechanisms responsible for the acoustic emission. Microstructural investigation is thus crucial to an understanding of the origin of the acoustic emission.

An example from previous work (10) can illustrate this point. We consider the effect of increasing the product qa upon melt depth and microstructure for commercially pure 1100 aluminum and an Al-6.5% Cu alloy (Figs. 2-4). The melt depth increased with increasing qa; the alloy, because of its lower melting temperatures, had a greater depth of melt than the 1100 aluminum. A comparison between the calculated and experimentally measured melt depths was made (10). Agreement was good considering various assumptions of the heat flow model including constant thermophysical properties. The alloy had a less clear transition from unmelted to melted regions, exhibiting a "mushy" zone of partially melted material (Fig.3). It was possible to tentatively link the source of emission to crack production in the alloy, since much greater signals were obtained in the alloy, which cracked, than the pure Al, which did not crack. There were, however, shortcomings with these initial experiments which are addresed below.

Acoustic emission. In this phenomenon, stress waves are emitted during rapid defect motion or phase transformations. The elastic waves then propagate throughout the structure to be detected by remotely located transducers. For a recent review of the potential for process monitoring see Ref. 4. The technique has a real-time capability (limited only by the time for elastic waves to travel from the source to receiver, usually a few microseconds) and has been successfully applied to monitoring commercial welding (5,6,7). In the present work, we report the use of acoustic emission to monitor rapid solidification of electron beam surface melted pure Al and Al-4.5 wt.% Cu. Heat flow theory has been used to design controlled experiments in which shallow melts were formed at the surface of aluminum plates. After attainment of steady-state temperatures the electron beam was switched off, resulting in rapid solidification. Acoustic emission was measured separately during both heating and cooling since the origin of the emission was different in these stages. It is important to understand both since they jointly contribute to the instaneous emission during (continuous) surface modification treatments or welding. The effects of alloying, cold work, and heat treatment were systematically studied as a function of heat flux level (i.e. ultimate steady-state temperature). Post mortem metallography and heat flow theory were correlated with acoustic emission to make tentative identifications of the defect mechanisms accompanying heating and cooling.

The present study was undertaken to gain a better understanding of the relative contributions to the detected acoustic emission of dislocations and cracks and how this is affected by microstructure. High purity Al and a binary Al-4.5% Cu alloy were used, rather than the commercial materials used in a previous study because these are less complex and maybe better charac-

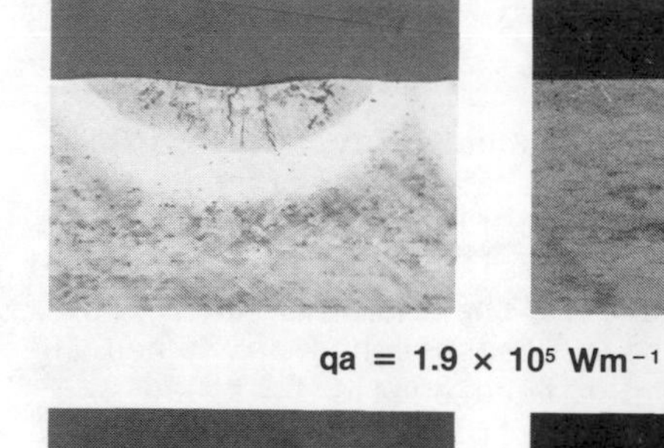

$qa = 1.9 \times 10^5\ Wm^{-1}$

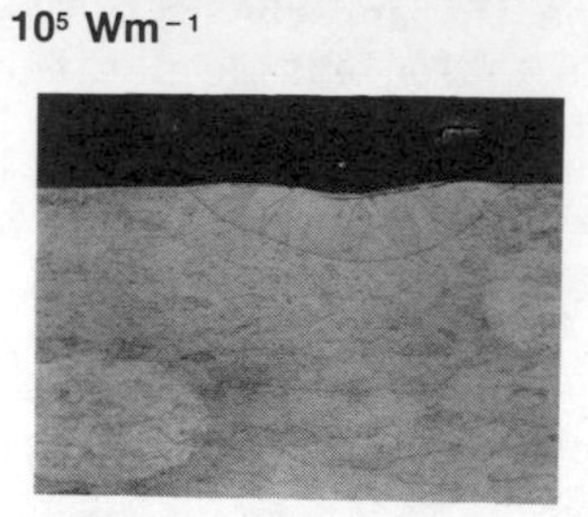

$qa = 2.7 \times 10^5\ Wm^{-1}$

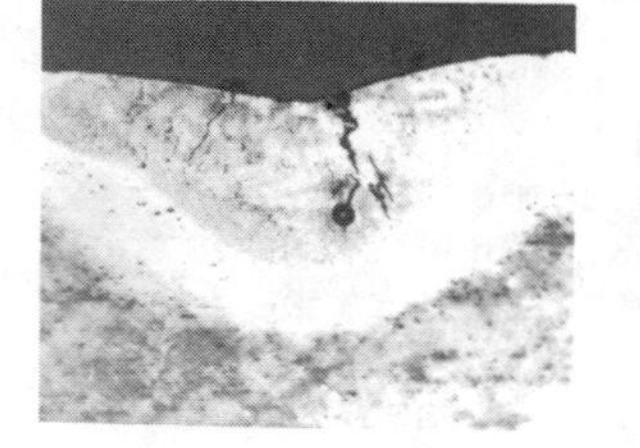

$qa = 3.3 \times 10^5\ Wm^{-1}$

FIG.2. Optical micrographs of 1100 Al and 2219 Al alloy (6.5% Cu) showing the effect of qa on the melt profile and ensuing microstructures in the "steady-state" region.

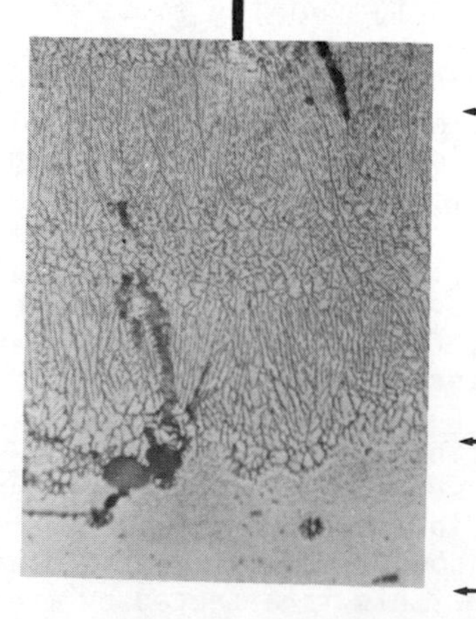

FIG.3. Optical micrographs of 2219 Al alloy (6.5 % Cu) melted at qa= 2.7×10^5 W /m showing detailed microstructure of the "mushy" zone at the bottom of the melt pool.

terized. In addition, previously published heat flow calculations (8) could be more rigorously applied. Also previous work (10) over a wider qa range (temperature range) encountered problems associated with vaporization. This complicated the analysis and has been excluded here. The measurement system has been improved reducing signal distortion and improving the signal-to-noise ratio, and a new specimen/ transducer holder design has improved the measurement reproducibility.

EXPERIMENTAL PROCEDURE

Materials. In this study, effects of substrate composition and condition were examined by using pure Al in an annealed and 20% cold-worked state and an Al-4.5 wt.% Cu alloy in two heat treated conditions.

Both materials were at least 99.99% pure and after casting were homogenized at 535°C for 1.5 hours, hot rolled from that temperature, solution treated at 535°C for 75 min. and ice water quenched. The alloy composition was verified to be 4.5$\pm$0.1 wt.% Cu (at four locations) by wet chemical analysis. The specimens were then machined to 0.4x2.5x2.5 cm squares from the annealed rolled plate . Half of the pure Al specimens were further cold rolled 20% in thickness. The alloy specimens were then given low temperature heat treatments: Group 1) aged to peak hardness (8 hours at 179 C) and Group 2) underaged (1 hour at 170 C). The final grain size was approximately 450 microns in the alloy and 350 microns in the pure Al (at the surface, whose 1 mm depths had recrystallized). The specimens were stored in liquid nitrogen to avoid the possibility of precipitation between stages of the above heat treatment and testing.

Electron Beam Operation. Figure 5 is a schematic diagram of the electron beam facility, together with the acoustic emission apparatus and associated instrumentation. The sample and holder were placed in the electron beam apparatus and evacuated to 10^{-5} Torr. The beam was accelerated across 22 kV and focused to 1 mm radius spot (with an approximately rectangular intensity profile). The absorbed heat flux was determined at each flux level by measuring the current flowing through a resistor between the specimen and ground. The error in qa was estimated to be approximately $\pm$ 15 %. A series of single 77 ms duration pulses of increasing qa level were applied to the surface of the samples. The beam and sample were stationary so that acoustic emission during heating was clearly separable from that of cooling.

Acoustic Emission Measurements. The acoustic emission measurement system is shown schematically in Fig. 5. A piezoelectric, 140 kHz resonant acoustic emission transducer was coupled to the back of the specimen with a low viscosity oil. To minimize ringing, which prolongs the period of emission and may cause obliteration of closely following signals, the specimen was covered on the underside with a plastic adhesive tape and held down by an O-ring. Spring loading insured constant coupling for improved measurement reproducibility. Bulk specimen heating was less than 5 degrees C per pulse, and since generally no more than eight pulses were applied to a sample, its bulk temperature rise was <40°C. An acoustic emission parameter analogous to energy was measured by summing the squared unbiased voltages of digitally recorded signals for periods of 77 ms during both heating and cooling.

RESULTS

A typical electron beam pulse and superimposed AE signal are shown in Fig. 6. Different AE signals had several common characteristics. There was a delay time after the beam was switched on before detectable emission occurred. This delay decreased with increasing qa (i.e. increasing temperature). Acoustic emission was then produced during heating. After the beam was switched off, there was a delay time prior to emission provided that T(0,0), the temperature at the center of the heated region, approached or exceeded melting (12). Then "cooling" acoustic emission occurred.

Acoustic emission on heating was relatively small for $T(0,0) < T_S$, the solidus temperature. Above T_S it increased dramatically (Figs. 7 and 8) for both materials. Acoustic emission on cooling was smaller than that produced during heating at temperatures below T_S. The "cooling" acoustic emission had three characteristic temperature (qa) regions indicated in Table 2 and Figures 7 and 8. In region I, the emission value increased with temperature (qa) up to a maximum value. It then decreased in region II to a vanishing value at the qa level at which melting was initiated. It then increased again in region III, during general melting.

Typical microstructures of the heated zones are shown in Figures 8-10. At the lower temperatures, corresponding to region I, all of the specimen surfaces showed coarse slip bands (Figs. 8A and 9A). The amount of deformation (number and width of slip bands) increased with increasing temperature (qa).

Although some recovery, as evidenced by subgrain boundary development, is seen at the lowest heat flux levels in pure Al (Fig. 8A), it predominanted in region II. Region II corresponds to temperatures in which recovery processes occurred, a typical example of which is shown in Fig. 9A for the Al-4.5 % Cu alloy.

No cracks were observed at any level of qa (temperature) in the pure Al samples (although the grain boundaries were thermally etched, especially at the higher temperatures, giving the appearance of cracks, Fig. 8B). Fig.8C shows a typical cross-section of a melt zone in pure Al. There are two grains visible, but no fractures.

In region III, where general melting

FIG.4. The effect of qa upon melt depth for 1100 and 2219 aluminum alloy. Melt depth is measured from the liquidus isotherm to the substrate surface.

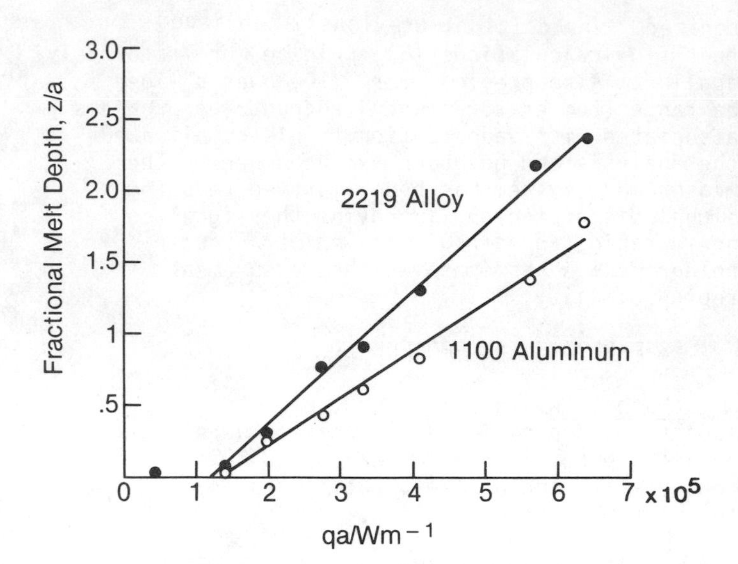

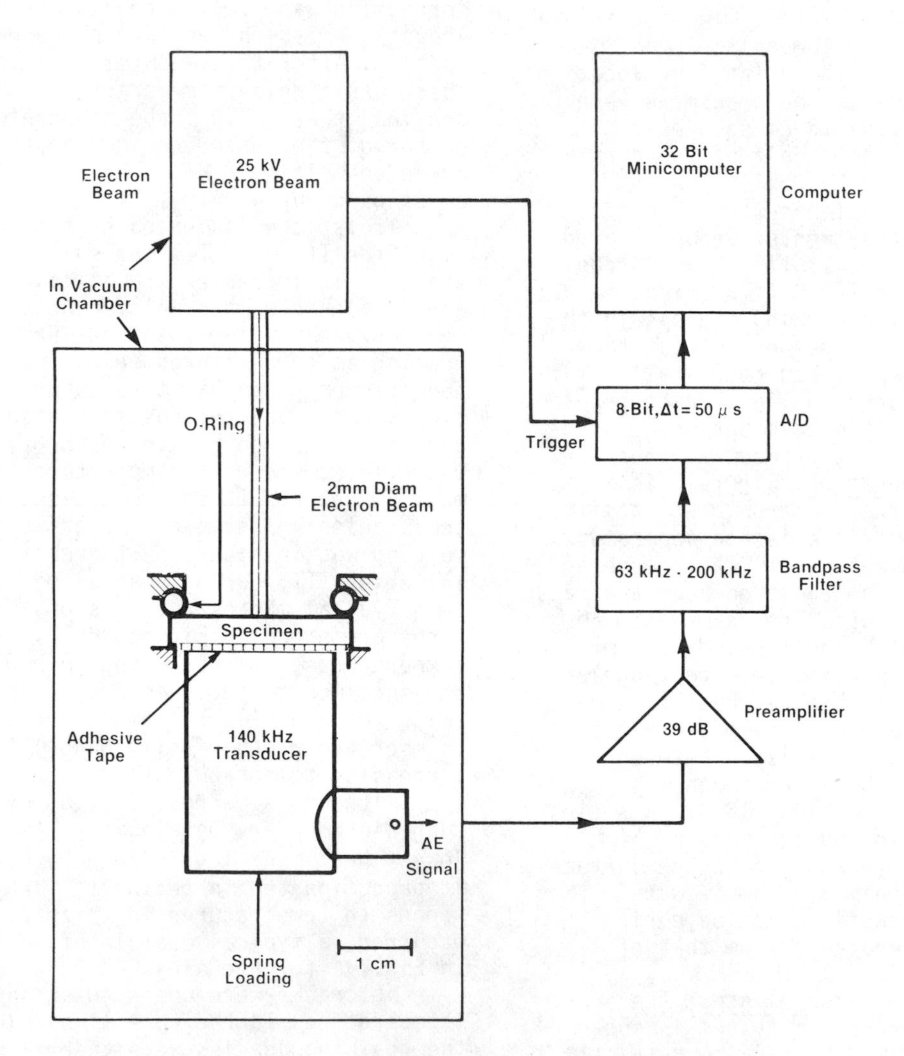

FIG.5. Schematic diagram of electron beam apparatus and acoustic emission instrumentation.

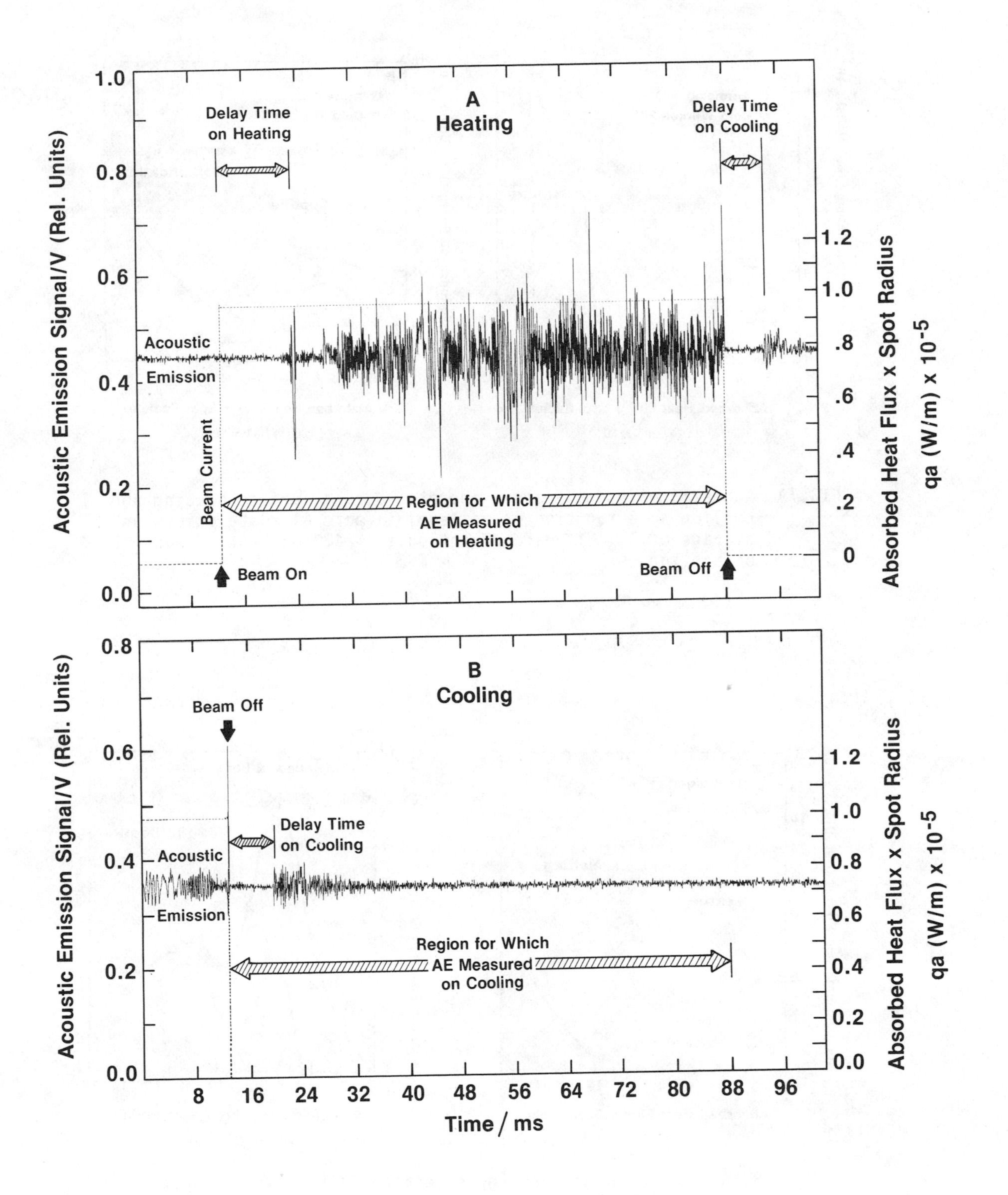

FIG.6. Typical acoustic emission signal produced by melting and resolidification of surface melt in peak-aged Al-4.5 %Cu alloy. (A) During heating at qa= $8x10^4$ W/m and (B) During cooling after beam is switched off.

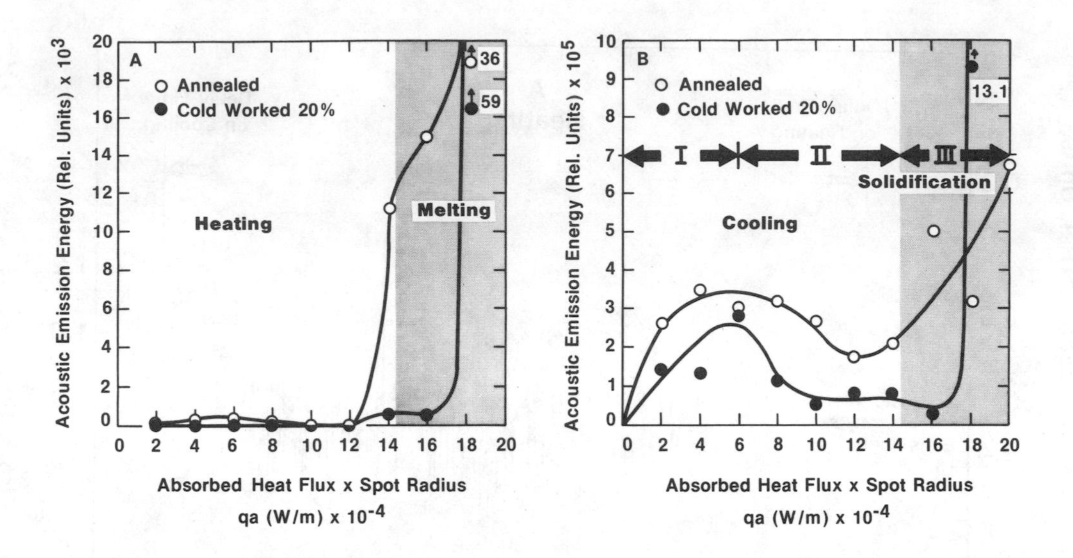

FIG.7A. Acoustic emission energy generated during surface heating and cooling as a function of qa level in pure Al. Data points an average of 8. Standard deviation is ± 43% on heating and ± 22 % on cooling.

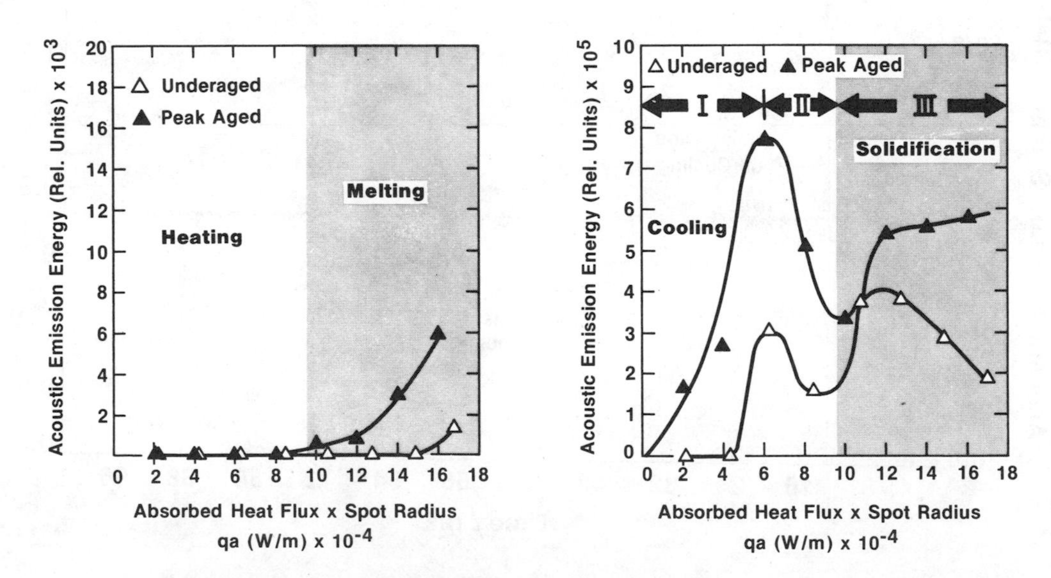

FIG.7B. Acoustic emission energy generated during surface heating and cooling at given qa level in Al-4.5% Cu alloy. Data points an average of 8. Standard deviation is ±43% on heating and ± 43% on cooling.

FIG.8A. Annealed pure Al. Optical micrograph on surface at center of irradiated area (qa=4×10^4 W/m). Surface polished prior to radiation; not etched. Notice recovery: sub-boundary development at center surrounded by coarse slip bands confined to grains.

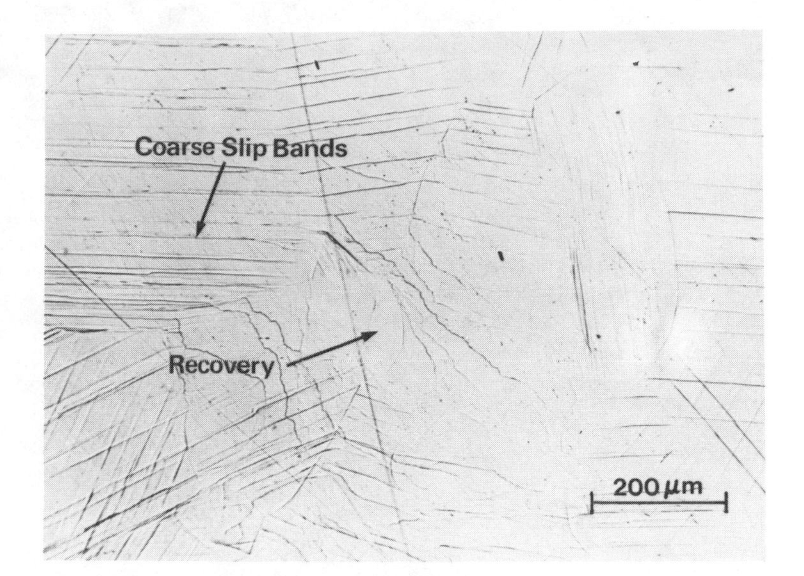

FIG.8B. Annealed pure Al. Optical micrograph of surface melt (qa=1.4×10^5 W/m); surface polished prior to irradiation; unetched. Gray crescent to right of melt zone is surface upheaval caused by melting. Note radial thermal grain boundary grooves in resolidified melt, frozen-in surface waves, recovery subboundaries, coarse slip.

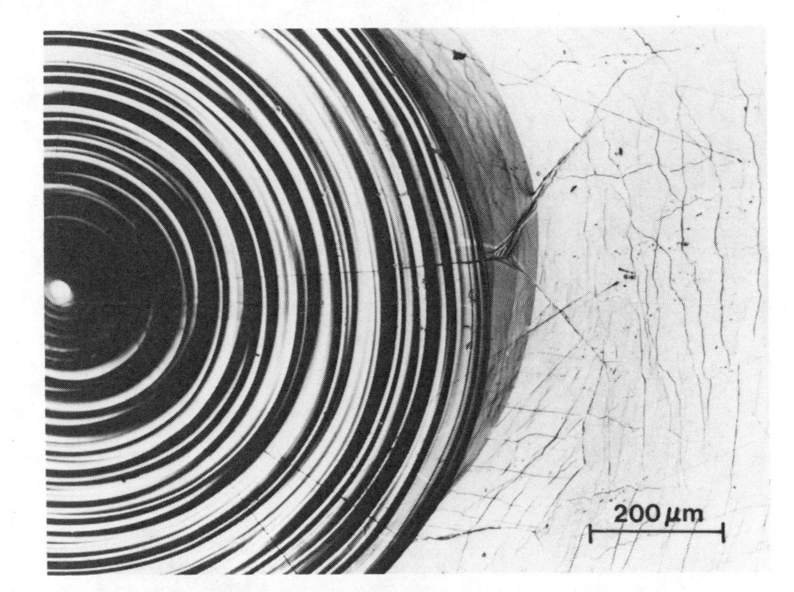

FIG.8C. Annealed pure Al. Optical micrograph section through surface melt (qa=1.4×10^5 W/m) similar to that above. Note that there are no cracks. Electropolished and Keller's etch.

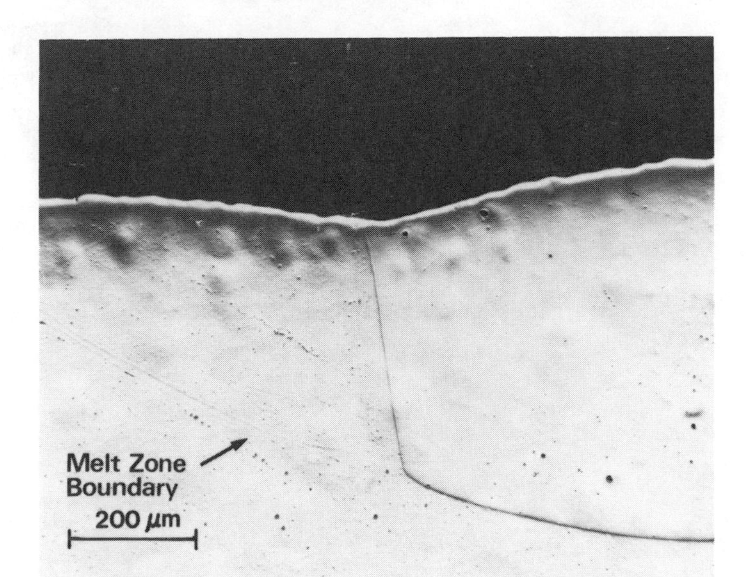

FIG.9A. Underaged Al-4.5% Cu. Optical micrograph of surface (polished prior to irradiation) at center of irradiated area (qa= $6x10^4$ W/m) showing coarse slip bands confined to grains.

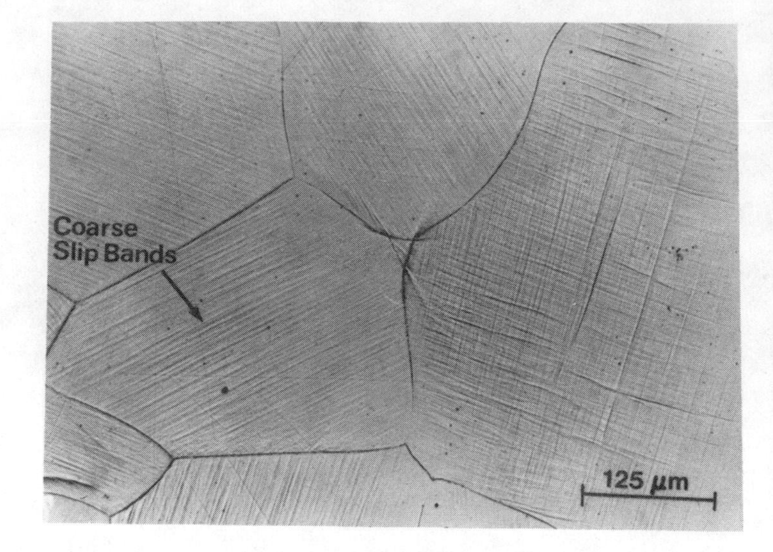

FIG.9B. Underaged Al-4.5% Cu. Optical micrograph of surface (polished prior to irradiation) (qa=$8x10^4$ W/m) showing partial melt at center of irradiated area, surrounded by circle of recovery, in turn surrounded coarse slip bands.

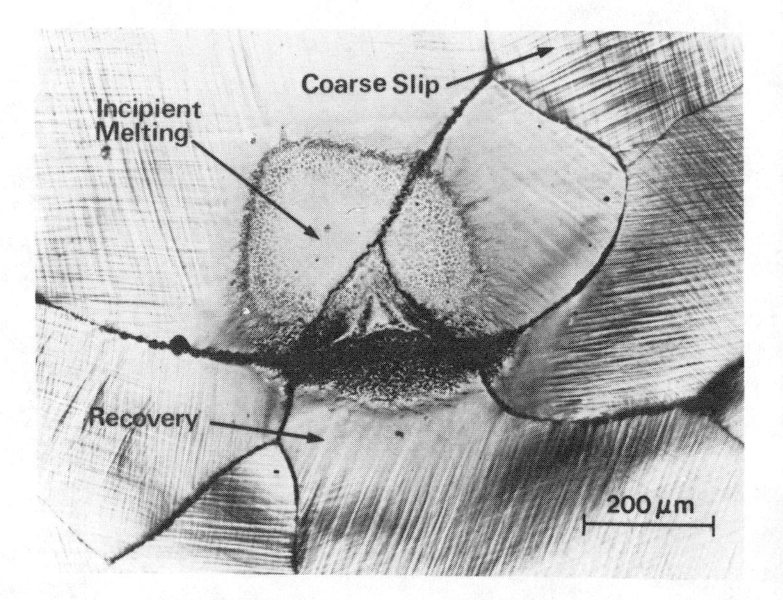

FIG.9C. Underaged Al-4.5% Cu. Optical micrograph of section through resolidified melt (qa=$1.6x10^5$ W/m) showing "hot tearing" cracks there. Keller's etch.

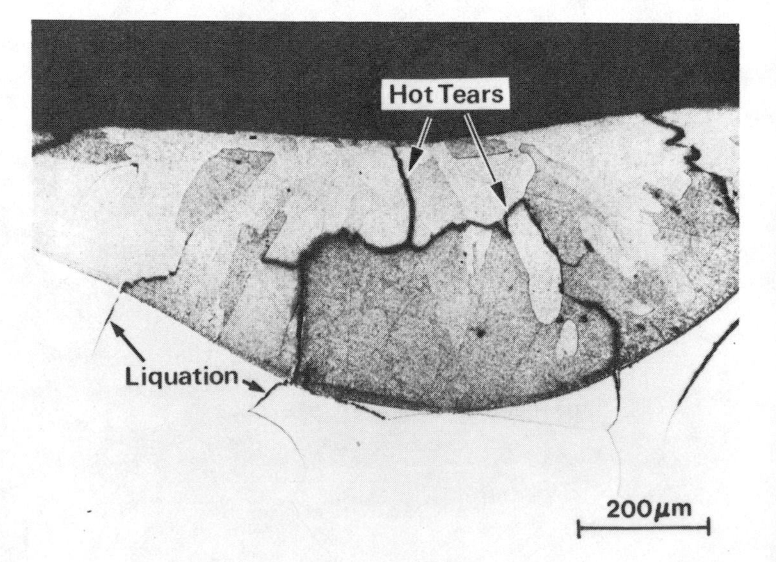

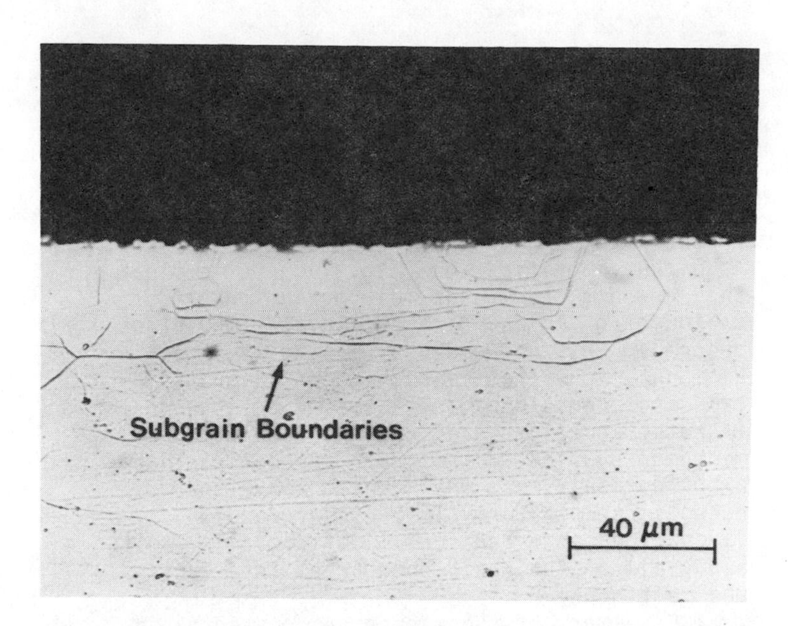

Fig.10A. Peak-aged Al-4.5% Cu. Optical micrograph of section through center of irradiated region (qa=8x10^4 W/m) showing sub-boundary development. Keller's etch.

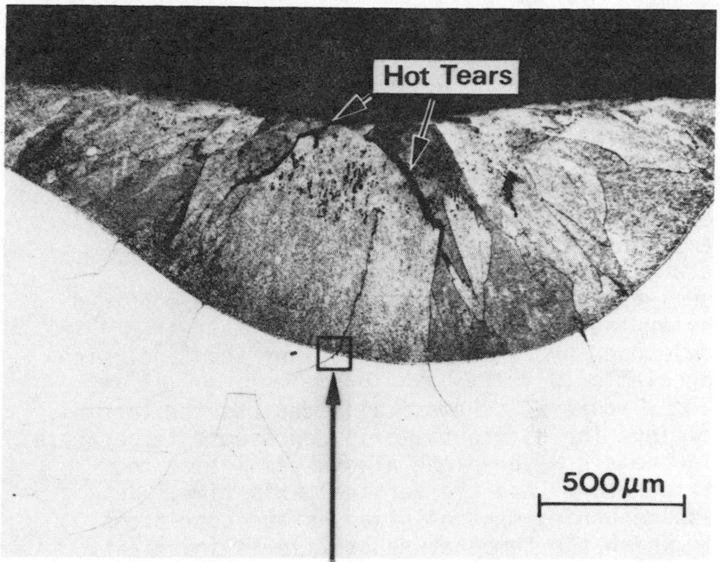

Fig.10B. Peak-aged Al-4.5% Cu. Optical micrograph through section of resolidified melt (qa=2x10^5 W/m). Keller's etch.

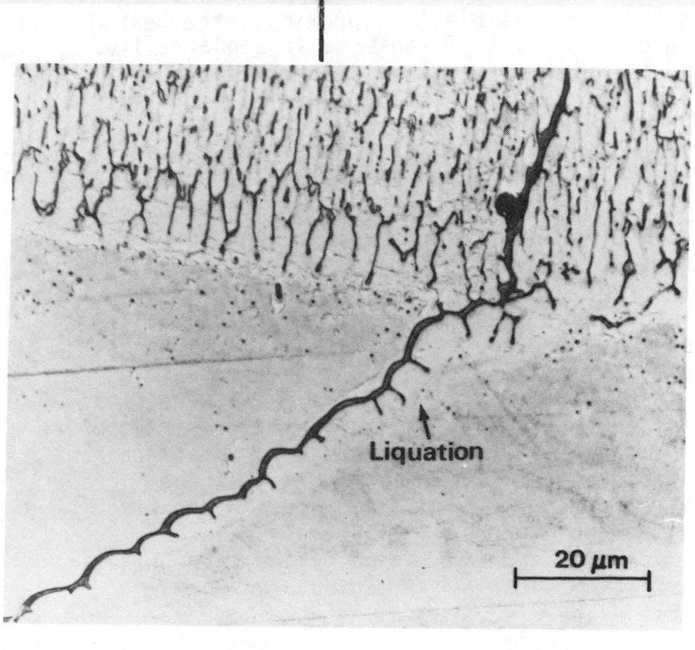

Fig.10C. Peak-aged Al-4.5% Cu. Expanded optical micrograph of the above showing resolidified intergranular "liquation" in substrate. Keller's etch.

occurred, the difference in microstructures of the resolidified melt zones of pure Al and the Al-4.5% Cu alloy is dramatic. In the resolidified melt zone of the alloy material, numerous "hot tear" fractures are observed in the melt zone cross section (Figs. 9C, 10B). These appear, at this magnification, to be intergranular cracks extending into the substrate. In fact, at higher magnification (Fig. 10C) it is clear they are not cracks but melting and resolidification of grain boundaries. Savage (16) first described this phenomenon, calling it "constitutional liquation", and the term "liquation" has since come into general usage. It occurs in eutectic alloys and is associated with microsegregation at the grain boundaries that preferentially melt when heated above the eutectic temperature (16). Intergranular "liquation" occurred for all of the alloy melts along all of the grain boundaries interfacing the melt zone.

A summary of these observations is given in Table 2.

DISCUSSION

Thermal Stress Cycle. When the circular region of the beam spot is heated and then cooled, material is subjected to a thermal stress cycle. Phase transitions, plastic flow and cracking can further complicate this stress state. To determine the origin of the acoustic emission it is essential to have some understanding of this sequence of stresses. A semi-quantitative understanding can be obtained using the concept of "stress-free" strain developed by Eshelby for study of the "inclusion problem" (13).

Figure 11 schematically depicts the thermal cycle. The horizontal axis represents temperature in the disc (the plate always remains at room temperature) and the vertical axis time. We assume achievement of steady-state conditions, in which the temperature profile is invariant. Then it is possible to approximate the heated region as a disc of radius a in a plate, i.e. the temperature profile is approximated as a "top hat" distribution of width 2a. In step 1 the disc is imagined to be cut out of the plate and heated, causing a "stress-free" strain. In step 2, forces are applied to the disc so that it once again fits into the original hole. The interface is welded, and the forces removed, causing "thermal" stresses (depicted by arrows) in the disc and adjacent matrix. It can be shown that the radial stresses are compressive everywhere. Tangential stresses are compressive within the disc and tensile outside (13). The compressive radial stresses inside the disc may exceed the yield strength of the hot material, resulting in plastic deformation (coarse slip) and rumpling of the surface. While there will be no tendency for cracking inside the disc during heating (because of compressive stresses), cracks (or intergranular separation aided by liquation) can form outside the disc.

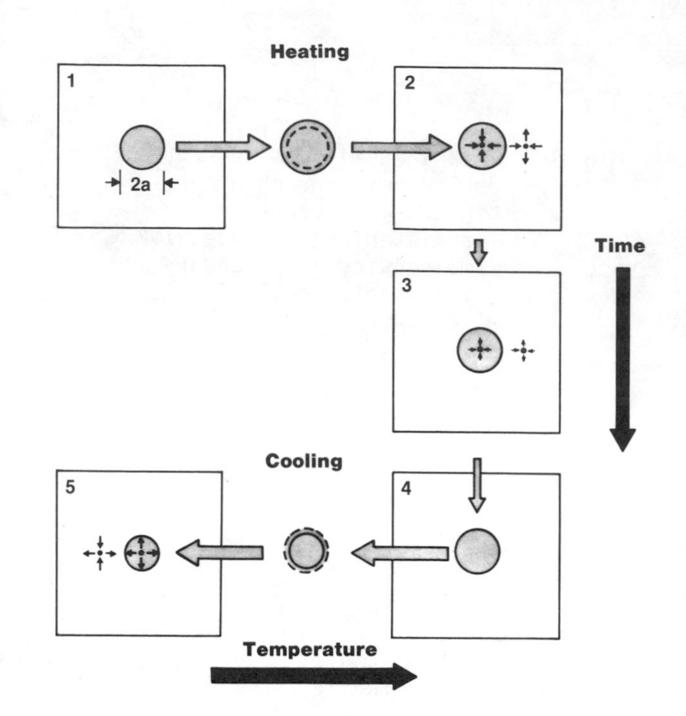

FIG.11. Two-dimensional representation of thermal stresses generated by pulsed electron beam irradiation of circular region of radius a.

Steps 3 and 4 correspond to stress relaxation in the steady-state condition by a number of mechanisms including melting, recovery (subgrain boundary formation), creep, crack growth and intergranular liquation. In the limit the disc and plate ultimately become once more stress-free.

In step 5 (cooling) we imagine the now stress-free but hot disc of radius a to again be cut from the plate. The disc is cooled, resulting in a stress-free strain, and then reinserted into the plate as before. The resulting thermally-induced stresses produced by cooling are opposite to those produced by heating. Accordingly there are tensile stresses in the radial directions everywhere, and the tangential stresses are tensile inside the disc and compressive outside. This biaxial tensile stress state produced inside the disc during cooling is conducive to crack formation there during and following solidification. Only one tensile stress (in the radial direction) is induced outside the disc during cooling. This could promote tangential cracks (i.e. cracks containing the tangential direction) during cooling, but these are not seen. These cracking processes and plastic deformation partially relax the stress during cooling.

Acoustic Emission Mechanisms. The heat flow theory discussed above indicates that, for electron beam heating, the rise and decay time of the temperature is on the order of 10 and 100 ms, respectively. As a result, the rise time of the thermal expansion stress corresponds to a frequency of <100 Hz, far too low to be detected as acoustic emission. Hence the electron beam heating itself can not be the source of the emission we observe. (This is not generally true of Q-switched laser pulses, which are well-known generators AE signals (17)). This is experimentally supported by the existence of a delay time for acoustic emission on heating. Heat flow theory likewise indicates motion of the liquid-solid interface occurs typically over 10-100 ms, so that it too can be eliminated as a direct AE source. Other processes, such as radiation pressure and ablation (vaporization) are calculated to be far below the detection sensitivity of the acoustic emission system (12).

Microstructural examination, as summarized in Table 2, suggests that plastic flow and/or fracture may be the principal sources of AE. All of the material underwent some amount of coarse slip, which is known to produce significant emission. Wadley et al.(4,14) have shown that acoustic emission can be detected provided that the product nav > 10^{-3} m^2s^{-1}, where n is the number of moving dislocation segments, a is the slip distance, and v is the velocity. Consider the early stages of slip where dislocation loops can expand across the entire grain before being arrested at the boundaries. If we take n=1, a=200 μm (the grain radius), and estimate v=200 ms^{-1}, we obtain nav=.040, which is over 10 times the detection threshold. Thus a single high speed dislocation loop could give a detectable signal.

During heating, as the steady-state temperature increases, the microstructure remains relatively fixed, so that a and n will be relatively constant. However, increasing thermal stress (increasing qa) and enhanced thermal activation are expected to increase the dislocation velocity. The volume of stressed material also increases with temperature (qa). Thus, we can expect the acoustic emission to increase with qa in region I in accordance with observations. For heating, the same processes continue in regions II and III. In addition, bulk and intergranular melting, with associated stress relief, also occur. This could account for the increasing AE with increasing qa on heating and its dramatic increase above the solidus temperature (Figs. 7A and 7B).

The picture is different during cooling because of resolidification and the annealing (recovery) of deformed material in the heat affected zone. One tentative explanation is as follows. In region I, where the temperature is lowest, these effects will be minimized, and deformation processes will be somewhat reversible, so that the AE on cooling increases in region I through the increased stress and volume of material cooling. In region II, recovery processes will be enhanced with increasing qa. Recovery results in a relaxation of stress and development of subgrain boundaries. The subgrains limit the distance dislocations move so that AE on cooling in region II decreases with increasing qa. In region III, however, the amount of material affected by the melting and resolidification process (which increases with qa) predominates over other effects to bring

Table 2

Temperature Ranges of Heating Phenomena

	Region	qa (W/m)	T(0,0) (°C)	Final Microstructure
Al	I	<$0.6x10^5$	<250	Coarse slip, sub-boundary development, no melting.
	II	$0.6-1.4x10^5$	250-660	Heavier coarse slip, sub-boundary development, no melting
	III	>$1.4x10^5$	>660	Melting, coarse slip, sub-boundary development.
Al-4.5%Cu	I	<$0.6x10^5$	26-250	Coarse slip, no melting
	II	$0.6-1.0x10^5$	250-560	Coarse slip, sub-boundary development no melting.
	III	>$1.0x10^5$	>560	Melting, coarse slip, sub-boundary development, intergranular liquation, hot tears in resolidified melt.

about an increase in acoustic emission.

The condition of the substrate will modify these processes. Cold work limits the distance dislocations may move during subsequent plastic flow. This, according to our model, is expected to reduce the AE intensity as observed in these experiments. In Al-4.5% Cu underaging produces only partially developed GP zones (15) which promote coarser slip compared with that of a solid solution or peak-aged structure (14). This results in an increase in acoustic emission (over that of a solid solution) in tensile tests at room temperature. The present tests, however, were (in effect) at elevated temperatures, during which the metastable GP zones are expected to begin to dissolve. Thus coarse slip may be suppressed and with it intense AE.

CONCLUSIONS

Table 2 summarizes many of the results responsible for AE at various levels of heat flux (Regions I, II and III). We can draw some general conclusions about the mechanisms involved in acoustic emission during rapid surface melting and resolidification.

1. Processes that are directly associated with electron beam heating generally vary relatively slowly with time. Their elastic wave radiation is thus at frequencies too low to directly cause observable acoustic emission. Thus, electron beam heating, radiation pressure, substrate quenching, and motion of the liquid-solid interface itself are not direct causes of detected acoustic emission.

2. However, relaxation of these stresses by defect motion (i.e. slip or cracking) can occur with sufficient speed to generate high frequency elastic waves and is postulated to be the origin of AE.

3. The condition of the substrate (its composition, heat treatment or degree of cold work) strongly affects the amount of emission by controlling the micromechanisms of slip.

4. The amount of acoustic emission is greater on heating than on cooling, at least for the materials studied here, increasing exponentially above the solidus.

If generally true, it will be necessary to spacially filter data during AE monitoring of welding or surface modification in which the heat source is moving, so that the emission which occurs on cooling may be unmasked from that of the region being heated.

ACKNOWLEDGMENTS

We acknowledge helpful advice from R. Schaefer and R. Mehrabian, metallographic assistance from C. Brady, and S.Levy of Reynolds Metals in obtaining materials. Financial support by the Defense Advanced Research Projects agency under DARPA Contract No. 4275 is gratefully acknowledged.

REFERENCES

1. Mehrabian, R., International Metals Reviews, Vol. 27, pp 185-208, 1982.
2. Hou, S. C., S. Chakravorty, and R. Mehrabian, Met. trans. B, Vol. 9B, pp 221-229, 1978.
3. Hsu, S. C., S. Kou, and R. Mehrabian, Met. Trans. B, Vol. 11B, pp 29-38, 1980.
4. Wadley, H. N. G., and R. Mehrabian, "Acoustic Emission Sensors for Materials Processing: A Review", in Process Monitoring and Microstructure Control, Proc. of the Symposium on Nondestructive Testing, Johns Hopkins Univ. and Univ. of PA., Hershey PA, April 1983.
5. Dickhaut, E., and J. Eisenblatter, Journal of Engineering for Power (Trans. of the ASME), pp 47-52, Jan. 1975.
6. Crostack, H. A., "Weld Quality Control Measurements of EB-Weldments," Proc. Third Meeting of EWGAE in Ispra, Italy, pp 28-47, Sept. 1974.
7. Jon, M. C., C. A. Keskimaki, and S. J. Vahaviolos, Materials Evaluation, 51, pp 41-44, March 1978.
8. Sekhar, J. A., S. Kou, and R. Mehrabian, Met Trans. B, Vol. 14B, pp 1169, 1983.
9. Sekhar, J. A., R. Mehrabian, and H. L. Fraser, Met. Trans B, Vol. 12B, pp 411, 1981.
10. Clough, R. B., H. Wadley, and R. Mehrabian, "Heat Flow-Acoustic Emission-Microstructure Correlations in Rapid Surface Solidification," Proc. of Second International Conference on Applications of Lasers in Materials Processing, E. Metzbower and S. Copley, eds., Amer. Soc. for Metals, Metals Park, Ohio, 1983.
11. Clough, R. B., and H. Wadley, "Laser Calibration of Acoustic Emission Source Energy," to be published.
12. Clough, R. B., H. Wadley, and R. Mehrabian, "In Situ Studies of High Intensity Electron Beam Interactions with Aluminum", to be published.
13. Eshelby, J. D., Proc. Roy. Soc., Vol. A241, pp 376, 1957; also Vol. A252, pp 561, 1959.
14. Scruby, C. B., H. Wadley, K. Rusbridge and D. Stockham-Jones, Metal Science, Vol. 15, pp 599, 1981.
15. Kelly, A., R. B. Nicholson, "Precipitation Hardening," pp 190, Pergamon Press (Distr. by MacMillan) 1963.
16. Pepe, J. J., and W. F. Savage, Weld, J., Vol. 46, pp 411s-422s, 1967.
17. Hutchins, D. A., R. J. Dewhurst, S. B. Palmer, and C. B. Scruby, Appl. Phys. Lett., Vol. 38, pp. 677-679, 1981.

AN ACOUSTIC EMISSION TECHNIQUE FOR CHARACTERIZATION OF SOLIDIFICATION CRACKING IN LASER-MELTED STAINLESS STEELS

R. J. Coyle, Jr.
Western Electric Corporation
Engineering Research Center
Princeton, NJ 08540

D. G. Cruickshank
Western Electric Corporation
Engineering Research Center
Princeton, NJ 08540

M. C. Jon
Western Electric Corporation
Engineering Research Center
Princeton, NJ 08540

V. Palazzo
Western Electric Corporation
Engineering Research Center
Princeton, NJ 08540

ABSTRACT

Acoustic Emission (AE) and quantitative optical metallography were used to detect and evaluate solidification cracking in several commercial austenitic stainless steels. Cracking was induced by high peak energy, pulsed laser irradiation. A photo detector was used to initiate a pre-set electronic gate incorporated with a delay function to capture the AE information. Four AE zero-crossing counts were obtained by using four thresholds to discriminate different AE amplitudes. These AE zero-crossing counts and AE burst counts were then correlated with the severity of cracking using a statistical method called the discriminant analysis function. Attempts were made to correlate the AE data to the size and frequency of the cracks for both single and multiple pulsed exposures.

INTRODUCTION

Solidification or hot cracking of austenitic stainless steels has been the subject of a considerable amount of research over the years (1-5). This type of defect is initiated within the weld fusion zone prior to the complete solidification of the weld metal. Solidification cracking has been attributed to the formation of weak or brittle low melting point compounds at interstices and boundaries in the fusion zone (6). It is at these sites that intergranular fracture can be inititated by the thermal stresses generated during the weld cooling cycle.

This work is concerned with the detection of solidification cracking in welds generated by pulsed laser irradiation. One of the unique characteristics of pulsed laser welding is extremely rapid solidification and cooling rates (up to 10^6 deg/s), which can produce steep temperature gradients and high thermal stresses (7). In addition, these cooling rates may produce welds which contain microstructural features not found in conventional welds, such as extremely fine grain size, unusual precipitate morphology, size, or distribution, or non-equilibrium structures (7-8). Any of the aforementioned factors can either exacerbate cracking or cause cracking to occur in materials which are not susceptible to cracking under conventional welding conditions.

Solidification cracking commonly is detected and characterized through mechanical tests and metallographic examinations (9-10). These are time-consuming, destructive evaluations which are not always feasible for routine weld inspection. In recent years an alternative, non-destructive technique, acoustic emission (AE) detection, has gained increased acceptance in the area of weld defect analysis (11-15). In particular, AE techniques have been applied to detect crack growth and propagation during both hot (solidification) and cold (delayed) cracking processes in austenitic stainless steel welds (16-17). These efforts have concentrated primarily on conventional weld processes with relatively slow cooling rates. AE monitoring, however, has the potential to analyze extremely fast localized events, such as those typical of pulsed laser welding. Furthermore, AE monitoring offers the added advantage of in situ, real-time detection of defects. This investigation was undertaken to develop an AE technique to detect and characterize solidification cracking generated by the rapid cooling rates during laser welding.

EXPERIMENTAL

In these experiments an AE monitoring system is synchronized with the pulsed laser output in order to sample the elastic waves

Table 1 - Analyzed Chemical Composition of Austenitic Stainless Steel Weld Metals

	Cr	Ni	Mo	Mn	Si	C	S	P	N	Bal
316L	17.0	11.0	2.16	1.49	0.35	0.02	0.007	0.030	-	Fe
NITRONIC 60	16.7	8.5	0.23	8.51	3.96	0.08	0.005	0.024	0.137	Fe

generated during and immediately after solidification. AE information is recorded for single and multiple overlaid melt spots in order to simulate the range of solidification conditions in a pulsed overlap seam weld. Optical metallography then is used to determine the extent of cracking for correlation to the AE data.

The chemical compositions of the alloys under investigation are given in Table 1. Both austenitic stainless alloys were tested in the annealed condition, in nominally 0.75 mm (0.030 in.) thick sheets. Type 316L, which is regarded as readily weldable (immune to solidification or hot cracking) by conventional or laser techniques (19-20), served as a control sample. Less is known about the weldability of Nitronic 60 (21), however, other stainless alloys with similarly high N, Mn, or Si contents often are prone to cracking, particularly with laser or electron beam welding (22-25). Thus, tests were conducted on two stainless steels with potentially distinct response to solidification cracking during laser welding.

SAMPLE PREPARATION - A pulsed Nd:Glass laser welder (1.06 μm wavelength) was used to produce melt spots for the cracking study. The laser was operated with a four inch focal length lens, at a pulse duration of 5 ms and a power density of approximately 3.5 KW mm^2. The beam was slightly defocused (optical spot estimated to be 0.5 - 0.6 mm) to promote a more uniform distribution of energy across the melt region (26).

As a sample was pulsed with the laser, the AE signal was recorded and the sample was examined macroscopically for damage. This sequence was performed for two, three and four overlaid pulses. Each pulse train was repeated to ascertain reproducibility as shown in the typical test piece in Fig. 1. This test sequence was used in lieu of monitoring actual seam welds for three reasons. First, the initial two pulses represent a reasonable simulation of pulsed overlap welding. Second, the latter pulses which are overlaid may produce cumulative damage or healing (repair) which could be of practical interest. And third, it is easier to isolate cracking metallographically in the single melt spot than in the seam weld.

Representative samples were sectioned and molded in transparent thermoplastic mounts. After routine grinding and polishing through 0.05 micron cerium chromium oxide, the samples were etched electrolytically in oxalic acid and examined for cracking. Several random sections were examined on each sample. However, the most significant data came from the center region of the weld, which was the last to solidfy and thus was affected most by the thermal stresses.

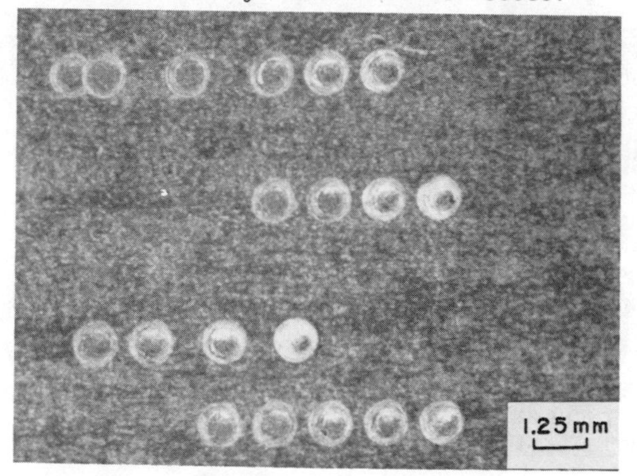

Fig. 1 Top view of a typical series of laser melt spots in an austenitic stainless steel.

ACOUSTIC EMISSION - The standard AE monitoring system shown schematically in Fig. 2, included a piezoelectric transducer bonded by an epoxy to the sample to detect AE signals, a preamplifier to amplify low level AE signals, and an electronic analyzer to process and display AE information. In this study, the signal from a resonant transducer was fed into a low noise preamplifier which in turn was amplified without further filtration by a post amplifier in the AE analyzer. The AE monitoring system used in this study is similar to that described in a previous publication (27).

This analyzer contained a multiple threshold detector designed to produce AE threshold crossing counts as follows. The voltage generated by a photo-detector was used to initiate a gate for capturing AE information. The gate had a pre-set delay

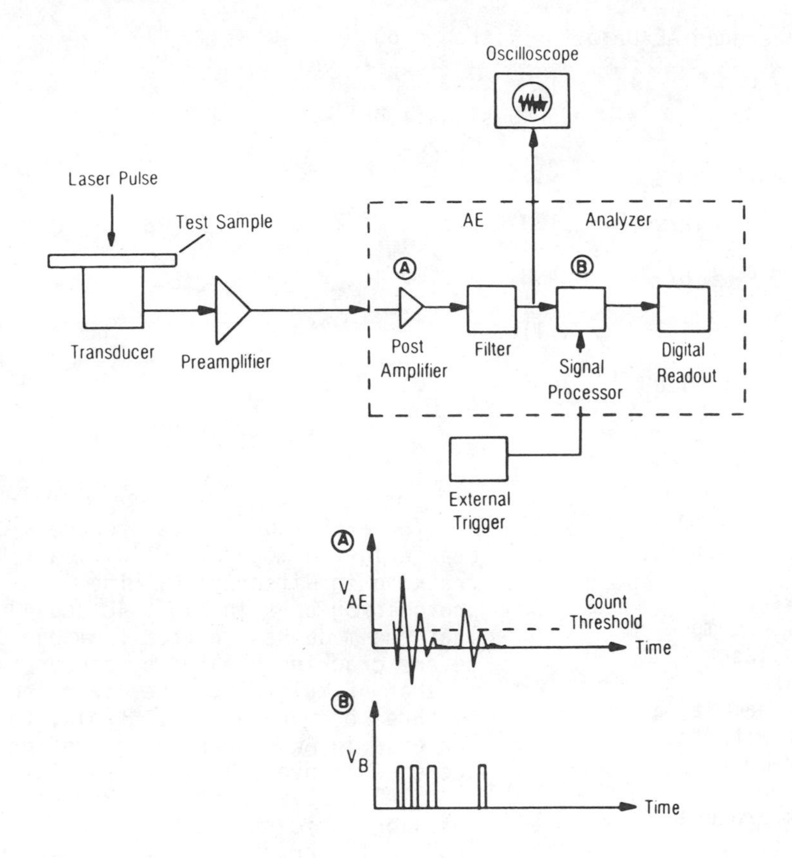

Fig. 2 Schematic representation of the acoustic emission weld monitor system.

equal to the laser pulse duration so that the AE analyzer monitored only the immediate past weld interval. During the time the gate was open, AE signals were counted as threshold crossing counts for each of four predetermined dc thresholds. The analyzer also contained a detector referred to as an event or envelope monitor. This was designed to count discrete resonant bursts or decay envelopes.

The count data were generated in the described manner to provide two types of semi-quantitative information. First, the amplitude thresholds were designed to discriminate between cracks of different sizes, which should have correspondingly different acoustic amplitudes. Second, the event monitor was designed to detect the quantity of discrete decay packets, which is caused by the individual crack events. Thus, the amplitude and event counters could be used to provide information about both the relative crack sizes and the number of cracks present.

RESULTS

ACOUSTIC EMISSION - Type 316L stainless steel, which is considered to be immune to solidification cracking, shows virtually no AE activity as a result of laser pulsing. On the other hand Nitronic 60, which may be susceptible to cracking, shows a considerable amount of AE activity. The AE data for this alloy are summarized in Table 2. The data presented are the average number of counts in each "window", for a given number of overlaid laser pulses. These "windows" consist of the event monitor (envelope detector) plus four amplitude brackets determined by the pre-set threshold voltages.

At first glance, there appear to be three distinct groupings of data. A single laser pulse generates clearly the highest number of average counts in all windows. Two overlaid pulses result in roughly one-half to one-third the number of counts observed after a single pulse. With subsequent pulses there is a significant reduction in the count level, particularly in the amplitude-dependent thresholds. Unfortunately, the three-level discrimination based on average count values is not as distinct as it may seem. A close examination reveals considerable scatter and overlap amount the individual data sets.

Table 2. Averaged AE Data for Nitronic 60

No. of Laser Puless	No. of Tests	Events	Window Designation 1	2	3	4
1	9	36	12296	9133	5583	2713
2	6	13	6797	4498	2009	937
3	5	7	1616	1111	664	263
4	5	10	1588	1088	701	348

Therefore, metallographic examinations were made on individual samples in an attempt to improve the correlation or, to explain the discrepancies in the data.

METALLOGRAPHY - Fig. 3 shows a photomicrograph of a typical 316L sample after four consecutive overlaid laser pulses. There is no evidence of solidification cracking in this sample, an observation which is consistent with the absence of AE activity in laser-melted 316L. This demonstrates that the normal solidification process does not produce a significant amount of AE activity.

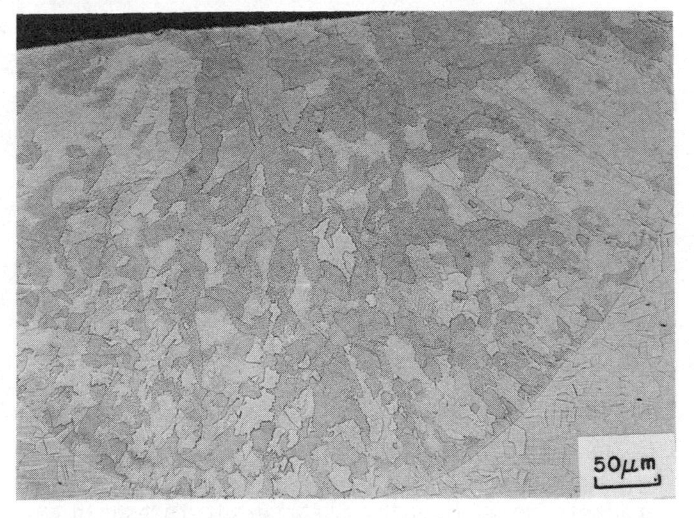

Fig. 3 Photomicrograph of the fusion zone in 316L stainless steel after four overlaid laser pulses.

Fig. 4 shows a composite of typical metallographic and AE data for one to four overlaid pulses in Nitronic 60. The first pulse (Fig. 4a) produces severe solidification cracking and high AE counts in all five windows. Two overlaid pulses create several large cracks along with some smaller cracks (Fig. 4b). This metallographic evidence seems to be consistent with the reduced AE counts for the case of two pulses (see also Table 2).

Three or four pulses produce a complete change in the mode of solidification cracking in Nitronic 60. The photomicrographs in Figs. 4c and 4d show that the mode has shifted from one of large severe cracking created by one or two pulses to that of multiple, fine, hairline cracking by three or more pulses. Again, there is a decrease in AE counts corresponding to the decrease in overall crack severity.

STATISTICAL ANALYSIS - Numerous metallographic samples were examined to confirm the cracking patterns presented in Fig. 4. In a general way this metallographic data corroborated trends in the AE data which were identified in Table 2. However, attempts to correlate AE results to crack size or distribution in individual samples were not successful. This is illustrated by a comparison of the AE data for Figs. 4b and 4c. The number of AE events are essentially the same for both although there is clearly more cracking in Fig. 4c. Also, the counts in the lowest threshold (smallest amplitude cracks) are considerably greater for the case of Fig. 4b, despite the fact that it is obvious that there are fewer (small) cracks present. The scatter in the AE data seems to limit any direct, one-to-one comparison of cracked samples. For this reason, a statisitcal technique referred to as "discriminant analysis" was employed (28).

Discriminant analysis, as it is applied here, is a multi-variant correlation coefficient. The basic technique is adapted to discriminate failure modes on the basis of a set of test variables or parameters. For our specific problem we have identified (by metallographic inspection) three solidification failure modes (FM):

FM1 - Severe solidification cracking characterized by large defects (Fig. 4a).

FM2 - A less severe type of cracking characterized by a few large defects along with some smaller defects (Fig. 4b).

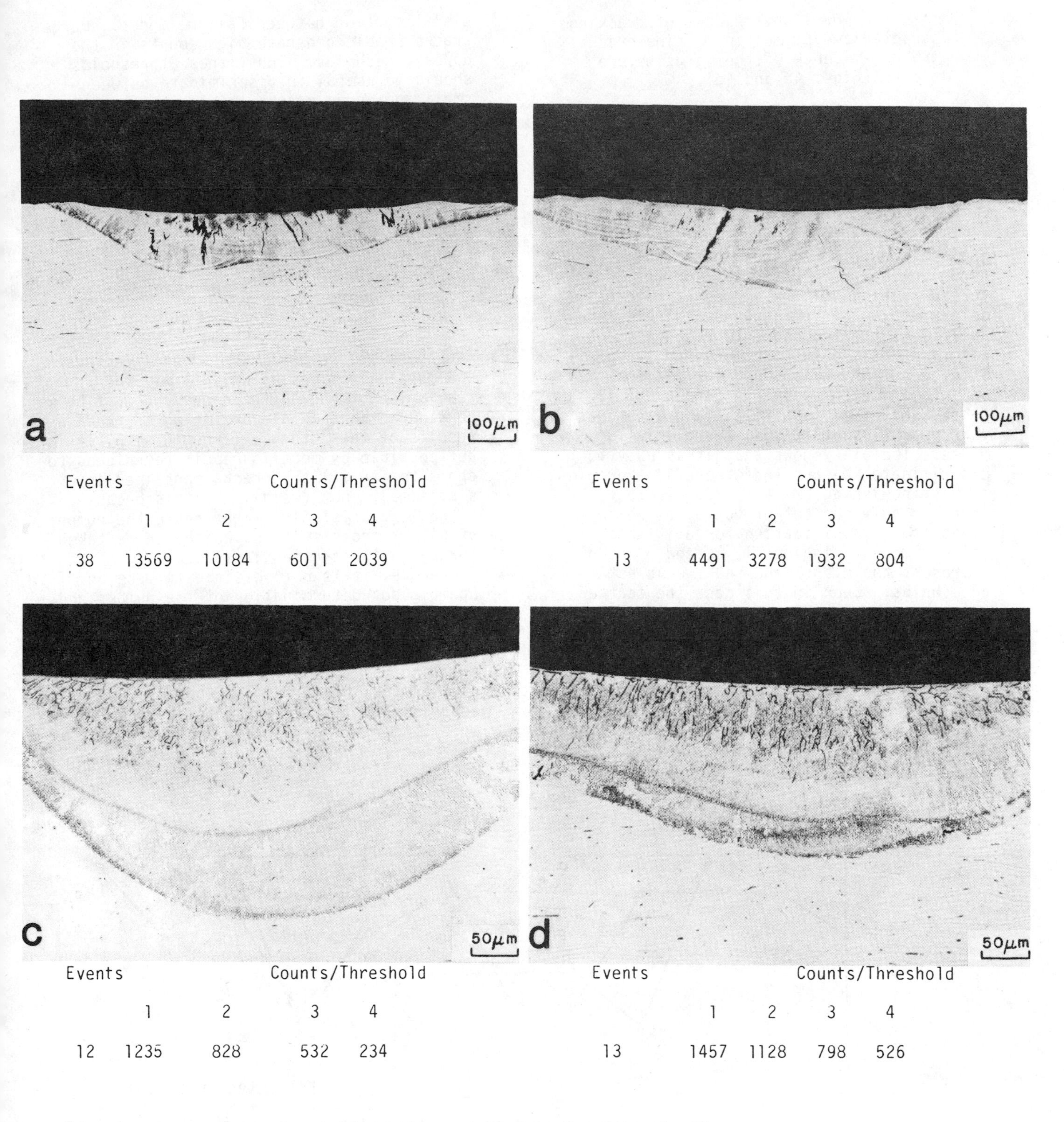

(a)

Events	Counts/Threshold 1	2	3	4
38	13569	10184	6011	2039

(b)

Events	Counts/Threshold 1	2	3	4
13	4491	3278	1932	804

(c)

Events	Counts/Threshold 1	2	3	4
12	1235	828	532	234

(d)

Events	Counts/Threshold 1	2	3	4
13	1457	1128	798	526

Fig. 4 Typical metallographic and AE data for Nitronic 60, (a) large, severe solidification cracks caused by the initial laser pulse, (b) less severe cracking with two overlaid pulses, (c and d) multiple, fine, cracking caused by three or four overlaid laser pulses.

FM3 - The least severe case of cracking characterized by multiple, fine cracks, with no evidence of the large severe defects (Figs. 4c and 4d).

The test variables, of course, consist of the five AE windows used in the experiments. The objective is to determine which variable, or set of variables, can classify accurately each test sample into its predicted failure mode.

The plot in Fig. 5 contains results from application of the discriminant analysis technique. For each variable or set of variables a classification percentage is determined. This percentage indicates the ability of the technique to recognize and discriminate the failure modes on the basis of each set of variables. For example, if only the AE event variable (index No. 1) is considered, the classification percentage is very poor (less than 80%) for all failure modes. The plot shows that it is easy to discriminate FM3 but that there is a problem in distinguishing FM1 and FM2. In fact, there is only one set of variables which gives 100% classification for all three failure modes: indices 1, 2, and 3 which represent the events and the two lowest thresholds. Even in this case the scatter in the data may weaken the interpretation. The data for individual samples presented in Table 3 shows that the probability of correct classification can be low (50 - 60%) even when the classification percentage is 100%.

Nevertheless, it is somewhat surprising which set of indices provides the best discrimination between failure modes. Large cracks should generate high counts in the upper thresholds. Thus, these thresholds should be useful to discriminate between small and large cracks. This is not the case since there seems to be no real advantage to the use of indices 4 and 5 in the analysis. Moreover, a good discrimination of the smaller cracks (FM3) can be achieved with nearly any combination of indices. Also, the data for the event detector (index 1) seem anomalous since the failure mode with the greatest number of cracks (FM3) has the smallest number of recorded events.

DISCUSSION

There are several experimental factors and physical processes which limit the resolution of the AE technique and must be considered in the interpretation of the results of this study. To begin with, it is not possible by metallographic techniques to characterize all the cracks contained in a single melt spot (weld). This is because there is a physical limitation to the number of planar cross sections which can be taken through a three-dimensional weld. Therefore, it is not possible to make an unambiguous determination of the number and size of all cracks present. It is possible to section a sufficient number of welds in order to perform a quantitative analysis of cracking but this is beyond the scope of the present investigation. Furthermore, this approach has limited value for an individual weld since it provides only a probability of the type of cracks present.

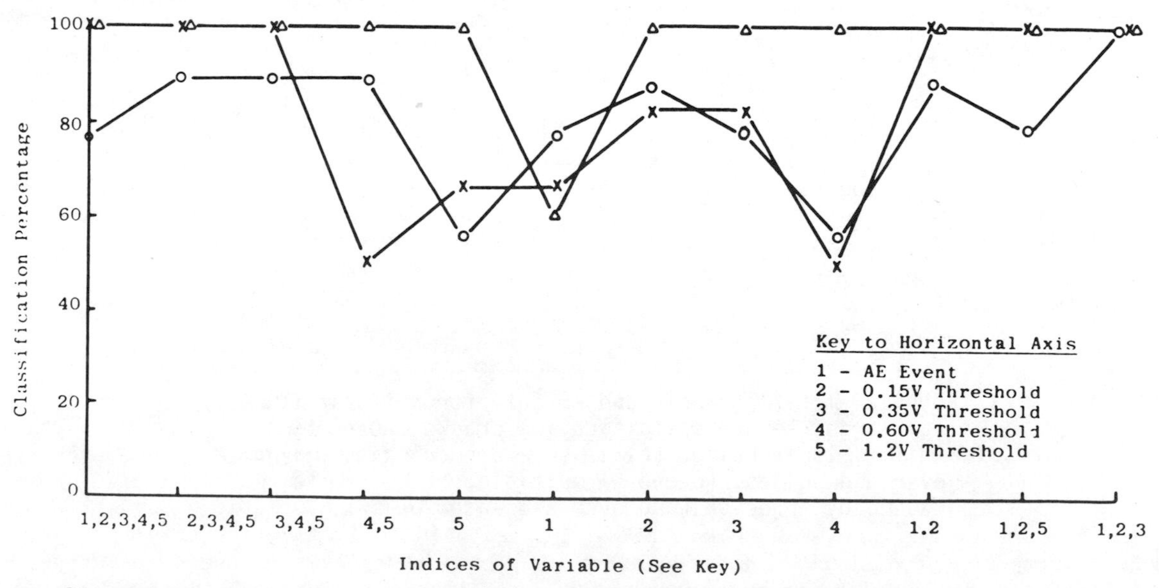

Fig. 5 Plot of classification percentage as a function of various sets of indices for discriminant analysis of AE data.

Table 3. Detailed Results of Discriminant Analysis for Nitronic 60
Using the Event Detector and the Two Lowest Thresholds (Indices 1, 2, & 3)

Prob. of Correct Class.	Class. As Pop#N	Original Pop#N	Data Var 1	Var 2	Var 3
1.000	1	1	38.000	1357.000	1318.000
0.997	1	1	49.000	905.000	648.000
0.581	1	1	24.000	999.000	727.000
0.581	1	1	25.000	974.000	689.000
1.000	1	1	42.000	1496.000	1213.000
0.999	1	1	41.000	1084.000	926.000
0.996	1	1	36.000	1489.000	845.000
0.629	1	1	16.000	1046.000	903.000
1.000	1	1	54.000	1717.000	1251.000
0.968	2	2	12.000	725.000	543.000
0.571	2	2	13.000	449.000	328.000
0.971	2	2	7.000	703.000	344.000
0.842	2	2	21.000	576.000	342.000
0.986	2	2	7.000	1125.000	711.000
0.669	2	2	18.000	494.000	380.000
0.971	3	3	12.000	124.000	83.000
0.936	3	3	9.000	189.000	127.000
0.885	3	3	7.000	241.000	165.000
0.954	3	3	4.000	152.000	111.000
0.974	3	3	3.000	102.000	70.000
0.949	3	3	11.000	171.000	112.000
0.959	3	3	10.000	150.000	81.000
0.956	3	3	11.000	160.000	116.000
0.964	3	3	13.000	146.000	113.000
0.948	3	3	7.000	168.000	124.000

CLASSIFICATION PERCENTAGE

	1	2	3
POP #1 [9]	1.000	.000	.000
POP #2 [6]	.000	1.000	.000
POT #3 [10]	.000	.000	1.000

Another important factor is that rapid solidification and simultaneous crack propagation make it difficult to separate individual events. This could explain the relatively low event counts recorded for FM3, which clearly contains the highest incidence of crack initiation.

A second experimental problem could be created by use of a resonant transducer. The extremely large cracks, characteristic of FM1, may generate excessive "ring down" or long-time resonant decay. This could result in erroneously high counts in the lower thresholds, which could mask the difference between the large severe cracking (FM1) and the smaller more subtle cracking (FM3).

There are several other factors which potentially could contribute to scatter in the data. The laser output can vary from pulse to pulse due to thermal instability in the laser rod. The surface condition of the sample can vary with each overlaid pulse. Both the surface roughness or color (due to oxidation) can alter the absorption of laser energy into the sample. The microstructure of the weld may also change with each pulse. This is illustrated in Fig. 6, which shows the variation of ferrite (body centered cubic phase) content and distribution as a function of the number of overlaid pulses.

The point to be emphasized is that any of the aforementioned factors may influence the weld size, geometry, or structure. Random variations in the fusion zone could influence cracking in a random manner and hence contribute to scatter in the data. A final observation is that transducer coupling (epoxy bond) may be altered by shock or heat from the laser pulse. Since the bond failures were not always catastrophic, it is conceivable that transducer de-bonding may have contributed to some of the observed scatter in the data.

CONCLUSIONS

1. An AE technique was used to provide real-time detection of solidification cracking in rapidly quenched pulsed laser melted samples. The technique was able to discriminate readily between a stainless alloy which was immune to cracking (316L) and one which was susceptible to cracking (Nitronic 60)

2. Both AE and metallographic data independently identified three distinct failure modes in Nitronic 60. The failure modes were (a) large, severe solidification cracking after the initial laser pulse, (b) less severe cracking, characterized by a mixture of large and small cracks after a second overlaid pulse, and (c) multiple fine cracking with three or more overlaid laser pulses.

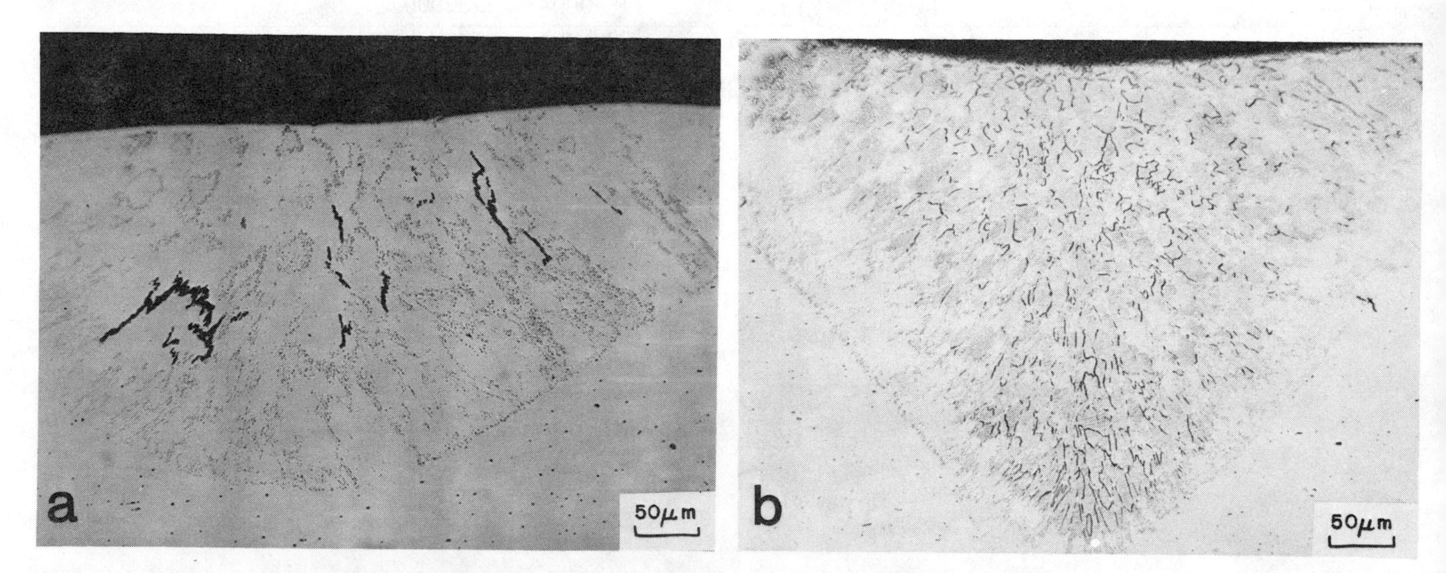

Fig. 6 Effect of the number of overlaid pulses on the ferrite (magnetic phase) content and distribution in Nitronic 60 etched with magnetic colloidal suspension, (a) single laser pulse (b) four overlaid pulses.

3. Two methods were used to analyze the AE data. The first was based on a simple arithmetic average of count levels in each window. The second was based on the discriminant analysis statistical method. Each was successful in identification of the three distinct failure modes.

4. For a single unknown example, discriminant analysis can be used to identify the failure mode on the basis of the information contained in the first three AE indices.

5. Several physical processes and experimental factors generate scatter in the data or limit the resolution of the technique. Some of these factors such as rapid solidification, rapid crack initiation and growth, and microstructural variations in the fusion zone, cannot be controlled. Other factors such as variations in laser output and transducer coupling can be controlled. These factors will be taken into consideration in future experiments.

REFERENCES

1. Cieslak, M. J., Ritter, A. M., and Savage, W. F., Weld. J., 61, 1s, (1982).

2. Cieslak, M. J. and Savage, W. F. Weld. J., 59, 136s, (1980).

3. Jolley, G and Geraghty, J. E., "Solidification and Casting of Metals", The Metals Society, 411, (1979).

4. Hull, F. C., Weld. J., 46, 399s, (1967).

5. Borland, J. C. and Younger, R. N., Brit. Weld. J., 7, 22, (1960).

6. Lippold, J. C., and Savage, W. F., Weld. J., 61, 388s, (1982).

7. David, S. A. and Vitek, J. M., in "Lasers in Metallurgy", K. Mukherjee and J. Mazumder, eds., AIME, New York, 1981, p. 247.

8. Coyle, R. J., "Lasers in Materials Processing", ASM, Metals Park, Ohio, to be published December, 1983.

9. Savage, W. F. and Lundin, C. D., Weld. J., 44, 433s, (1965).

10. Lundin, C. D., Chou, C. P. D., and Sullivan, C. J., Weld. J., 59, 226s, (1980).

11. Vasudevan, R., Stout, R. D., and Pense, A. W., Weld. J., 60, 155s, (1981).

12. Wadley, H. N. G., Scruby, C. B., and Speake, J. H., Int. Met. Rev., 25, 41, (1980).

13. Spanner, J. C., "Elastic Waves and Non-Destructive Testing of Materials", Y. H. Pao, ed., ASME, 71, (1978).

14. Spanner, J. C. and McElroy, J. W., "Monitoring Structural Integrity by Acoustic Emission", ASTM STP 571, ASTM, (1975).

15. Harris, D. O. and Dunegan, H. L., Non-Destructive Testing, 7, 137, (1974).

16. Romrell, D. M. Mater. Eval., 30, 254, (1972).

17. Jolly, W. D., Mater. Eval., 28, 135, (1970).

18. Jolly, W. D., Weld. J., 48, 21, (1969).

19. Hensley, W. E., "Handbook of Stainless Steels", D. Peckner and I. M. Bernstein, eds., 26-1, McGraw-Hill, (1977).

20. Willgoss, R. A., Megaw, J. H. P. C. and Clark, J. N., Weld. Met. Fabr., 47, 117, (1979).

21. Espy, R. H., Weld. J., 61, 149s, (1982).

22. Brooks, J. A., Thompson, A. W. and Williams, J. C., "Physical Met. of Metal Joining", TMS-AIME, 117, (1980).

23. Ogawa, T. and Tsunetomi, Weld. J., 61, 82s, (1982).

24. Brooks, J. A., Weld, J., 54, 189s, (1975).

25. Janson, B., J. Iron Steel Inst/. 209, 826, (1971).

26. Longfellow, J., "Lasers in Industry", S. S. Charschan, Ed., Van Nostrand Reinhold, N.Y., 302, (1972).

27. Jon, M. C. and Lord, H. A., Mater. Eval., 40, 663, (1982).

28. Jon, M. C. and Sturm, G., Int. J. Non-Dest. Test., 15, 185, (1982).

ULTRASONIC NDE OF RAPIDLY SOLIDIFIED PRODUCTS AND ITS APPLICATION TO PROCESS CONTROL

C. L. Friant
Materials Science and Engineering Department
The Johns Hopkins University
Baltimore, MD 21218

M. Rosen
Materials Science and Engineering Department
The Johns Hopkins University
Baltimore, MD 21218

R. E. Green, Jr.
Materials Science and Engineering Department
The Johns Hopkins University
Baltimore, MD 21218

R. Mehrabian
Metallurgy Division
National Bureau of Standards
Washington, DC 20234

ABSTRACT

The enhancement of certain material properties obtainable through rapid solidification techniques makes the large-scale commercial application of such techniques desirable. However, the properties and morphology of rapidly solidified alloys are very sensitive to variations in process parameters. Hence, nondestructive evaluation during or after manufacture is necessary to provide feedback for process control. Several end-products of rapid solidification processing techniques, including sheets, ribbons and surface modified layers on conventional substrates readily lend themselves to ultrasonic NDE. Ultrasonic techniques are also attractive for process control because the volume of material inspected in a single rapid test can be large enough to give an accurate indication of the product's aggregate properties. Guided longitudinal and shear waves were used in the present study to determine the elastic properties of rapidly solidified alloys in ribbon and sheet geometries.

INTRODUCTION

Until recently most compositions of rapidly solidified materials have been produced in only laboratory quantities. Of the alloys for which large quantities of material have been produced, the predominant forms have been powder and wire. However, the mix appears to be changing as more and more rapidly solidified material is being fabricated and utilized in wide ribbon and sheet geometries. The numerous process variables associated with production of these materials require that reliable nondestructive evaluation be implemented. To this end, techniques were developed for ultrasonic inspection of ribbon and sheet geometry rapidly solidified materials. Of particular interest was the use of ultrasonic wave velocities to determine the elastic properties of these materials. The method described in this paper is unique because guided waves of both longitudinal and shear polarization are used to directly determine all the elastic moduli. In previous ultrasonic studies on rapidly solidified ribbon samples, only one elastic modulus could be determined because only one wave polarization was used.[1-9]

For rapidly solidified materials the critical parameter from the viewpoint of heat flow is the smallest dimension of the end product, i.e., the diameter for powders and wires, and the thickness for ribbons or sheets. The smallest dimension is critical because it determines the solidification rate for the material. The directionality of solidification is determined by the method used for cooling the molten metal. Heat flow during the solidification of spherical powder is generally uniform, while for wire it is radially symmetric in the plane normal to the wire axis, and for ribbon it is unidirectional.

Unidirectional solidification through the ribbon's thickness can often lead to internal (volume) and external (surface) inhomogeneities. This is because most ribbons are solidified during contact with a single chill surface. The side of the ribbon in contact with the chill surface replicates that surface and generally experiences the highest cooling rate. The opposite side of the ribbon is free to solidify without the constraint of a chill surface and generally experiences the slowest cooling rate. Internal inhomogeneities are generally a result of this variability in cooling rate, whereas external inhomogeneities are attributed to the behavior of the liquid-solid and liquid-gas interfaces during solidification.

For alloys that can be made amorphous, the most prevalent (and important) internal inhomogeneity is the presence of microscopic crystallites imbedded in the amorphous matrix. For rapidly solidified alloys (amorphous or crystalline) porosity is often a problem. It can occur within the volume or on either surface of the ribbon forming craters where pores break the surface. The presence of crystallites and porosity produces corresponding changes in the sound wave velocity and the ultrasonic attenuation values for the ribbon. A velocity change due to the presence of a crystalline phase will be observed because of the differences between the elastic moduli of this phase and those of the amorphous phase. The moduli of the crystalline phase are always greater than those of the amorphous phase; therefore, the sound wave velocity will increase with increasing volume fraction of the crystalline phase. It is plausible that ultrasonic velocity measurements may prove valuable for qualitatively assessing the degree of crystallinity for as-spun and annealed metallic glasses and, hence, provide feedback for process control during the manufacture and heat treatment of such ribbons. The ultrasonic method has one potential advantage over the currently used x-ray diffraction method in that the volume of material inspected can be thousands of times greater because the entire thickness of the ribbon is interrogated.

Determining the extent of defects such as porosity is a potential application for the ultrasonic attenuation measurement. In particular, porosity and surface irregularities associated with the solidification process act as scatterers of ultrasonic energy in the megahertz frequency range. The cumulative effect of these defects is to diminish or attenuate the ultrasonic energy that is propagated through the ribbon. This measurement in practice is difficult because other factors such as nominal ribbon thickness, ultrasonic frequency and reproducibility of the transducer-specimen bond also have very strong influence on the apparent attenuation measured.

The application of techniques for measuring the sound velocity and ultrasonic attenuation in rapidly solidified ribbons has been limited to laboratory investigations. The two dimensional geometry and the 25 to 75 μm thickness range have been the main drawbacks to ultrasonic investigations associated with processing because this sample size precludes the use of conventional bulk waves. Instead, guided waves must be implemented for property determination. Unlike bulk waves, the velocity of guided waves usually shows dispersion; velocity is a function of extrinsic factors such as specimen thickness and ultrasonic frequency. The attenuation of guided ultrasonic waves is also a function of these extrinsic factors but is a much more difficult analytical problem; hence, no attempt was made in this study to interpret attenuation data. However, it was possible to use the velocity of guided waves to characterize the elastic moduli of rapidly solidified ribbon specimens, because the effect of extrinsic factors on the velocity can be predicted and quantified. In the present study this characterization and subsequent calculation of isotropic elastic moduli were performed for ribbon or foil samples of amorphous 2605-SC Metglas, Pd-Cu-Si, Cu-Zr and commercial purity aluminum using conventional ultrasonic wave velocity measuring apparatus. The assumption of elastic isotropy is valid when the product obtained by melt spinning is amorphous. Elastic isotropy for microcrystalline products, however, does not always result.

THEORY OF GUIDED WAVES-ELASTIC MODULI

Guided waves are the type of waves which propagate in bounded elastic media such as plates, sheets, foils, ribbons, cylinders, rods and wires. For some of these geometries, the exact solutions to the equations of motion can be derived from classical elasticity theory.[10] These solutions must satisfy specific boundary conditions. For example, the two plane parallel surfaces of plates, sheets, foils and ribbons are assumed to be stress free. Similarly, in the case of cylinders, rods and wires the single curved surface is assumed to be stress free. The primary types of guided waves that exist in solids fall into three categories: longitudinal, flexural and shear (or torsional). The longitudinal and flexural waves that travel in plate-type materials are often referred to as symmetric and anti-symmetric Lamb waves, respectively.

Several characteristics of guided waves distinguish them from bulk waves. In particular, these waves show dispersion of wave velocity with frequency and plate thickness or rod diameter. Whenever dispersion exists, the group velocity deviates from the phase velocity. Also, higher order modes of each type of wave exist, each exhibiting its own unique dispersion relation. Regions of high dispersion have also been shown to correlate with regions of high attenuation. Due to the multiplicity of modes for each type of wave, coupling of modes is often observed at certain values of thickness (or diameter) to wavelength ratio.

For the case of symmetric Lamb waves the relation between wavelength and phase velocity is given by the equation:[11]

$$4\,pq\,\tan((\pi t/\lambda)q)+(p^2-1)^2\tan((\pi t/\lambda)p)=0 \quad (1)$$

where: $p=[(V/V_S)^2-1]^{1/2}$; $q=[(V/V_D)^2-1]^{1/2}$

V= phase velocity of symmetric Lamb wave
V_D= longitudinal wave velocity in an infinite three-dimensional medium
V_S= shear wave velocity in an infinite three-dimensional medium
t= plate thickness
λ= wavelength

Numerical solutions are obtained by first choosing a value of Poisson's ratio to fix V_D in terms of V_S, and by selecting values of $\pi t/\lambda$. The necessary V/V_S is then obtained by a simple iteration process using a digital computer. The solutions for the fundamental mode, i.e., the case for which no nodal planes exists, are provided by solving for the lowest root. The group velocity, V_G, may be expressed in terms of phase velocity, V, and thickness-to-wavelength ratio, as follows:[12]

$$V_G = V + (t/\lambda)\frac{dV}{d(t/\lambda)} \quad (2)$$

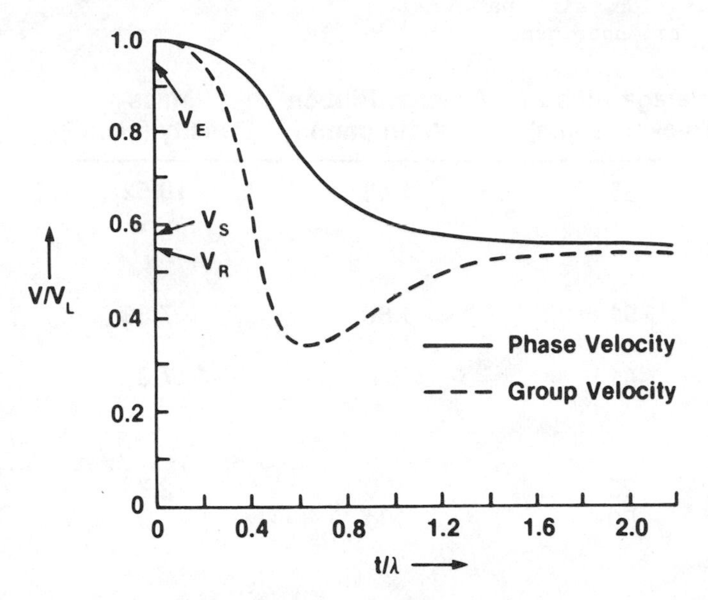

$$V_L = \left[\frac{E}{\varrho(1 - \nu^2)} \right]^{1/2}$$

$$V_E = \left[\frac{E}{\varrho} \right]^{1/2}$$

$$V_S = \left[\frac{G}{\varrho} \right]^{1/2}$$

V_R = Rayleigh Wave Velocity

$V_R = 0.928\ V_s$ for $\nu = 0.29$

ν = Poisson's Ratio

Fig. 1 - Theoretical phase and group velocity dispersion curves for an infinite sheet or plate (ν=0.29, after Kolsky[13]) and equations for V_L, V_E and V_S in terms of elastic moduli and density

A typical theoretical phase and group velocity dispersion curve for the fundamental longitudinal (symmetric) Lamb wave is shown in Fig. 1. The velocity is plotted versus thickness (t) to wavelength (λ) ratio to facilitate comparison of experimental measurements on specimens of various thicknesses at a number of frequencies. At the zero value for the thickness to wavelength ratio, the phase and group velocity converge to the value, V_L, which is related to the isotropic Young's modulus (E), density (ρ) and Poisson's ratio (ν) by the equation that appears at the top right of the figure. Underneath is the equation for the longitudinal extensional velocity, V_E, which applies to cylinders, wires and rods. As the thickness to wavelength ratio increases the phase velocity decreases monotonically from V_L to V_R while the group velocity passes through a minimum. To the right the Rayleigh or surface wave velocity, V_R, is given in terms of the shear wave velocity, V_S, for a Poisson's ratio, ν, of 0.29. In Fig. 1 an equation is also given for the velocity of a bulk shear wave, V_S, this is identical to the equation for the velocity of the fundamental horizontally polarized shear wave, $V_{SH(0)}$, the only nondispersive guided wave. The SH(0) mode is useful for determining the shear modulus directly because no other elastic moduli appear in the equation. The only limitation on this wave is that the plate, sheet, foil or ribbon must be several wavelengths in width for the wave to be truly nondispersive.

Measurement of both the fundamental SH-type velocity ($V_{SH(0)}$) and the velocity of the fundamental longitudinal Lamb wave at zero frequency (V_L) for any ribbon, foil or sheet specimen is sufficient for the complete determination of all the isotropic elastic moduli of the material if the mass density is known. A summary of the equations required to calculate the more important of these moduli is presented in Table 1.

Table 1 - Equations for isotropic elastic moduli in terms of measured guided wave velocities

$$G = \rho V_S^{\,2} \qquad E = 4G \left[1 - \left(\frac{V_S}{V_L} \right)^2 \right]$$

$$\nu = 1 - 2 \left(\frac{V_S}{V_L} \right)^2 \qquad k = \frac{E}{3(1-2\nu)}$$

Where:
G= Shear Modulus
ρ= Mass Density
E= Young's Modulus
ν= Poisson's Ratio
k= Bulk Modulus
V_S= $V_{SH(0)}$; Velocity of fundamental horizontally polarized shear wave
V_L= Velocity of fundamental longitudinal Lamb wave at zero frequency

Table 2 - Chemical composition, size and density of ribbon and foil specimens

Specimen	Atomic Composition (%)	Average Ribbon Thickness (μm)	Average Ribbon Width (mm)	Mass Density (g/cm³)
Amorphous Pd-Cu-Si	$Pd_{77.5}\ Cu_6\ Si_{16,5}$	53	1.48	10.52
Amorphous Cu-Zr	$Cu_{50}\ Zr_{50}$	51	1.68	7.33
Metglas 2605 · SC	$Fe_{81}\ B_{13.5}\ Si_{3.5}\ C_2$	41	$\leqslant 51$	7.3
Aluminum Foil	>99% AL			
Regular		25	100	2.7
Heavy Duty		50	100	

EXPERIMENTAL TECHNIQUE

The ribbon specimens used in this study were produced by either melt-spinning or melt-extraction techniques. Metglas 2605-SC was provided by Allied Chemical Corporation. Pd-Cu-Si and Cu-Zr samples were supplied by the Metallurgy Division of the National Bureau of Standards. Commercially available aluminum foil and sheet of greater than 99% purity were also used. In Table 2 the atomic composition, ribbon cross-sectional dimensions and mass density are listed for each specimen. All specimens were at least 4 cm long. In addition to these ribbon and foil specimens an amorphous Pd-Cu-Si rod 3.5 mm in diameter and 2.1 cm long was available for bulk longitudinal and shear wave velocity measurements. Unlike most metallic glasses, amorphous Pd-Cu-Si can be produced in relatively large cross-sections. Metglas 2605-SC and Cu-Zr were only available in ribbon form. Mass density values for ribbons were obtained using a submerged balance technique similar to that described by Davis.[14]

In this study the group velocities of the fundamental longitudinal Lamb wave and the fundamental horizontally polarized shear wave (SH(0)) were measured. Conventional longitudinal and shear polarized piezoelectric transducers were used to generate and receive these two types of waves. Commercial longitudinal transducers with nominal frequencies of 1, 2.25, 5 and 10 MHz were used for generation and detection of the fundamental longitudinal Lamb wave. AC-cut quartz operating at 5 MHz was used to generate and receive the SH(0) wave. This type of transducer required impedance matching networks whereas the commercial longitudinal transducers did not. For both wave types a viscous silicone fluid was used as an acoustic couplant between the transducer and the ribbon specimen. A two transducer pitch-catch arrangement was used for the time-of-flight measurements instead of the single transducer pulse-echo method due to the absence of at least two echoes of appreciable amplitude.

The orientations of the generating and receiving transducers relative to the ribbon specimen proved to be an extremely important consideration. The transducer-specimen configurations used in this study are based on those reported by Gelles[15] in his work on films, foils, thin plates, whiskers, fibers and fine wires. In particular, two transducer positions for each type of wave were found to yield the largest received signal amplitudes. The longitudinal Lamb wave transducer-specimen configuration is shown in Fig. 2 and the SH(0) wave transducer-specimen configuration is shown in Fig. 3. In both cases the "a" configuration was used for samples that could not support their own weight but could be easily bent around a sharp corner. The "b" configuration was used for stiffer samples that could support their own weight but could not be easily bent around a sharp corner. For generation and detection of the SH(0) type wave the transducers were positioned in such a manner as to orient the shear particle displacements parallel to the width direction in the plane of the ribbon or sheet specimen.

A schematic diagram of the experimental apparatus used for velocity measurements is shown in Fig. 4. The apparatus is essentially equivalent to that used for the pulse-echo overlap method.[16] The continuous wave (cw) oscillator provided a variable frequency sync signal which was fed into the decade-divider and dual-delay strobe generator. The decade-divider divided the frequency of the sync signal by selected powers of 10. This divided sync signal was used to trigger the rf pulse modulator and the monitoring oscilloscope. The dual delay strobe generator provided strobed intensification of the attenuated rf pulse and a selected portion of the signal received by the second transducer. The rf pulse was fed into an attenuator so that it did not overload the oscilloscope's vertical

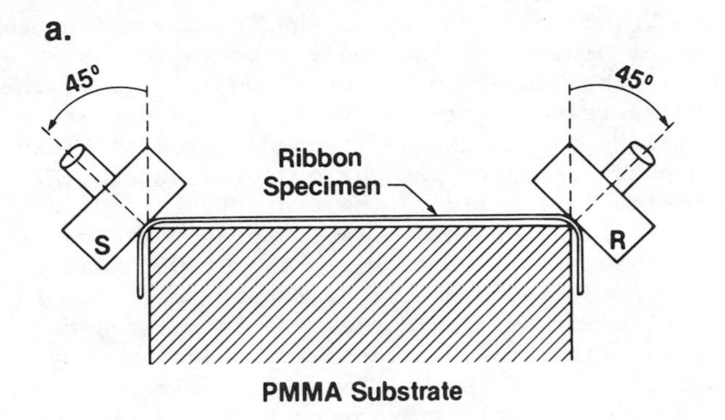

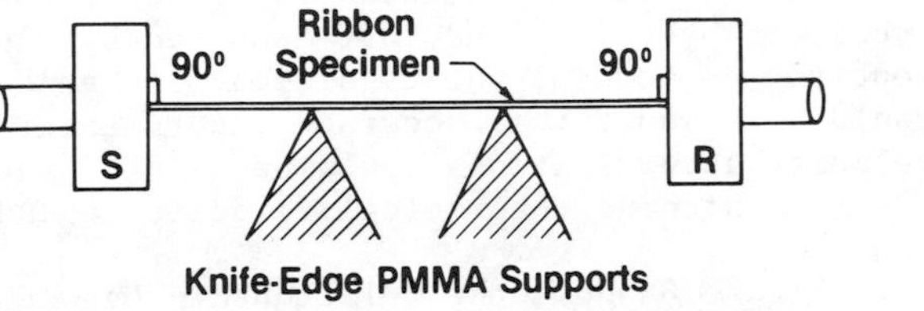

Fig. 2 - Sending (S) and receiving (R) transducer configurations relative to ribbon position for longitudinal Lamb wave generation and detection

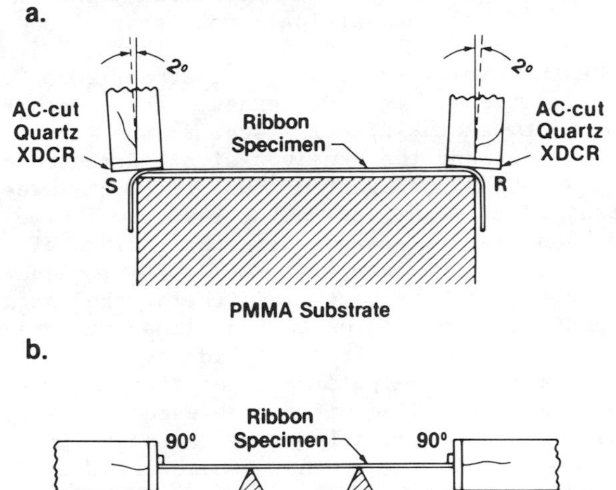

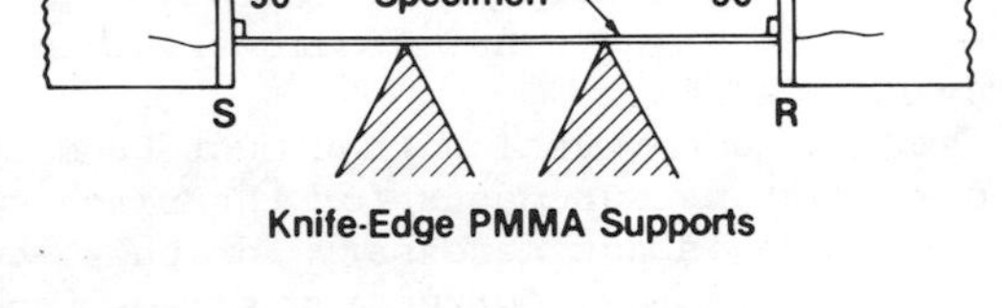

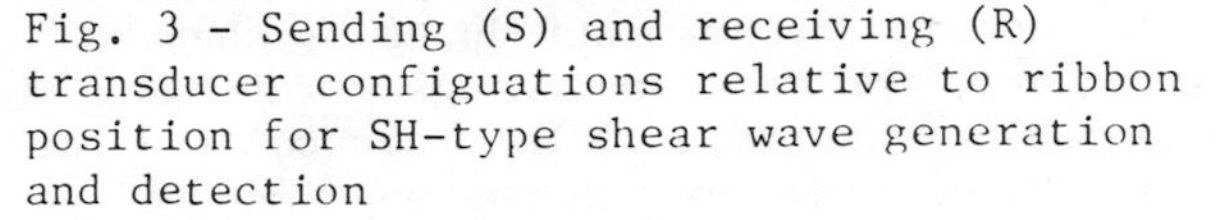

Fig. 3 - Sending (S) and receiving (R) transducer configuations relative to ribbon position for SH-type shear wave generation and detection

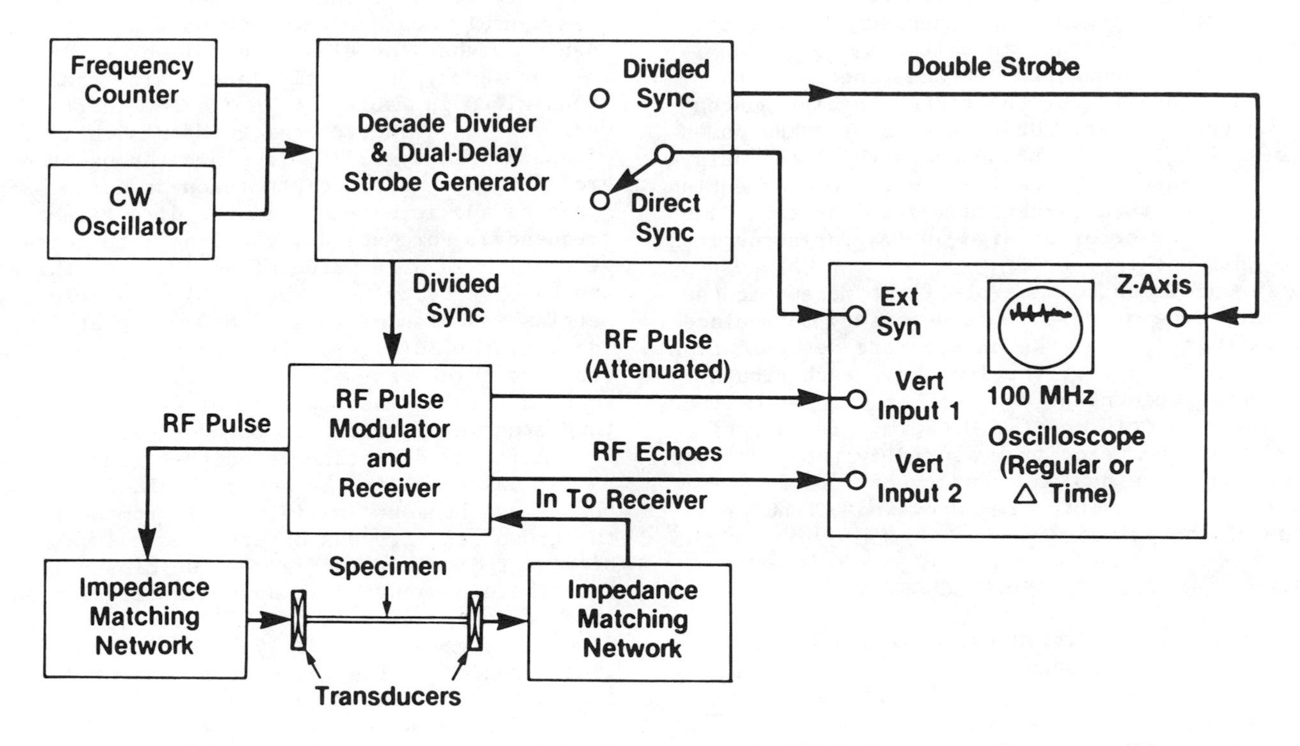

Fig. 4 - Experimental apparatus for measuring the sound velocity in melt spun ribbons

input. When the "divided" sync was used to trigger the oscilloscope sweep it was possible to view and tune the signals received by the second transducer on one vertical input channel and the attenuated rf pulse on another. In the "direct" mode the oscilloscope is swept at a frequency whose period is equal to exactly the time difference between the rf pulse sent to the first transducer and the received signal detected by the second transducer. In practice this sweep is obtained by first estimating the travel time between signals, calculating the reciprocal, then adjusting the cw oscillator to this frequency value. In the "direct" mode the strobed signals appear superimposed on the oscilloscope screen. At this point fine frequency adjustment is made until a cycle-to-cycle match of these signals is obtained. The time difference used to determine the velocity is exactly equal to the reciprocal of the sync frequency that provides the correct overlap of signals.

There is a correction that must be made to time-of-flight measurements for the pitch-catch technique that is not necessary for pulse-echo overlap method. This correction arises from the sum of the time delays introduced by the two transducers, impedance matching networks (if used) and receiving amplifier. In order to find the actual travel time through the specimen the appropriate time delay must be subtracted from the total travel time measured. The major contribution to the time delay was that due to the transducers. Commercial transducers contain both piezoelectric elements and matched wear plates, both of which increase in thickness as the nominal frequency decreases. Hence, the lower the frequency the greater the time delay introduced by the transducers. Using a pair of 1 MHz commercial longitudinal transducers the total delay time was measured to be 1.4 μsec. Not correcting for a delay of this magnitude can lead to a Lamb wave velocity error as high as 17% for a specimen 4 cm in length.

In addition to time-of-flight measurements for each type of wave it is necessary for velocity calculation to make an accurate determination of specimen length. The length of each ribbon or sheet specimen was made with a vernier caliper that had a precision of ± 0.02 mm. It is estimated that the velocity measured by the technique described above has an accuracy of 1 to 2% depending on ultrasonic frequency and attenuation value of the material.

RESULTS AND DISCUSSION

Commercial purity aluminum was selected as a reference material because it was readily available in bulk, plate, sheet and foil geometries. Velocity measurements were taken on various thickness specimens at several frequencies to check the agreement between experimentally measured values of group velocity and the theoretically derived dispersion curve. In Fig. 5 the group velocity of the fundamental longitudinal mode, V, divided by the shear wave velocity, V_S, is plotted versus the thickness-frequency product divided by the shear wave velocity. Unlike the dispersion curve shown in Fig. 1, the nondispersive shear wave velocity was used to normalize the thickness-frequency product because the phase velocity was not measured and the group velocity does not give the proper value for the ultrasonic wavelength. The theoretical curve was calculated for a Poisson's ratio of 0.34 using the method discussed in the Theory section. In practice, dispersion curves are calculated for several values of Poisson's ratio and the one that fits the data the closest is selected.

Experimental data points taken on aluminum foil and sheet at various frequencies fell very near the theoretical curve. Agreement to better than 1% was obtained between the group velocity predicted by theory and that measured by experiment for commercial purity aluminum. Hence, it could be assumed that accurate experimental group velocity determinations could be made on materials for which no theoretical velocity dependence was known.

Figure 6 shows the relationship between group velocity of the fundamental longitudinal mode and frequency for Metglas sheet 5 cm in width, 41 μm in thickness and 10 cm in length. In this case, the group velocity was plotted versus frequency because all samples were the same thickness. For reference, 10 MHz corresponds to $t \cdot f/V_S=0.14$ this is well within the linear portion of the dispersion curve in Fig. 5. It is necessary to determine V_L, the velocity of the fundamental longitudinal Lamb wave at zero frequency so that the elastic moduli of the ribbon, foil or sheet can be calculated using the equations given in Table 1. In Fig. 6 a dotted line, representing a theoretical Poisson's ratio value of approximately 0.315, is drawn through the group velocity data points taken at various frequencies and is extended in the direction of zero frequency. For Metglas, the line intersects the vertical axis at a value of 4895 m/sec, which is the Lamb velocity V_L. The SH(0) wave velocity of Metglas was also measured (2860 m/sec at 5 MHz). Ideally, the SH(0) mode is nondispersive for a constant thickness. In practice, it was found that the ribbon must be several wavelengths wide to prevent dispersion due to edge effects.

A rather different velocity-frequency behavior was exhibited by amorphous Pd-Cu-Si ribbon as can be seen in Fig. 7. A prominent dip in the group velocity was observed that looks similar to the dip in Fig. 1. On closer inspection the dip occurred at approximately 2.5 MHz which corresponds to a $t \cdot f/V_S$ ratio of 0.07. Since this value definitely lies within the linear region of Fig. 5, some factor other than ribbon thickness must be responsible.

It was shown earlier (Figs. 1 and 5) that dispersion took place in a region where the thickness was on the same order as the acoustic wavelength. At high values of thickness-to-wavelength ratio the velocity approached that of a Rayleigh wave and at low values it approached V_L, the value for a thickness-to-wavelength ratio

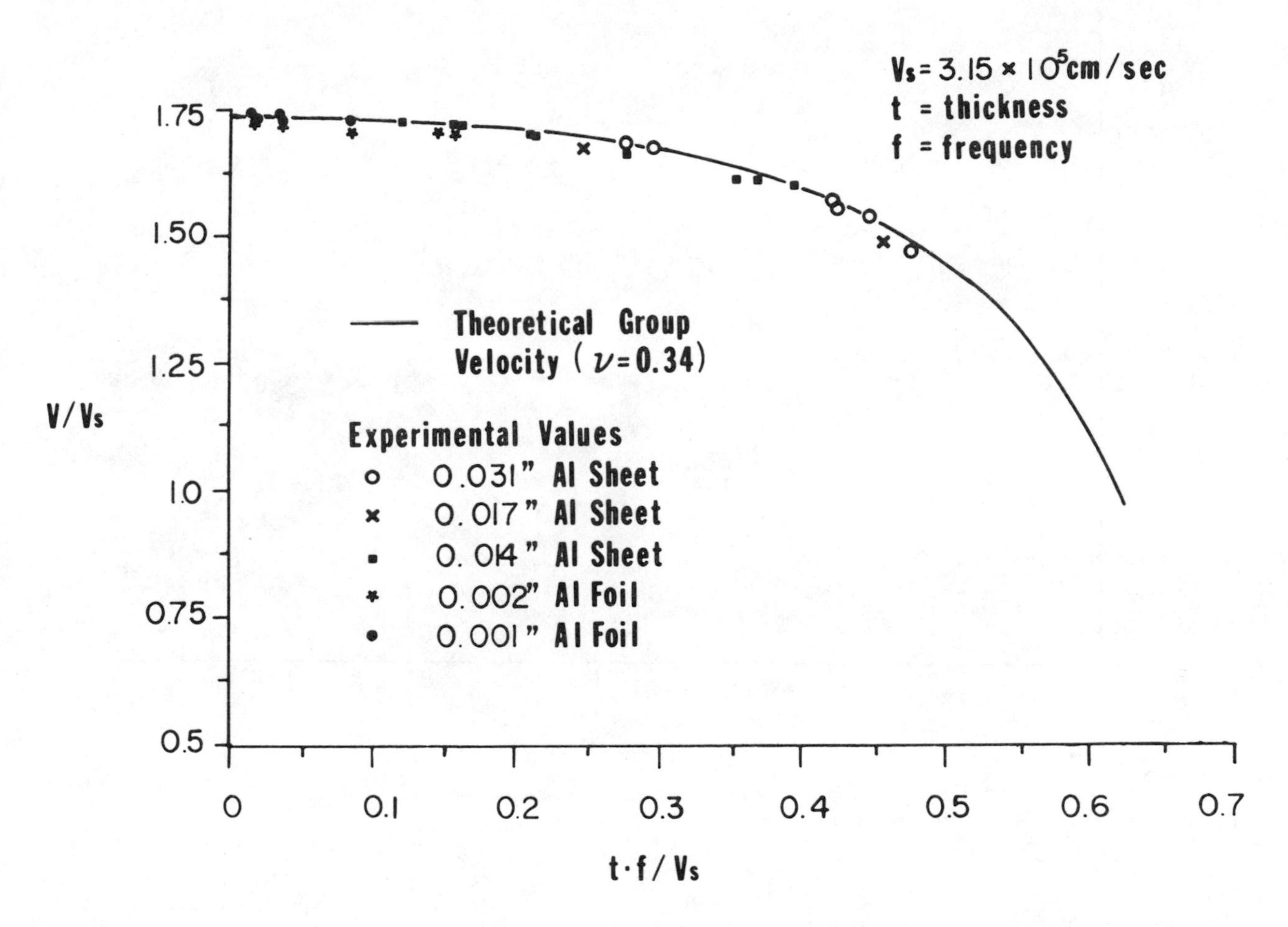

Fig. 5 - Normalized velocity of fundamental longitudinal Lamb wave in aluminum versus normalized thickness·frequency (ν=0.34)

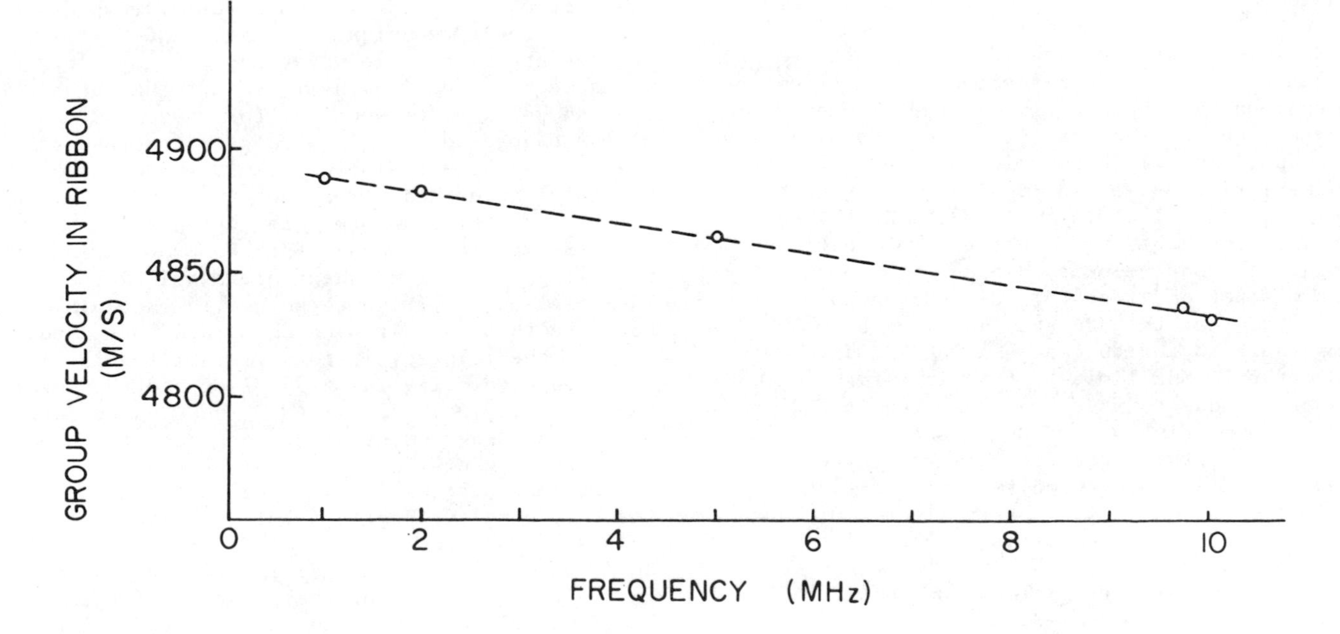

Fig. 6 - Group velocity of fundamental longitudinal Lamb wave versus frequency for Metglas ribbon (ν=0.315)

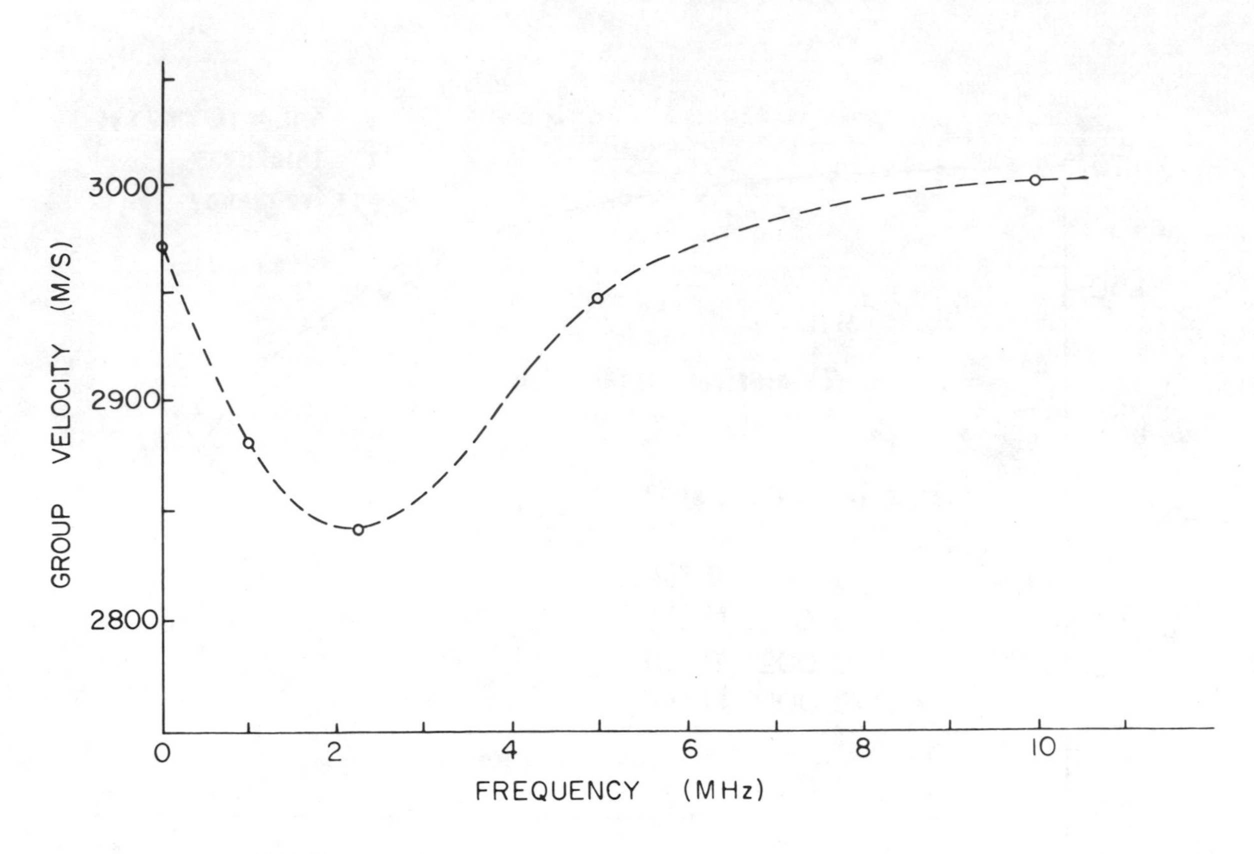

Fig. 7 - Group velocity of fundamental longitudinal Lamb wave versus frequency for amorphous Pd-Cu-Si ribbon

of zero. It is also possible for dispersion to occur due to the width of the ribbon.[11] In this case the dispersion due to ribbon width is superimposed on that due to ribbon thickness. This yields a very complicated dispersion relation in which the character of the wave is initially extensional (rod-type) then with increasing frequency changes to Lamb (plate-type) and finally to Rayleigh (surface-type). If the specimen has a square or round cross-section the wave remains extensional in character until the frequency is high enough for the transition to a Rayleigh wave to occur. The width to thickness ratio for the Pd-Cu-Si ribbons was 28 to 1. In the region of the group velocity minimum at 2.5 MHz the acoustic wavelength became comparable to the ribbon width. A group velocity dip due to ribbon thickness was not observed for Pd-Cu-Si because it is estimated that the acoustic wavelength becomes comparable to the thickness at 50 MHz which was above the 11 MHz limit of this investigation. No dispersion due to width was observed in the cases of aluminum (Fig. 5) and Metglas (Fig. 6) because the width to thickness ratios were greater than 1000 to 1. A group velocity dependence similar to that of Pd-Cu-Si was observed for amorphous Cu-Zr ribbon which had almost identical thickness and width dimensions.

A group velocity minimum due to width can, as in the case of Pd-Cu-Si and Cu-Zr ribbon, obscure the low frequency value for the velocity of the fundamental longitudinal Lamb wave, V_L. For these specimen geometries the ribbon width is small enough for the longitudinal extensional wave velocity, V_E, to be determined at the low frequency end of the dispersion curve. In this regime the acoustic wavelength is much greater than both the thickness and the width of the ribbon specimen. The low frequency extensional wave velocities were measured by Rosen et al.[1] and Chang et al.[2] for Pd-Cu-Si and Cu-Zr, respectively. Their technique utilized a Nd-Yag laser to generate ultrasonic waves and a piezoelectric transducer to receive them. This technique has been shown to be very useful for the measurement of the change in Young's modulus of amorphous Pd-Cu-Si and Cu-Zr ribbons with time during crystallization heat treatment.

The 2.1 cm by 3.5 mm dia. cylinder of Pd-Cu-Si was large enough for the propagation of bulk longitudinal and shear ultrasonic waves. The pulse-echo overlap technique,[16] which is more accurate than the pitch-catch method, was used to measure the velocity in both cases. The longitudinal wave velocity was 4495 m/sec and the shear wave velocity was 1787 m/sec. These values are in very good agreement with those reported by other researchers for the same material.[17]

Using the equations listed in Table 1 values for the Young's, shear and bulk moduli and Poisson's ratio were calculated for Pd-Cu-Si, Cu-Zr, Metglas and commercial purity aluminum. In addition, values for the velocity of the bulk longitudinal wave, V_ℓ, and the longitudinal type extensional wave, V_E, were calculated for Metglas from V_L and V_S. For Pd-Cu-Si and aluminum, V_ℓ, V_S and V_E were obtained from direct measurements Finally, for Cu-Zr, V_E and V_S were determined by

Table 3 - Summary of experimental and calculated values for sound velocities and isotropic elastic moduli of ribbon and foil specimens

Alloy	V_E(cm/s)	E(dynes/cm^2)	V_S(cm/s)	G(dynes/cm^2)	V_l(cm/s)	K(dynes/cm^2)	ν
Pd-Cu-Si	2.97×10^5	9.45×10^{11}	1.787×10^5	3.36×10^{11}	4.495×10^5	16.78×10^{11}	0.41
Cu-Zr	3.30×10^5	7.98×10^{11}	1.98×10^5	2.87×10^{11}	4.67×10^5*	12.15×10^{11}	0.39
Metglas 2605-SC	4.64×10^5	15.72×10^{11}	2.86×10^5	5.97×10^{11}	5.56×10^5*	14.56×10^{11}	0.32
Aluminum Foil	5.12×10^5	7.09×10^{11}	3.13×10^5	2.65×10^{11}	6.36×10^5	7.19×10^{11}	0.34

***Longitudinal velocities calculated from experimentally observed extensional and shear velocities.**

direct measurements and V_ℓ was calculated. Table 3 summarizes the results obtained for elastic moduli, Poisson's ratios and sound wave velocities.

SUMMARY AND CONCLUSIONS

The present investigation showed that it is possible to determine all the elastic moduli of melt spun metallic ribbons 50 μm or less in thickness. This was achieved by determination of the wave velocities V_L, V_E, and V_S along with the mass density of the materials and insertion of these values into the proper equations.

A method for extrapolating the velocity of the fundamental longitudinal Lamb wave, V_L, from measurement of the group velocity at several frequencies was utilized successfully for ribbons and foils which were many acoustic wavelengths in width.

The low frequency Lamb wave velocity, V_L, could not be determined for narrow ribbons whose width was on the same order as the acoustic wavelength. Instead, the longitudinal extensional wave velocity, V_E, was used for elastic modulus determination.

Conventional ultrasonic velocity measurement techniques and apparatus can be adapted to make accurate group velocity measurements of guided waves that propagate in ribbon and foil geometries.

ACKNOWLEDGEMENTS

This research was supported by the Defense Advanced Research Projects Agency (DARPA) under DARPA Order No. 4175. The authors wish to gratefully acknowledge Dr. Louis Testardi, and Dr. H.N.G Wadley of the National Bureau of Standards (Gaithersburg, Md.), and Dr. Emanuel Horowitz, Director of the Center for Materials Research, The Johns Hopkins University for their encouragement and support. The technical assistance of F. Biancaniello of the National Bureau of Standards, S. Fink and B. Elkind of Johns Hopkins is appreciated.

REFERENCES

1. Rosen, M., H.N.G. Wadley and R. Mehrabian, Scripta Met. 15, 1231-6 (1981)
2. Chang, J.C., F. Nadeau, M. Rosen and R. Mehrabian, Scripta Met. 16, 1073-8 (1982)
3. Davis, L.A., Y.T. Yeow and P.M. Anderson, J. Appl. Phys. 53 (7), 4834-7 (1982)
4. Chen, H.S. and J.T. Krause, Scripta Met. 11, 761-4 (1977)
5. Chou, C.-P., Phys. Rev. Letters 37, 1004-7 (1976)
6. Davis, L.A., R. Ray, C.-P. Chou and R.C. O'Handley, Scripta Met. 10, 541-5 (1976)
7. Davis, L.A., C.-P. Chou, L.E. Tanner and R. Ray, Scripta Met. 10, 937-40 (1976)
8. Krause, J.T., and H.S. Chen, "Rapidly Quenched Metals, Sec. I", pp. 425-7 (1976)
9. Testardi, L.R., J.T. Krause and H.S. Chen, Phys. Rev. B 8, 4464-9 (1973)
10. Meeker, T.R., and A.H. Meitzler, "Physical Acoustics, Vol. I, Pt. A", pp. 112-66, Academic Press, New York (1963)
11. Morse, R.W., J. Acoust. Soc. Amer. 20, 833-8 (1948)
12. Grigsby, T.N. and E.J. Tajchman, I.R.E. Trans. Ult. Eng. 8, 26-33 (1961)
13. Kolsky, H., "Stress Wave in Solids", p. 83, Dover Publications, New York (1963)
14. Davis, R.S., Metrologia 18, 193-201 (1982)
15. Gelles, I.L., J. Acoust. Soc. Amer. 40, 138-47 (1966)
16. May, J.E., 1958 I.R.E. Natl. Conv. Rec. 6 (pt. 2), 134-42 (1958)
17. L.A. Davis, "Rapidly Quenched Metals, Sec. I", p. 378, MIT Press, Cambridge (1976)

NONDESTRUCTIVE CHARACTERIZATION OF RAPIDLY SOLIDIFIED Al-Mn ALLOYS BY ULTRASONIC AND ELECTRICAL METHODS

John J. Smith
The Johns Hopkins University
Baltimore, Maryland 21218

Moshe Rosen
The Johns Hopkins University
Baltimore, Maryland 21218

Haydn N. G. Wadley
National Bureau of Standards
Washington, DC 20234

ABSTRACT

The effect of supersaturation induced by rapid solidification on the ultrasonic velocity and electrical resistivity of aluminum-manganese alloys (containing between 0.1 and 12 wt.% Mn) has been measured. Rapid solidification, achieved by melt spinning, extended the ambient temperature solubility of Mn in aluminum from the equilibrium value of 0.5 wt.% Mn to 9 wt.% Mn. The ultrasonic technique employed a Nd:YAG laser to generate ultrasonic pulses in a specimen that were subsequently detected by a piezoelectric transducer placed a known distance from the origin of the pulse. Young's moduli were calculated from the ultrasonic wave velocity and x-ray density measurements. An initial decrease of modulus was seen as the Mn concentration increased up to 3 wt.% Mn. This was followed by an increase in modulus for higher manganese concentrations. Annealing the specimens raised elastic moduli regardless of manganese concentration. Electrical resistivity, metallographic, and x-ray diffractometry were used to determine the cause of these modulus effects. It was concluded that a subtle structural transition existed in the supersaturated solid solution near the equilibrium eutectic composition. Both ultrasonic velocity and electrical resistivity measurements were shown to be suitable nondestructive evaluation techniques for on-line process monitoring of melt spun products.

INTRODUCTION

Rapid solidification processing is becoming an attractive route for producing high performance materials with improved mechanical properties. These improvements are achieved by more homogeneous alloy element distribution, reduced grain size, and non-equilibrium microstructural constituents. An area of research that has been attracting substantial interest is the production of alloys with enhanced concentrations of alloy elements in solution. Non-equilibrium extended solubility obtained through rapid solidification shows promise for the development of alloys with improved mechanical and physical properties.

The increasing technological importance of rapidly solidified alloys has created a demand for the development of nondestructive methods to assess both the microstructural state and the physical properties of the alloys produced. In the case of extended solubility, a means of ensuring that a sufficient concentration of alloy elements is retained in solid solution during solidification is essential. As important is a way of detecting unintentional losses of alloy elements from solution during subsequent thermo-mechanical treatments as components are prepared. These treatments are necessary because rapid solidification products are in powder or ribbon form and must be consolidated and sintered to form the blanks from which components are ultimately produced. In this paper we report on the use of ultrasonic velocity and electrical resistivity measurements for characterizing rapidly solidified aluminum-manganese alloys, both before and after thermal processing.

The aluminum-manganese system was chosen for investigation because significant extension of the equilibrium solubility limit is possible upon rapid solidification (1-9). Thus, the properties of aluminum-manganese solid solutions could be studied over a wide range of manganese concentrations. The equilibrium solubility of manganese in aluminum is 2 wt.% at the eutectic temperature of 657°C (930°K) as is shown in Fig. 1. At higher Mn concentrations an intermetallic compound (Al_6Mn) is formed by a peritectic transformation. Under conditions of rapid solidification this reaction is bypassed, and microcrystalline substitutional solid solutions containing up to 15 wt.% Mn reportedly have been formed (7). Increases of hardness (3,5,6,8) and ultimate tensile strength (9)

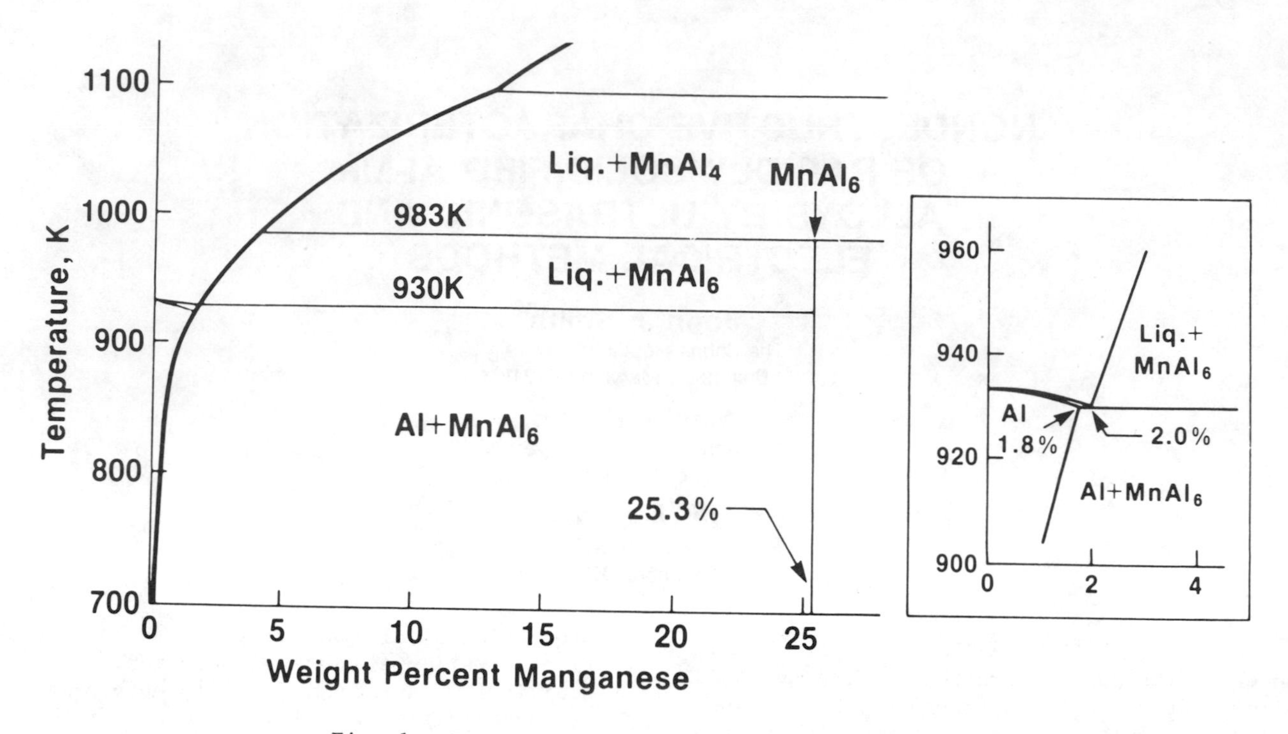

Fig. 1 - Partial phase diagram of the aluminum-manganese system with the region around the eutectic point enlarged

which are proportional to manganese content have been observed for these nonequilibrium solid solutions. Until this investigation, however, no measurements of the ultrasonic properties of rapidly solidified Al-Mn alloys have appeared in the literature.

Conventional ultrasonic testing techniques are often unsuitable for examining the powder or ribbon forms of rapidly solidified materials. In this study a laser ultrasonic technique was employed to systematically examine the use of ultrasonic velocity measurements for nondestructive evaluation of extended solubility in rapidly solidified alloys produced by melt spinning. To explore the physical basis of the ultrasonic phenomena observed x-ray diffraction, electrical resistivity, and optical microscopy measurements were made.

PRINCIPLE OF ULTRASONIC MEASUREMENTS - It is well known that elastic moduli are sensitive to alloy element concentration. Accurate values of the elastic moduli can be provided by measurements of ultrasonic velocity in a material. For a bulk specimen, both longitudinal and shear wave velocities must be known to calculate Young's modulus. Only the longitudinal velocity need be known, however, to calculate the Young's modulus of a bar with lateral dimensions <0.1 the longitudinal sound wavelength. In this case, transverse contractions (i. e., Poisson's ratio effects) occur in phase with longitudinal motions and a longitudinal wave is called an extensional wave. In an isotropic bar of constant cross-section the extensional velocity is related to Young's modulus by the equation

$$v = (E/\rho)^{1/2} \tag{1}$$

where E is Young's modulus, v is the extensional wave velocity, and ρ is the density. It is not necessary for the bar to have a circular cross-section for Eq. 1 to hold (10). Thus, Eq. 1 can be used to determine the Young's modulus of a melt spun specimen with a ribbon geometry if the sound wavelength is at least 10 times the largest dimension of the ribbon cross-section.

EXPERIMENTAL

SPECIMEN PREPARATION - Aluminum-manganese alloys were prepared by melting 5N purity aluminum and 4N purity manganese and casting under 0.5 atmospheres (380 torr) of argon. Rapidly solidified ribbons were then prepared from these alloys by the Pond melt spinning technique (11). A small charge of a casting was melt spun from a fused quartz tube (coated on the inside with a zirconia wash to prevent a reaction between the molten alloy and the quartz). The melt spinner was enclosed in a chamber under 1 atmosphere (760 torr) pressure of helium to minimize oxidation. The alloys were melted inductively and the molten metal was ejected with 10 psi (69 kPa) pressure from the fused quartz tube onto the outside of a 4 inch diameter copper wheel rotating at approximately 5000 RPM. Thin ribbon specimens, 1 to 1.5 mm wide and 30 to 50 μm thick with continuous lengths of >1 meter were produced.

For thermal stability experiments the as-spun specimens were annealed isothermally at 450°C (723°K) for varying lengths of time. Each

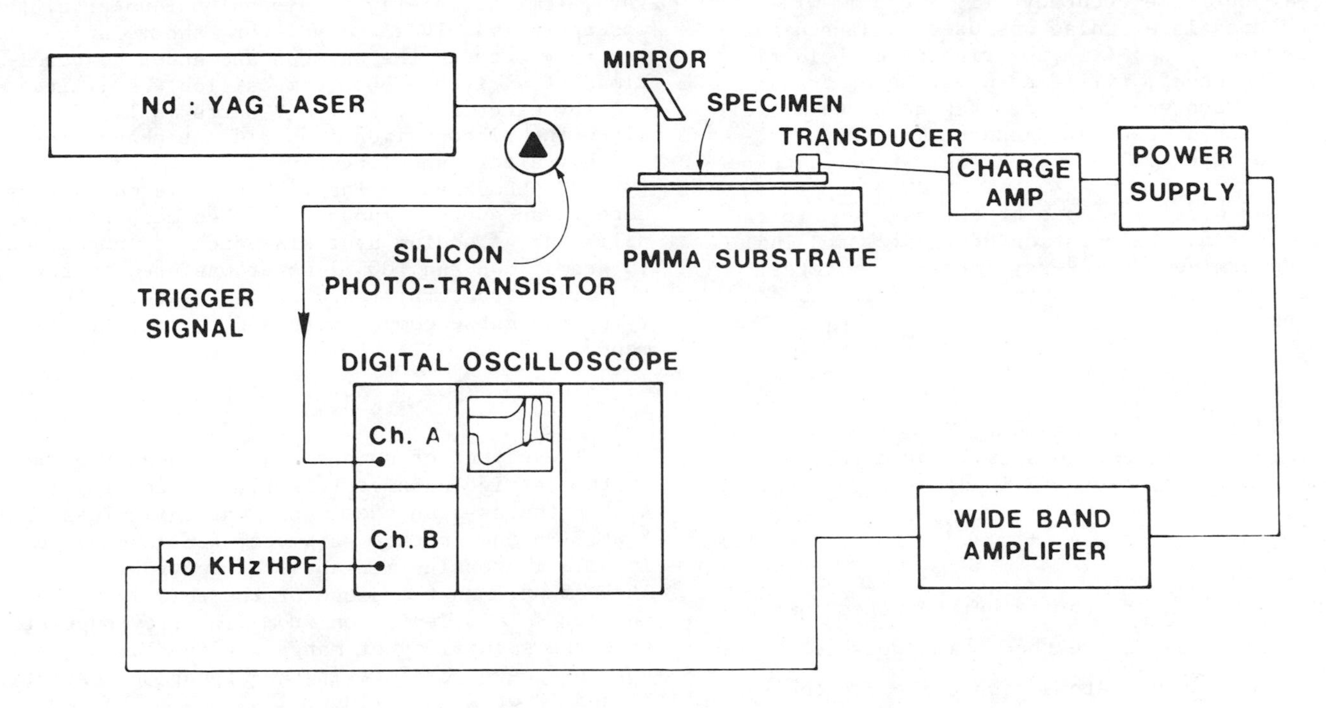

Fig. 2 - Ultrasonic velocity measuring system setup with laser generation and piezoelectric detection

ribbon was encapsulated in a borosilicate glass tube, which was then evacuated and backfilled with helium to a pressure of 1/2 atmosphere (380 torr). The encapsulated specimens were heated in a forced air convection furnace and air cooled.

LATTICE PARAMETER - Specimens 1 cm long were mounted on a glass slide and placed in an x-ray diffractometer. A correction factor was applied to compensate for systematic error due to displacement of the specimen from the diffractometer axis in a direction parallel to the reflecting plane normal. The correction was calculated by plotting the lattice parameter deduced from each diffraction peak against $\cos^2\theta/\sin\theta$ and extrapolating to a Bragg diffraction angle of 90° (i.e., $\cos^2\theta/\sin\theta=0$). The precision associated with diffractometer measurements corrected by this extrapolation function is 0.001 Å (12).

ULTRASONIC VELOCITY (YOUNG'S MODULUS) - The ultrasonic velocity measuring system is illustrated in Fig. 2. A single shot from a Q-switched Nd:YAG laser was directed at the ribbon surface. A fraction of the electromagnetic pulse was absorbed, heating the ribbon and thereby thermoelastically generating an ultrasonic pulse at one end of a ribbon. The pulse propagated the length of the ribbon and was detected by a 3 mm diameter acoustic emission transducer. The frequency of the detected waves was below 300 kHz, corresponding to an extensional wavelength of ∿1.7 cm in aluminum (11 to 17 times the specimen ribbon width), satisfying the dimensional requirement discussed above. The amplified transducer signal was digitally recorded at a 50 ns sampling rate using a 8-bit digitizer. A silicon photodiode, activated by the scattering of the light pulses from the sample was used to trigger the digitizer. The trigger and transducer signals were recorded and could be displayed on an oscilloscope screen. Transit time was determined by reading the time difference between the trigger signal and the arrival of the ultrasonic pulse at the transducer. A typical result is shown in Fig. 3.

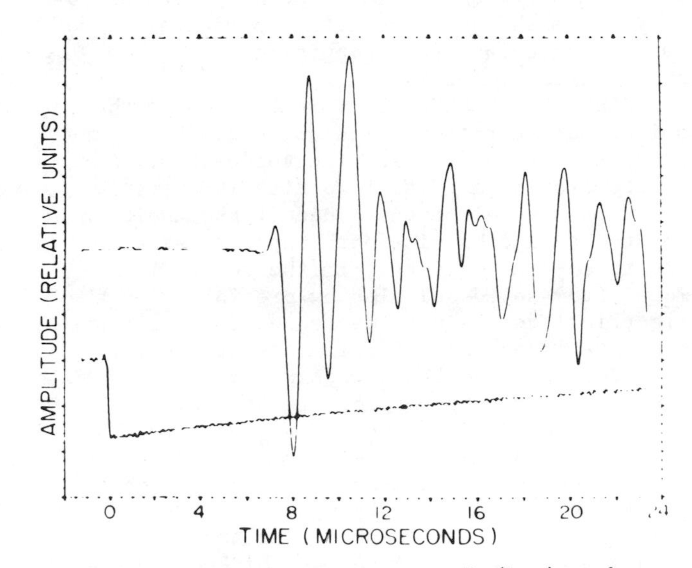

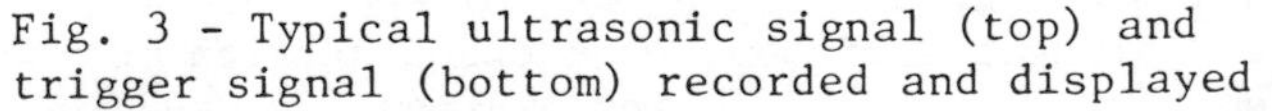

Fig. 3 - Typical ultrasonic signal (top) and trigger signal (bottom) recorded and displayed

To optimize accuracy, (i) a 0.5 mm wide line-shaped laser pulse was used to launch the ultrasonic pulse; (ii) the ribbon was held flat on a PMMA sheet; (iii) measurements of transit time for each specimen were taken at two differ-propagation distances. Accuracy of the ultrasonic velocity measurements was estimated to be 1%.

Mass density values of the as-spun ribbons (necessary for calculation of Young's modulus) were determined from x-ray lattice parameter measurements. The density can be calculated from the equation

$$\rho = \Sigma A / NV \tag{2}$$

where N is Avogadro's number, V is the unit cell volume, and ρ is the density. For a substitutional solid solution, ΣA is described by Eq. 3,

$$\Sigma A = n_1 A_1 + n_2 A_2 \tag{3}$$

where n_1 and n_2 are the numbers of atoms per unit cell and A_1 and A_2 are the atomic weights of the solvent and solute atoms respectively. The sum, $n_1 + n_2$, is a constant that equals the total number of atoms per unit cell (4 for the fcc aluminum lattice).

The x-ray technique fails in the presence of a second phase unless the volume fraction of that phase is known. This was not the case here. Therefore, the densities of several conventionally cast Al-Mn alloys, which were considered to approximate the precipitated state of the annealed specimens, were measured using a submerged balance technique described in reference (13). The differences between the calculated as-spun densities and the measured bulk densities were less than the experimental error of the velocity measurements. Thus the densities calculated from the as-spun lattice parameters were used to calculate the Young's moduli of the annealed specimens.

ELECTRICAL RESISTIVITY - Changes in the solute concentration of an alloy can be observed using electrical resistivity techniques. A solute atom in a crystal lattice is effective at scattering electrons. Hence, the addition of solute to an otherwise pure metal raises the electrical resistivity. If the solute is removed from solid solution by precipitation the electrical resistivity decreases because a precipitate is not as effective a scatterer.

The specimen resistance was measured with a four point probe potentiometric technique. A DC current of 200 milliamperes was conducted through a ribbon, and the voltage drop between two knife edges 5 cm apart was measured. The ribbon cross-section was determined by measuring the width with a travelling microscope and the thickness with a micrometer. Since the thickness and width varied along the ribbon, nine cross-section measurements were made, and the average was used to calculate the resistivity. The absolute accuracy, limited chiefly by dimensional uncertainty, was approximately ±5%. However, the relative accuracy between the as-spun and annealed conditions of a given ribbon composition was limited only by electrical errors (2 parts in 10^4), since the ribbon dimensions did not change measurably after annealing.

METALLOGRAPHY - The ribbons were mounted on edge in an epoxy compound and then mechanically polished. A sodium hydroxide etchant composed of 10 grams NaOH and 100 ml water was used to reveal the microstructural features associated with varying manganese composition and thermal treatment.

RESULTS

The effect of composition and heat treatment on the lattice parameter is illustrated in Fig. 4. In the as-spun specimens containing less than 5 wt.% Mn the lattice parameter decreased linearly with increasing manganese concentration. Above 5 wt.% Mn the slope of the relation changes. This deviation from linearity suggests that the solubility of manganese in aluminum may not have been complete in the higher Mn content (9 and 12 wt.% Mn) alloys.

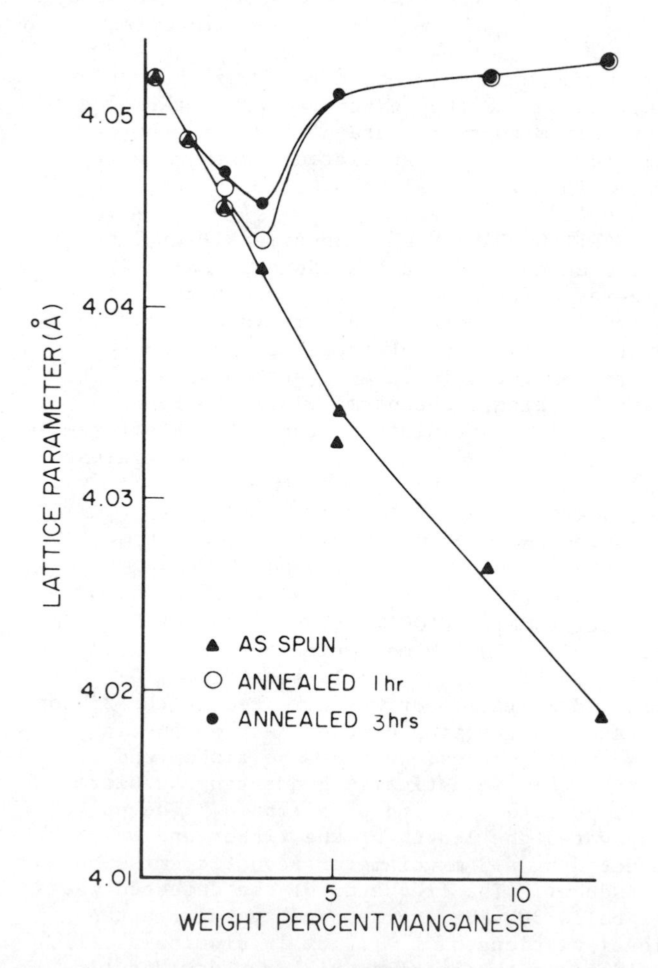

Fig. 4 - Lattice parameter variation of as-spun and annealed Al-Mn alloys

In Figs. 5 and 6 the ultrasonic velocity and Young's modulus as a function of manganese content are shown. The velocity and modulus curves have essentially the same shape, although the slope of the modulus curve is increased because Young's modulus is proportional to the square of the velocity, and the density increases linearly with manganese content (0.02 gm-cm^{-3}/wt.% Mn). It can be seen that manganese concentration had a much larger effect on velocity and Young's modulus than did the heat treatment of a particular alloy composition.

Pure aluminum ribbons had a modulus of 6.93×10^{11} dynes/cm^2 (69.3 GPa). Upon the addition of 0.1 wt.% Mn the modulus decreased sharply to 6.35×10^{11} dynes/cm^2 (63.5 GPa). The modulus decreased more slowly with increasing manganese content up to $\sim$2 wt.% Mn. Above $\sim$2 wt.% Mn the modulus increased. Above 9 wt.% Mn the rate of increase abated. The Young's modulus of the heat treated alloys was always higher than in the as-spun state. The difference due to annealing was not a simple function of concentration.

The electrical resistivity of the alloys in the as-spun state increased linearly from 0.1 wt.% to 9 wt.% Mn (Fig. 7). Above 9 wt.% Mn the resistivity leveled off, indicating that a complete solid solution did not exist in the as-spun 12 wt.% Mn specimen. Isothermal annealing at 450°C for periods as short as 5 minutes had a large effect on the electrical resistivity of the alloys with higher manganese content. After 3 hours at 450°C the resistivity of those alloys decreased to the resistivity of the annealed 2 wt.% Mn specimen, consistent with the view that the Al_6Mn precipitates, seen with x-ray diffraction, contribute little to the electrical resistivity.

Optical micrographs of ribbons with 1, 5, and 12 wt.% Mn are shown in Figs. 8, 9, and 10 respectively. In the micrographs the straight edge of each sample is the side of the ribbon that contacted the spinning wheel during solification.

A columnar microstructure, indicative of through thickness directional solidification, is seen in the 1 wt.% Mn sample. In the as-spun state, grain boundaries are slightly visible. Darkish spots seen in the structure are small pores, which may have caused the calculated densities to be somewhat overestimated. The grain boundaries are accentuated after annealing for 3

Fig. 5 - Extensional sound wave velocity of as-spun and annealed Al-Mn alloys

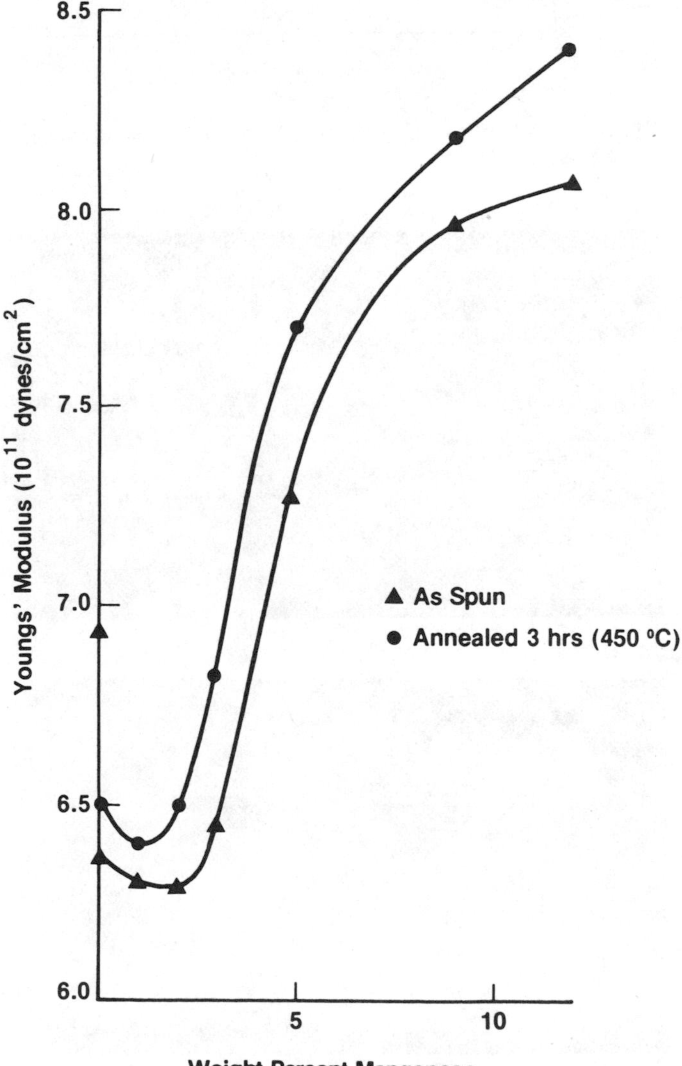

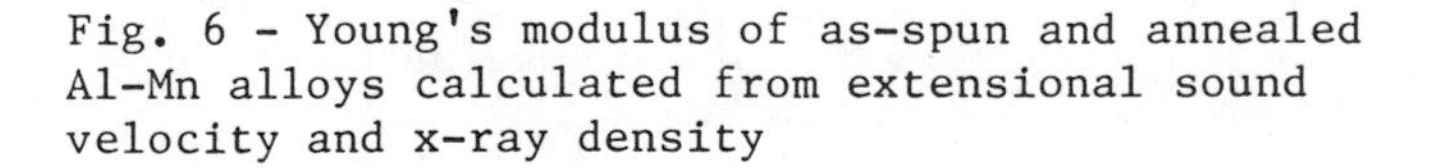
Fig. 6 - Young's modulus of as-spun and annealed Al-Mn alloys calculated from extensional sound velocity and x-ray density

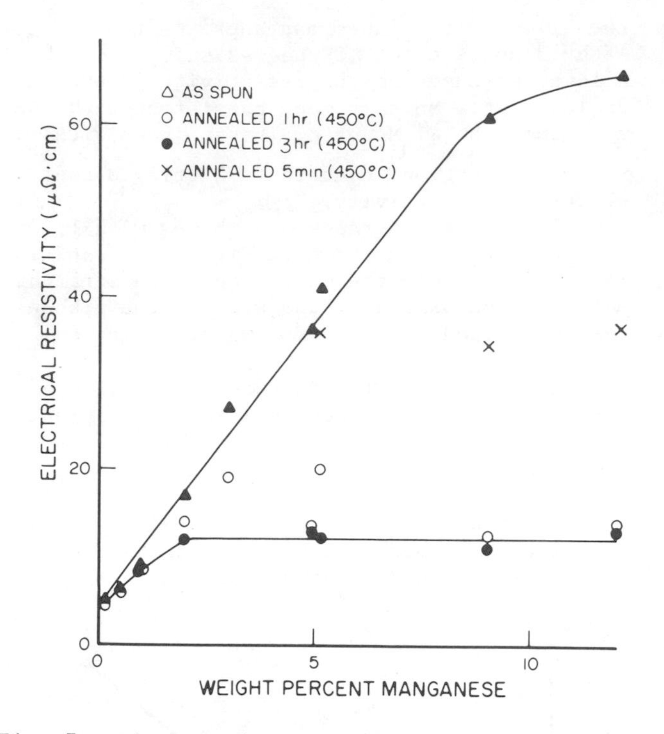

Fig. 7 - Electrical resistivity of as-spun and annealed Al-Mn alloys

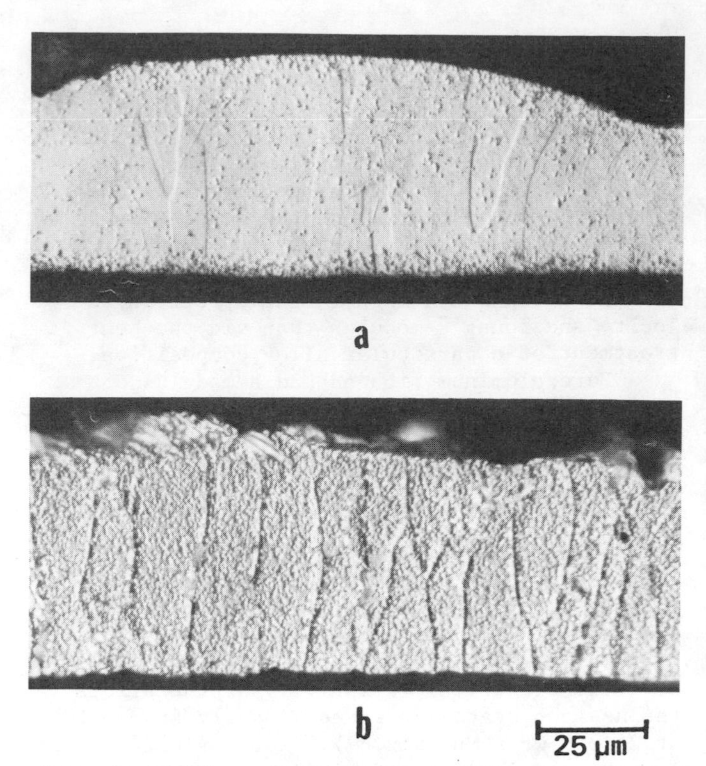

Fig. 8 - Al-1 wt.% Mn ribbon edge as quenched (a) and annealed 3 hours at 450°C (b)

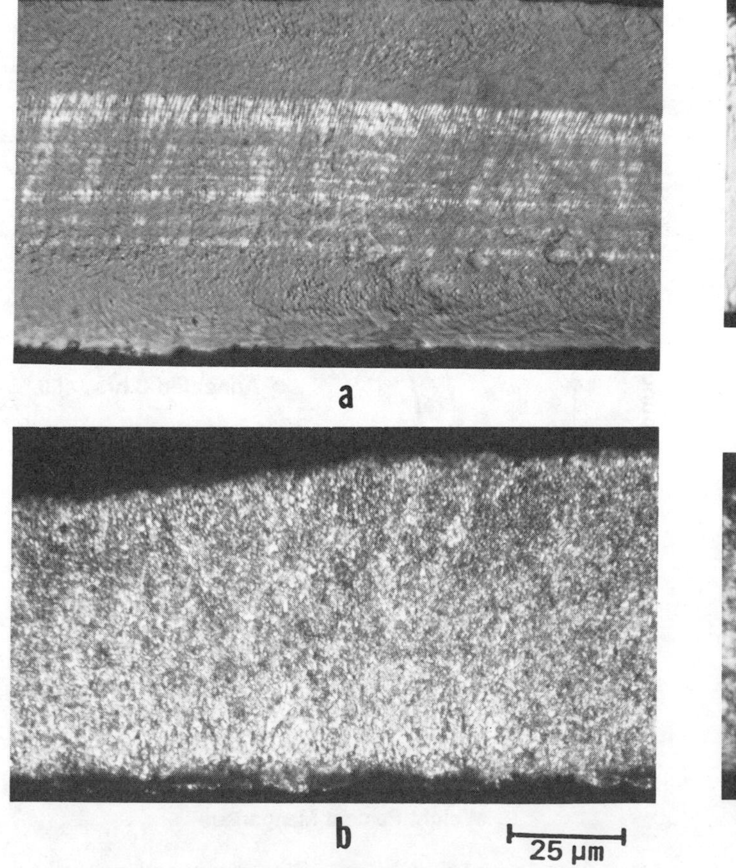

Fig. 9 - Al-5 wt.% Mn ribbon edge as quenched (a) and annealed 3 hours at 450°C (b)

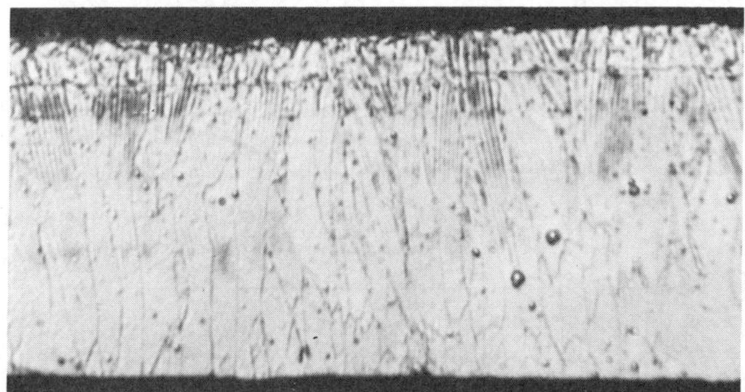

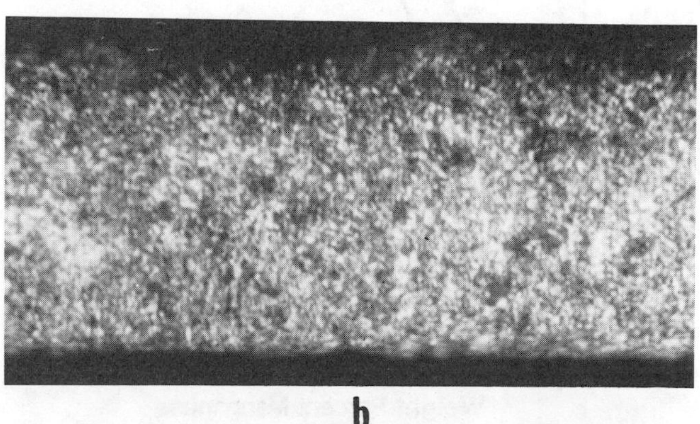

Fig. 10 - Al-12 wt.% Mn ribbon edge as quenched (a) and annealed 3 hours at 450°C (b)

hours at 450°C (Fig. 8b). In addition, a roughened surface is observed, which could be due to precipitates.

The as-spun 5 wt.% Mn specimen in Fig. 9a appears inhomogeneous only in the center, where the presence of a cellular solidification microstructure indicates a decreased temperature gradient (14). Substantial precipitation exists in the specimen annealed 3 hours at 450°C (Fig. 9b), consistent with the large resistivity decrease and lattice parameter increase reported above.

Figure 10a reveals the microstructure of the as-spun 12 wt.% Mn alloy. Some primary precipitation exists along the grain boundaries. Directional solidification is indicated by the columnar grain structure and by the solute segregation at the edge of the ribbon opposite the wheel side. Evidently, the quench rate was not sufficient to retain all 12 wt.% Mn in solution. A large amount of Al_6Mn precipitated after annealing at 450°C for 3 hours (Fig. 10b).

DISCUSSION

SOLUBILITY OF MANGANESE - The distribution of manganese in aluminum can be deduced by comparison of the different measurement results. In the as-spun state the linear changes of lattice parameter and electrical resistivity with manganese content indicate that Mn is completely in solution up to 5 wt.%. Primary precipitation in rapidly solidified specimens containing ≥ 5 wt.% Mn is revealed by microscopy (Fig. 9a) and reflected in a slope change of the lattice parameter vs. Mn concentration curve. The flattening of both the ultrasonic velocity and electrical resistivity curves above 9 wt.% Mn is linked to solute segregation in the as-spun material (Fig. 10).

Lattice parameter and electrical resistivity observations are consistent with the hypothesis that annealing at 450°C for 3 hours effectively reduces the concentration of manganese in solution to the equilibrium value of <0.5 wt.% Mn in alloys containing more than 3 wt.% Mn. In the alloys containing less than 3 wt.% Mn the small changes of resistivity and lattice parameter observed after annealing indicate that the extended solid solutions of these low composition alloys are relatively stable.

VELOCITY AND YOUNG'S MODULUS BEHAVIOR - The statically determined Young's modulus of conventionally cast Al-Mn alloys has been reported to increase linearly for compositions between 1.7 and 14.5 wt.% Mn (15). This suggests that the Al_6Mn phase has a substantially higher modulus than aluminum. Between manganese concentrations of 3 and 9 wt.% Mn the elastic modulus results of the present study confirm this finding, even though the rapidly solidified ribbons are expected to contain much higher concentrations of Mn in solution and, thus, a smaller volume fraction of Al_6Mn. The leveling of Young's modulus above 9 wt.% Mn could be a manifestation of manganese segregation (Fig. 10a). Due to the high concentration of manganese near the ribbon side not exposed to the melt spinning wheel, a substantial fraction of the specimen volume was depleted of manganese. Therefore, the bulk of the ultrasonic pulse may have propagated in material containing much less than 12 wt.% Mn.

The cause of the abrupt decrease of Young's modulus upon the addition of 0.1 wt.% Mn to pure aluminum is at this time unclear and is still under investigation. Several comments can be made, however, about the continued decrease of Young's modulus upon the addition of more Mn to aluminum. A similar decrease in Young's modulus with increasing solute content has been observed in substitutional solid solutions of some other alloy systems, notably ones based on copper and silver (16). Zener proposed a model that predicted, from thermoelastic considerations, an elastic modulus decrease in substitutional solid solutions because of solute induced lattice strain (17). Support for the notion that this can occur in Al-Mn alloys comes from a slight dip in the longitudinal sound velocity that was observed in hot pressed Al-Mn alloys at a very low Mn concentration corresponding to a solid solution (18). The extreme conditions of rapid solidification extend the solubility limit and, thus, would be expected to magnify any effect seen in equilibrium solid solutions.

According to Zener's model, Young's modulus would be expected to continue to decrease monotonically with increasing Mn solute in the as-spun ribbons. This was not observed here. Instead, the modulus rose sharply with the addition of manganese above 3 wt.% Mn. Although some precipitation was seen in as-spun alloys containing more than 3 wt.% Mn, the lattice parameter and resistivity results indicate that most of the manganese was quenched in solution. It seems surprising, therefore, that the small volume fraction of Al_6Mn that existed in the as-spun state could so strongly increase the modulus. Moreover, the precipitation of much more Al_6Mn during annealing treatments caused only a relatively small modulus increase.

This leads one to consider the possibility that the nature of the solid solution itself undergoes a transition around 2 wt.% Mn, perhaps to a semi-ordered solution. The critical composition of 2 wt.% Mn also corresponds to the onset of instability of the supersaturated solid solution, as indicated by the sharp changes of the lattice parameter and electrical resistivity upon annealing (Figs. 4 and 7). Furthermore, 2 wt.% Mn is the eutectic composition, at which the equilibrium solubility of manganese in aluminum is maximum.

The rise of Young's modulus caused by annealing at 450°C (Fig. 6) could be due to the elimination of disorder from the lattice by precipitation or clustering, thus removing the modulus-decreasing effect of lattice strain which Zener's theory addresses. Another effect of annealing could be precipitation hardening, which is known to increase slightly the Young's modulus of some Al-Cu

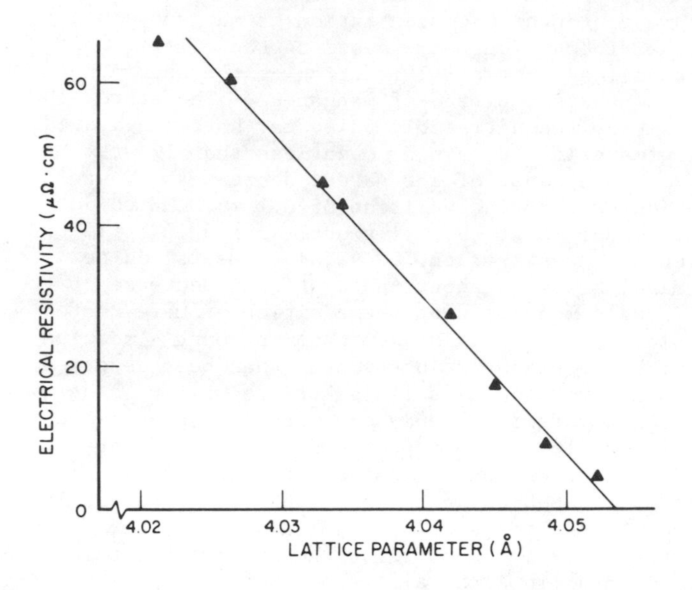

Fig. 11 - Linear relation between the electrical resistivity and lattice parameter of the as-spun ribbons

alloys (19). The contribution of precipitation to the elastic modulus increase of Al-Mn is most likely minor, however, because the increase of Young's modulus over the as-spun value is relatively independent of manganese content.

PROCESS CONTROL APPLICATIONS - Both ultrasonic velocity and electrical resistivity measurements appear to be well suited to the nondestructive evaluation of melt spun products. The two measurements complement each other. Resistivity is sensitive to the precipitation of solute from solution. A linear relationship exists between the resistivity and lattice parameter of the solid solution (Fig. 11). The ultrasonic velocity provides independent information about the elastic modulus changes that accompany microstructural transformations.

A practical approach to implementing these nondestructive techniques for on-line alloy monitoring is for measurements of electrical resistivity and ultrasonic velocity to be used in conjunction with chemical analysis to ensure that extended solubility is attained in the alloys produced. There is a need for reference resistivity and velocity data as a function of composition for this approach to be assessed more fully.

CONCLUSIONS

1. Young's modulus and electrical resistivity measurements have been made nondestructively on rapidly solidified ribbons of Al-Mn supersaturated solid solutions.

2. The measurements indicate it may be possible to infer the fraction of Mn that is dissolved in the Al matrix (forming a solid solution) provided the bulk concentration of Mn is known.

3. In both the as-spun and annealed alloys, Young's modulus has a minimum at a composition around 2 wt.% Mn.

4. The presence of manganese in Al-Mn alloys appears to have a greater effect on Young's modulus than the state in which manganese is distributed (solid solution or as part of a precipitate).

5. The physical processes that are responsible for the softening and stiffening of the modulus as the Mn concentration increases are at present not fully understood but point to the possibility of ordering in the solid solution.

ACKNOWLEDGEMENT

The authors wish to thank Mr. Charles Brady for technical assistance and Dr. Robert Schaefer for much of the x-ray data. This research was funded by the Defense Advanced Research Projects Agency under DARPA Order No. 4275. The technical monitor of this contract was Capt. S. Wax.

REFERENCES

1. Falkenhagen, G. and W. Hofmann, Z. Metallkd. 43, 69-81 (1952)
2. Salli, I. V. and G. V. Kudryumov, Dokl. Akad. Nauk SSSR (Engl. Transl.) 132 (6), 557-9 (1960)
3. Burov, L. M. and A. A. Yakunin, Russ. J. Phys. Chem. (Engl. Transl.) 39 (8), 1022-5.
4. Dobatkin, V. I., V. I. Elagin and V. M. Fedorov, Russ. Metall. (Engl. Transl.) 1970 (2), 122-7
5. Fridlyander, I. N., A. A. Kolpachev, T. V. Vukolova, and V. D. Sharabkova, Russ. Metall. (Engl. Transl.) 1979 (4), 132-5
6. Babic, E., E. Girt, R. Krsnik, B. Leontic, M. Ocko, Z. Vucic, and I. Zoric, Phys. Status Solidi A 16, K21-5 (1973)
7. Limina, L. P., Russ. Metall. (Engl. Transl.) 1968 (2), 95-9
8. Jones, H., J. Mater. Sci. Lett. 1, 405-6 (1982)
9. Yakunin, A. A., I. I. Osipov, V. I. Tkach, and A. B. Lysenko, Phys. Met. Metallogr. (Engl. Transl.) 43 (1) 120-4 (1977)
10. Kolsky, H., "Stress Waves in Solids," p. 74, Dover Publications Inc., New York (1963)
11. Pond, R. B., U. S. Patent 2,825,108
12. Cullity, B. D., "Elements of X-ray Diffraction," pp. 350-60, Addison-Wesley Publishing Company, Inc., Reading, Massachusetts (1978)
13. Davis, R., Metrologia 18, 193-201 (1981)
14. Cohen, M., B. H. Kear, and R. Mehrabian, "DARPA/AF Review, Progress in Quantitative Nondestructive Evaluation," Aug. 2-7, 1981, Boulder, Co., edited by D. O. Thompson and D. E. Chimenti, Plenum Press, New York pp. 421-32 (1982)
15. Dudzinski, N., J. R. Murray, B. W. Mott and B. Chalmers, J. Inst. Met. 74, 291-310 (1948)

16. Koster, W., Z. Metallkd. 39, 111-20 and 145-58 (1948)
17. Zener, C., Acta Crystallogr. 2, 163-6 (1949)
18. Rokhlin, L. L., Phys. Met. Metallogr. (Engl. Transl.) 28 (3), 206-8 (1969)
19. Friant, C. L., M. Rosen, and R. E. Green, Jr., "Proceedings of the 13th Symposium on Nondestructive Evaluation," April 21-23, 1981, San Antonio, Texas, edited by B. E. Leonard, Southwest Research Institute, San Antonio, Texas, pp. 123-34 (1981)

ULTRASONIC CHARACTERIZATION OF OXYGEN CONTAMINATION IN TITANIUM 6211 PLATES AND WELDMENTS

Sanford R. Buxbaum
The Johns Hopkins University
Baltimore, Maryland 21218

Robert E. Green, Jr.
The Johns Hopkins University
Baltimore, Maryland 21218

ABSTRACT

The mechanical properties of titanium alloys are known to be sensitive to oxygen content. Above a certain threshold concentration oxygen appears to have a severe embrittling effect on titanium alloys. Since gas-tungsten arc (GTA) and gas-metal arc (GMA) welding techniques are susceptible to dissolved gas contamination, it is of specific interest to investigate methods for non-destructively determining the oxygen content of a Ti-6211 welded joint. Accordingly, ultrasonic wave velocity and ultrasonic attenuation measurements were performed on a series of five plate specimens with nominal oxygen levels of 0.07, 0.14, 0.20, 0.24, and 0.29 percent by weight. Variations in the ultrasonic data were correlated with results from quantitative metallographic analysis and hardness testing. The specimens were subsequently annealed and the same battery of tests performed to gain further insight into the material properties responsible for the observed ultrasonic behavior. Similar data for welded Ti-6211 specimens will also be presented.

INTRODUCTION

Titanium alloys exhibit high strength-to-weight ratios and good corrosion resistance and are, therefore, desirable for use in structural applications. The safe, in-service use of titanium alloy weldments in critical, load-bearing parts requires the development of reliable nondestructive methods for inspecting the mechanical integrity of the weld region. Unfortunately, current nondestructive evaluation technology has proven inadequate for such alloys and applications. Since one of the primary nondestructive inspection techniques is ultrasonics, it is expedient to determine the usefulness and limitations of ultrasonic inspection for the materials in question. This requires careful acoustical characterization of these materials in order that appropriate accept/reject criteria be established. Hence, the first stage of this research was undertaken with the following objectives:

(1) Acoustically characterize weldments of Ti-6211 plate through ultrasonic wave velocity measurements.
(2) Calculate elastic moduli for base plate and weld metal from measured ultrasonic wave velocities and measured densities.
(3) Correlate ultrasonic wave velocity measurements with microstructural variations as observed through metallographic analysis and x-ray diffraction analysis.

The mechanical properties of titanium alloys are known to be sensitive to oxygen content. Above a certain threshold concentration oxygen appears to have a severe embrittling effect on titanium alloys. Since gas tungsten-arc (GTA) and gas metal-arc (GMA) welding techniques are susceptible to dissolved gas contamination, it is of specific interest to investigate methods for nondestructively determining the oxygen content of a Ti-6211 welded joint. The objective of the second stage of this research was to evaluate the feasibility of ultrasonic testing for detecting quantitatively the presence of interstitial gas contamination in weldments of Ti-6211 by:

(1) Performing ultrasonic wave velocity and ultrasonic attenuation measurements on a series of five specimens with varying oxygen contents in the as-received and annealed conditions.
(2) Comparing variations in ultrasonic data with results of scanning electron microscopy and hardness testing.

BACKGROUND

Pure titanium has a hexagonal close-packed structure (alpha phase) at room temperature, transforms to a body-centered cubic structure (beta phase) at 883°C, and melts at 1668°C. In this investigation the alloy used was titanium 6211, which has a nominal composition of 6 weight percent aluminum, 2 weight percent niobium, 1 weight percent molybdenum, and 1 weight percent tantalum with the balance ideally being pure titanium. This alloy of structural importance is classed as a near alpha alloy, because it contains primarily the alpha phase with small amounts of beta phase material interspersed.

The welded specimens evaluated in this work were joined either by a gas tungsten-arc (GTA) or a gas metal-arc (GMA) process. Gas tungsten-arc welding is a fusion welding process in which metals are joined by melting them with an electric arc which is generated between a nonconsumable tungsten electrode and the sections of metal to be welded. An inert gas or gas mixture shields the arc and the weld puddle during the welding process. Pressure may or may not be applied to the joint and filler metal may or may not be added. Plate section GTA welds in titanium and titanium alloys typically require the addition of filler metal. In gas metal-arc welding the metals are joined by melting them with an electric arc between a continuous, consumable filler metal and the sections to be welded. As in GTA welding, an inert gas or gas mixture shields the arc and the weld puddle during the welding process (1). In both GMA and GTA welding the filler metal used was Ti-6211. Contaminants such as oxygen, nitrogen, hydrogen, and carbon can be introduced into the weld by incomplete gas shielding and/or slight contaminant gas impurities in the inert cover gas mixture.

Ultrasonic wave propagation analysis of a material is a nondestructive evaluation technique that provides information about the elastic properties and absorption characteristics of the material in which the wave propagates. Absolute measurements of shear and longitudinal wave speed can be used to calculate useful material parameters, such as the effective Young's modulus and the effective shear modulus. The term "effective modulus" refers to the fact that the modulus is calculated for a material that is assumed to be linear elastic, homogeneous, and isotropic, which is a fair approximation for some fine-grained polycrystalline materials. The energy loss, or ultrasonic attenuation of an elastic wave propagating in a solid, may be divided into contributions from geometrical and intrinsic effects. Of interest here are the intrinsic effects which include scattering of the ultrasonic wave by inhomogeneities, conversion of sound energy to heat as a result of elastic deformation, interaction with thermal phonons, and dislocation damping (2).

Since the concentration of an alloying element, oxygen in this case, affects the same set of variables that determine ultrasonic wave propagation characteristics, it was expected that ultrasonic techniques would be suitable for determining oxygen concentration in contaminated Ti-6211 specimens. It has been shown that foreign solute atoms invariably change elastic moduli (3-6). Additionally, solute atoms can act as pinning points for dislocations and would, therefore, be expected to influence attenuation measurements. Some previous studies of the effects of interstitial gases on elastic wave propagation also indicate the feasibility of this method for determination of dissolved gas concentration. Hsu and Conrad (6) looked at the effect of oxygen on the elastic properties of titanium-oxygen alloys through ultrasonic wave velocity measurements performed on specimens with oxygen content in the range of 0.04 to 2 weight percent and controlled impurities. It was found that both density and longitudinal wave velocity increased with increasing amount of oxygen. Since the relative change in the wave velocity was greater than that in the density, the dynamic elastic modulus was increased by the oxygen in solid solution. Ultrasonic attenuation was also observed to increase with increasing oxygen content. The authors concluded that it was likely that the interstitial oxygen atoms in the titanium lattice actually increased the binding forces between atoms.

EXPERIMENTAL PROCEDURES

The material used in the present investigation, Ti-6211, was examined using various destructive and non-destructive techniques. These techniques included scanning electron microscopy, ultrasonic wave velocity and attenuation measurements, density measurements, Rockwell "C" hardness measurements and x-ray diffraction analysis.

METALLOGRAPHY - Metallographic specimens were taken from sections cut from the weld region, the base metal, and from the oxygen contaminated Ti samples. The specimens were ground on successively finer silicon carbide papers down to 600 grit, and then polished on lapidary wheels using 15 micron and 0.05 micron compounds. A two-step etching technique was used to reveal microscopic detail. In the first step, the polished specimen was briefly swabbed with a 2 ml HF, 98 ml water solution. This etched the alloy and stained the alpha phase. The second step, swabbing with a 1 ml HF, 2 ml HNO_3, 97 ml water solution, removed the stain, leaving a dark field of alpha phase material in which the beta phase appeared as finely dispersed white lines when viewed through a scanning electron microscope (SEM).

ULTRASONIC MEASUREMENTS - The specimens used for ultrasonic examination were fabricated from welded specimens and oxygen contaminated specimens provided by the Naval Research Laboratory in Washington, D.C. These plate-section weldments required many welding passes and were not suitable for ultrasonic inspection with contact transducers in the as-received condition due to surface irregularities resulting from welding. The oxygen contaminated specimens were also unsuitable in the as-received condition for contact ultrasonic inspection due to poor surface finish. Flat parallel faces were machined on each specimen prior to performing both ultrasonic measurements and hardness measurements. A high degree of parallelism of specimen faces was required to minimize diffraction errors caused by divergence of the ultrasonic pulse as it propagated back and forth through the specimen.

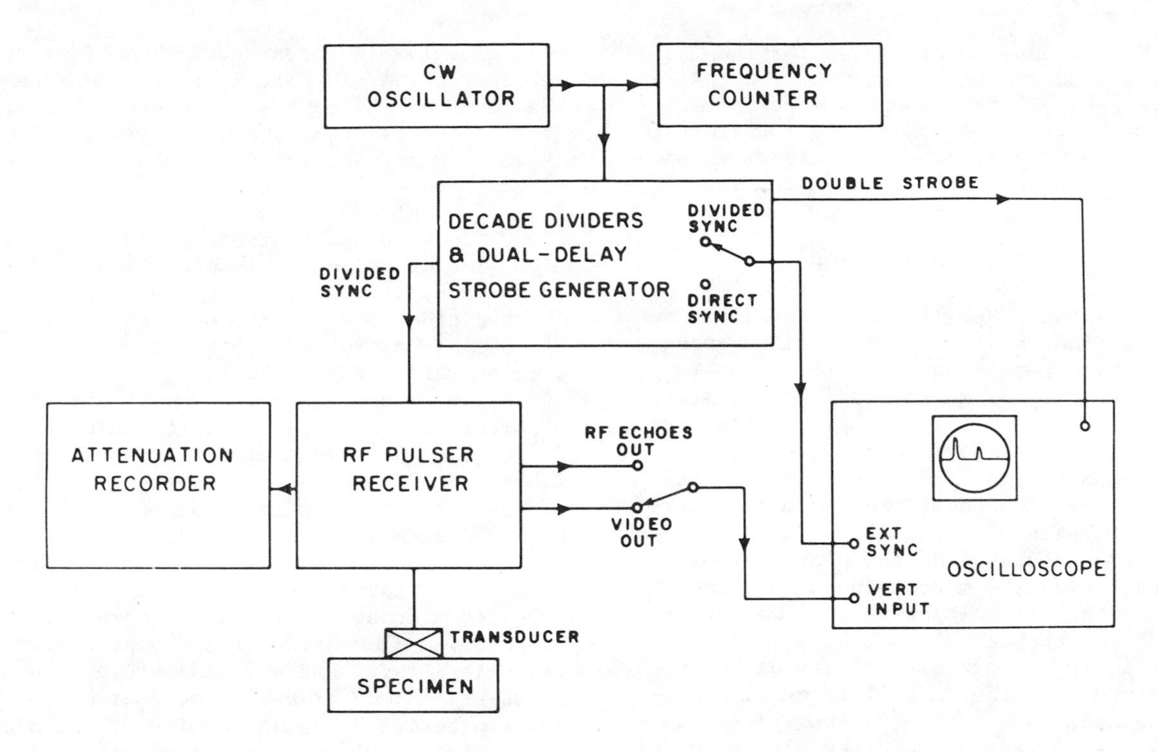

Fig. 1 - Ultrasonic Wave Velocity and Attenuation Measuring System

The ultrasonic wave velocity and attenuation measurement system used in the present research is shown schematically in Fig. 1. Conventional pulse-echo overlap techniques as described by Chung, Silversmith, and Chick (7) were used to measure the ultrasonic wave velocities. Attenuation measurements were made with an automatic attenuation recorder, which includes a time gate permitting selection of any two echoes from the received wave train, an automatic gain control to stabilize the amplitude of the first echo, and circuitry to obtain the logarithm of the ratio of two selected echo amplitudes and display the result in decibels.

Commercial, longitudinal wave, ceramic transducers along with appropriate couplants were used in this research. An Aerotech couplant (a light oil) was used for the wave velocity measurements. A 2 mm thick elastomer coating (developed by Martin Marietta Laboratories in Catonsville, MD) was applied to a 2.25 MHz, longitudinal wave, commercial transducer and was used to acoustically couple the transducer to the oxygen contaminated specimens for the attenuation measurements. The viscoelastic properties of this coating make it an ideal couplant for attenuation measurements. When subjected to low frequency stress cycles (low rated of deformation), the elastomer is flexible and can be forced down on a sample for a reproducible acoustical couple. When subjected to high frequency stress cycles (high rates of deformation) the coating behaves like a rigid body. The ultrasonic pulses effectively propagated through a rigid acoustical coupling media. This elastomer coating provided more reproducible attenautio measurements than was possible with conventional couplants. The thickness of the coating made it unsuitable for accurate velocity measurements.

HARDNESS MEASUREMENTS - Rockwell "C" hardness measurements were made on the oxygen contaminated Ti-6211 specimens in accordance with ASTM Standard E18. A diamond tipped "Brale" indentor with a 150 kg load was used. Since Ti-6211 continued to exhibit plastic flow after the application of the major load, the dial indicator continued to move after the operating lever stopped. For this reason the operating lever was brought to its latched position at an elasped time of 30 seconds between application and removal of load.

DENSITY MEASUREMENTS - Cubes nominally 1 inch on a side were machined from each of the oxygen contaminated Ti-6211 specimens prior to annealing. The volume was measured to an accuracy of 0.001 cm^3 and the mass was measured to an accuracy of 0.001 g. Two cubes were fabricated from each of the five oxygen contaminated specimens for mass density determination.

X-RAY DIFFRACTION ANALYSIS - In order to investigate differences in the relative amounts of alpha and beta phase material between the weld region and the base metal and changes in the relative amounts of alpha and beta phase material with increasing oxygen content in Ti-6211, x-ray diffractometer measurements were performed. A proportional counter was used to record intensities and angular positions of the diffracted beam filtered at the detector by a thin foil of nickel. The x-ray tube had a copper target (Cu K_α radiation, λ=1.54Å) and was generally operated at 44.5 kV and 33 mA. The diffractometer was equipped with a beam slit of 3° MR, a soller slit of medium resolution, and a detector slit of 0.2°. This produced a rectangular beam of primary radiation. The target angle was 4° with a scan rate of 4° per minute. Full scale for the detector was 5000 counts per second (100 cps = 1 milliroentgen/hr). The samples were initially scanned through 2θ angles of 10° to 100°; after appropriate peaks were chosen for closer examination, the samples were only scanned through angles of 33° to 45°.

RESULTS AND DISCUSSION

METALLOGRAPHY - Metallographic analysis revealed significant microstructural differences between the base metal and the weld region in the Ti-6211 alloy. At low magnifications the microstructures observed in the base plate metal are broken up and show texturing due to rolling. Prior beta phase grain boundaries are decorated by the alpha phase, because the grain boundaries are the first regions to transform during cooling at higher temperatures. Narrow platelet formations characterize the microstructure of the alpha phase; the remaining beta phase is interspersed between the acicular alpha platelets.

The weld region exhibits a very different microstructure from the base plate metal. Acicular alpha and alpha prime (martensitic-type structure) has formed on preferred planes in the prior beta crystals. The alpha platelets are narrower, straighter, and more sharply defined in the weld region than in the base metal. This is probably a result of the faster cooling rates present during welding. The boundaries of the large, equiaxed prior beta grains are decorated by alpha metal as in the base plate. Figure 2 illustrates the microstructural differences between weld metal and base metal.

Metallographic analysis performed on the five as-received oxygen contaminated specimens showed distinct changes in microstructure with increasing oxygen content (Fig. 3). As the oxygen content increased, the alpha platelets became wider; hence, the separation of the layers of beta increased. Additionally, the beta layers thickened until they began to break up and become more globular in appearance. This is especially evident in the SEM photomicrograph of specimen E, the most heavily contaminated specimen. It should be noted that the microstructures of the oxygen contaminated plates are not representative of those observed in the weld region of welded Ti-6211 specimens.

The oxygen was introduced into the molten Ti-6211 samples in the form of TiO_2 powder, and the samples were subsequently beta processed (heated and rolled in the beta phase field at 925°C in a controlled, inert atmosphere). The samples were then air-cooled. Chemical analysis of the specimens was performed by the RMI Company of Niles, Ohio, producers of the plate samples. The results of this analysis are presented in Table 1.

After annealing (one hour at 1050°C in inert atmosphere, helium cooled) the five oxygen contaminated specimens exhibited very similar acicular microstructures (see Fig. 4). The annealed microstructures were coarser than the as-received microstructures (compare magnifications in Figs. 3 and 4). As before prior beta phase grain boundaries are decorated by alpha phase material. The uniformity of the microstructures enables one to eliminate scattering as a potential cause of any wave velocity changes due to varying oxygen content.

DENSITY MEASUREMENTS - The results of density measurements performed on this series of five as-received oxygen contaminated specimens are presented in Fig. 5. Note that the mass density (ρ) increases with increasing oxygen content, because the oxygen enters the titanium lattice interstitially. These results are consistent with the results of Hsu and Conrad (6).

ULTRASONIC WAVE VELOCITY MEASUREMENTS - Longitudinal wave velocity was measured at various locations along several welded specimens. Data for such a scan are shown in Fig. 6. The sound wave travel path corresponds to the vertical path up and down the paper. Wave velocity was lowest

Base Metal
10 µm

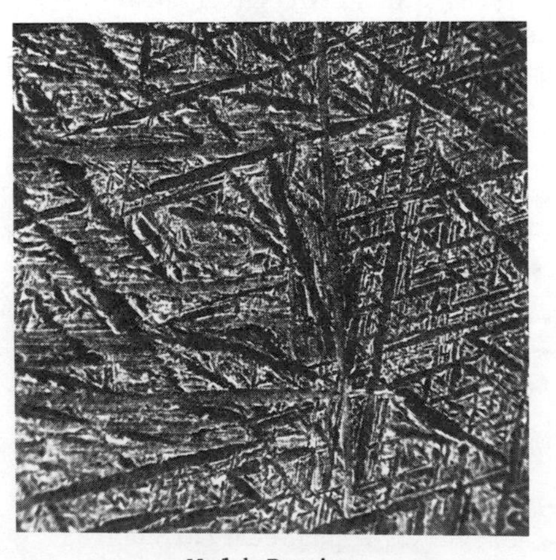

Weld Region
10 µm

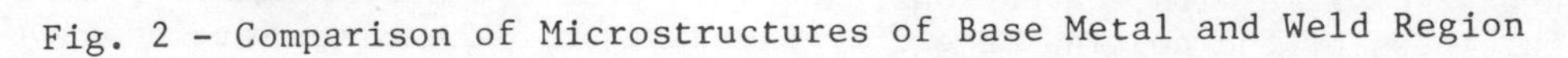

Fig. 2 - Comparison of Microstructures of Base Metal and Weld Region

SPECIMEN A
0.069 wt% OXYGEN

SPECIMEN B
0.136 wt% OXYGEN

SPECIMEN C
0.194 wt% OXYGEN

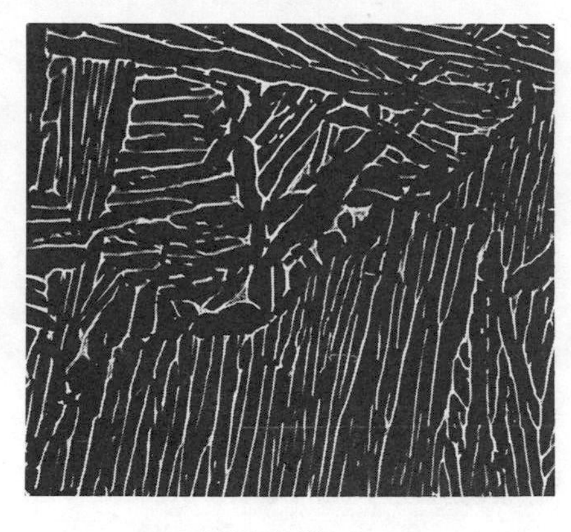

SPECIMEN D
0.238 wt% OXYGEN

Fig. 3 - Scanning Electron Photomicrographs of Ti-6211 Plate with Varying Oxygen Contents (⊢—⊣ 10 μm.)

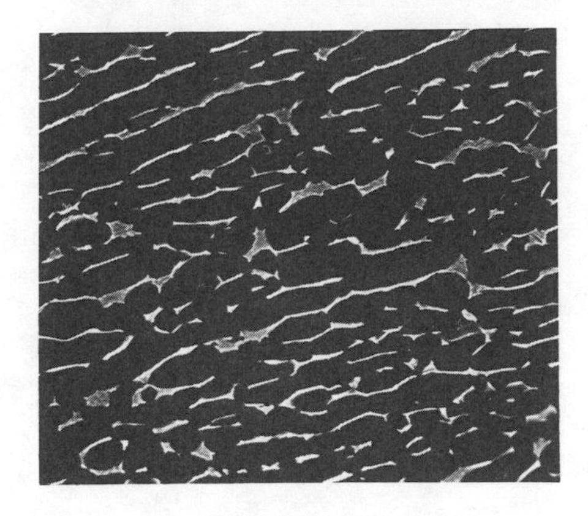

SPECIMEN E
0.290 wt% OXYGEN

Table 1 - Specimen Composition

SPECIMEN	COMPOSITION (weight percent)								
	Ti	Al	Nb	Mo	Ta	C	N	Fe	O
A	90.3	6.0	1.95	0.7	0.88	0.02	0.010	0.05	0.069
B	89.9	6.0	2.09	0.9	0.97	0.02	0.006	0.03	0.136
C	90.2	5.9	1.90	0.8	0.94	0.02	0.005	0.03	0.194
D	90.1	5.8	2.05	0.8	0.99	0.02	0.006	0.03	0.238
E	89.8	5.9	2.16	0.7	1.06	0.03	0.008	0.03	0.290

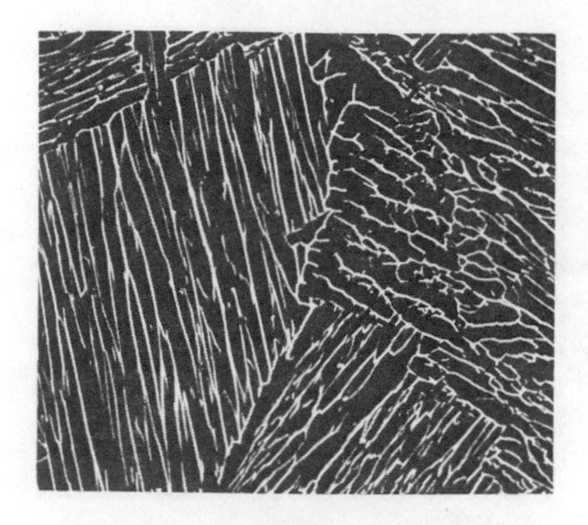

SPECIMEN A
0.069 wt% OXYGEN

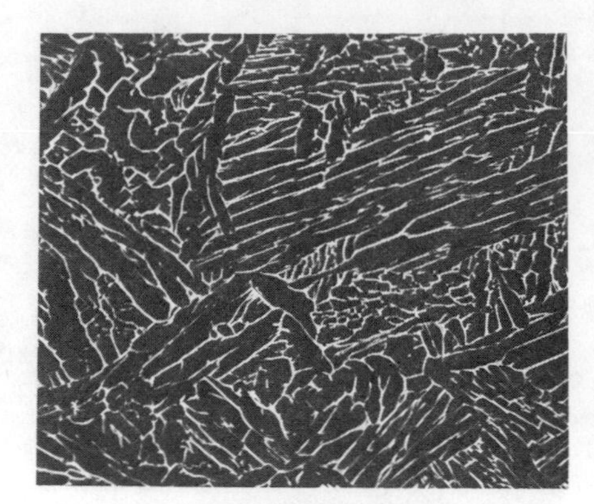

SPECIMEN B
0.136 wt% OXYGEN

SPECIMEN C
0.194 wt% OXYGEN

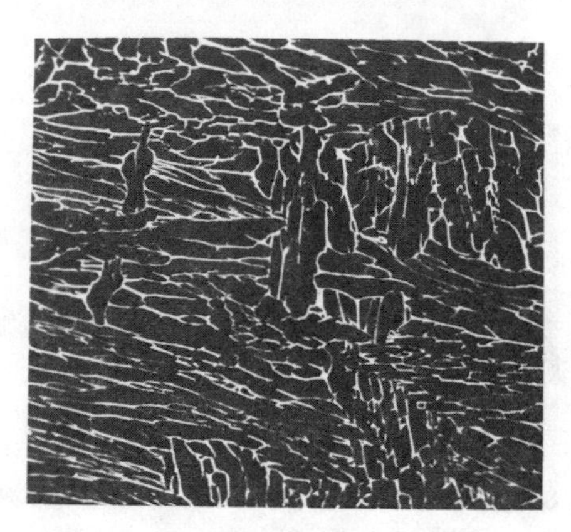

SPECIMEN D
0.238 wt% OXYGEN

Fig. 4 - Scanning Electron Photomicrographs of Annealed Ti-6211 Plate with Varying Oxygen Contents (|——| 50 μm.)

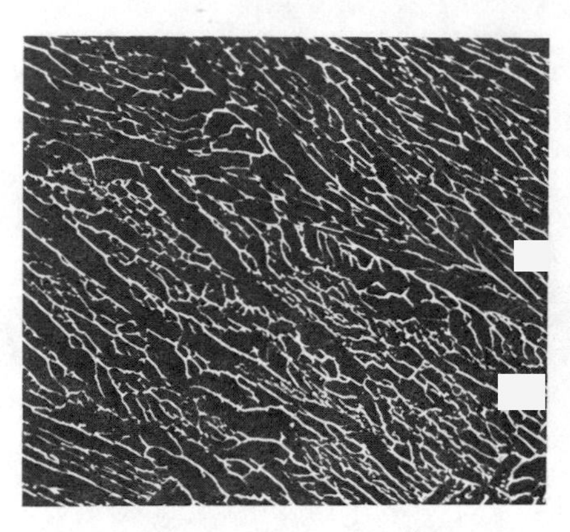

SPECIMEN D
0.290 wt% OXYGEN

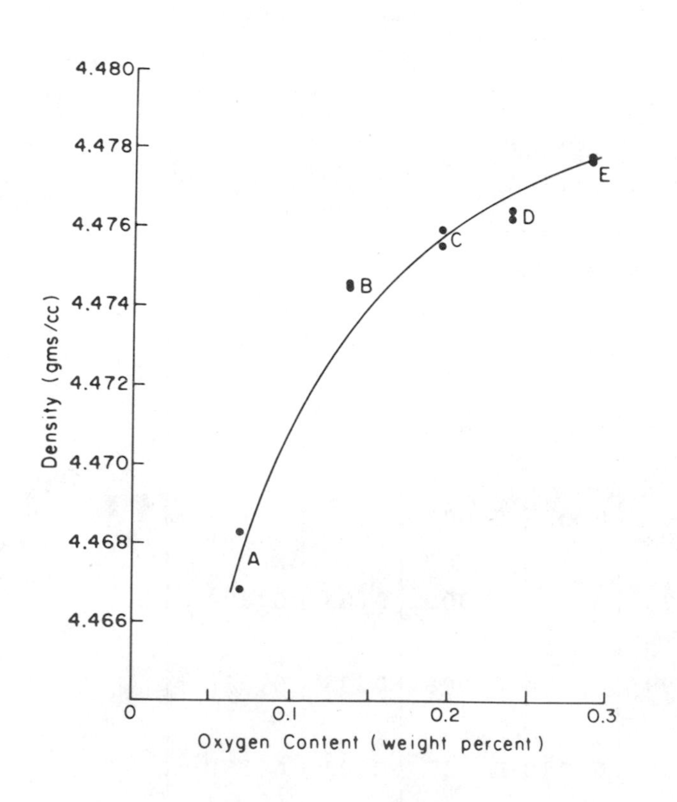

Fig. 5 - Mass Density vs. Oxygen Content

in the center of the weld where the ultrasound primarily passed through weld metal. The velocity difference from weld metal to base metal was about one percent. Peaks in the velocity versus distance curve occurred just outside the weld zone. Further from the weld the velocity decreased slightly as it returned to the unaffected base plate value.

Longitudinal wave velocity (V_L) was found to decrease with increasing oxygen content in the as-received plate specimens (see Fig. 7). Specimen C gave anomalous velocity results; it is suspected that this was due to improper thermomechanical preparation. Figure 8 is a plot of the average product ρV_L^2 (normalized) versus oxygen content for the five specimens. The behavior of the quantity ρV_L^2 is indicative of the behavior of the elastic moduli, so Fig. 7 shows that the elastic moduli tend to decrease with increasing oxygen content.

Discussion of the observation of a decrease in the longitudinal wave velocity requires consideration of the material properties that affect V_L in both the as-received oxygen contaminated Ti-6211 specimens and the welded specimens. In 1949 Zener (3) theoretically demonstrated that the residual stress introduced by slight interstitial alloying in any metal necessarily reduces the elastic moduli. However, his theory was based on purely geometric considerations and did

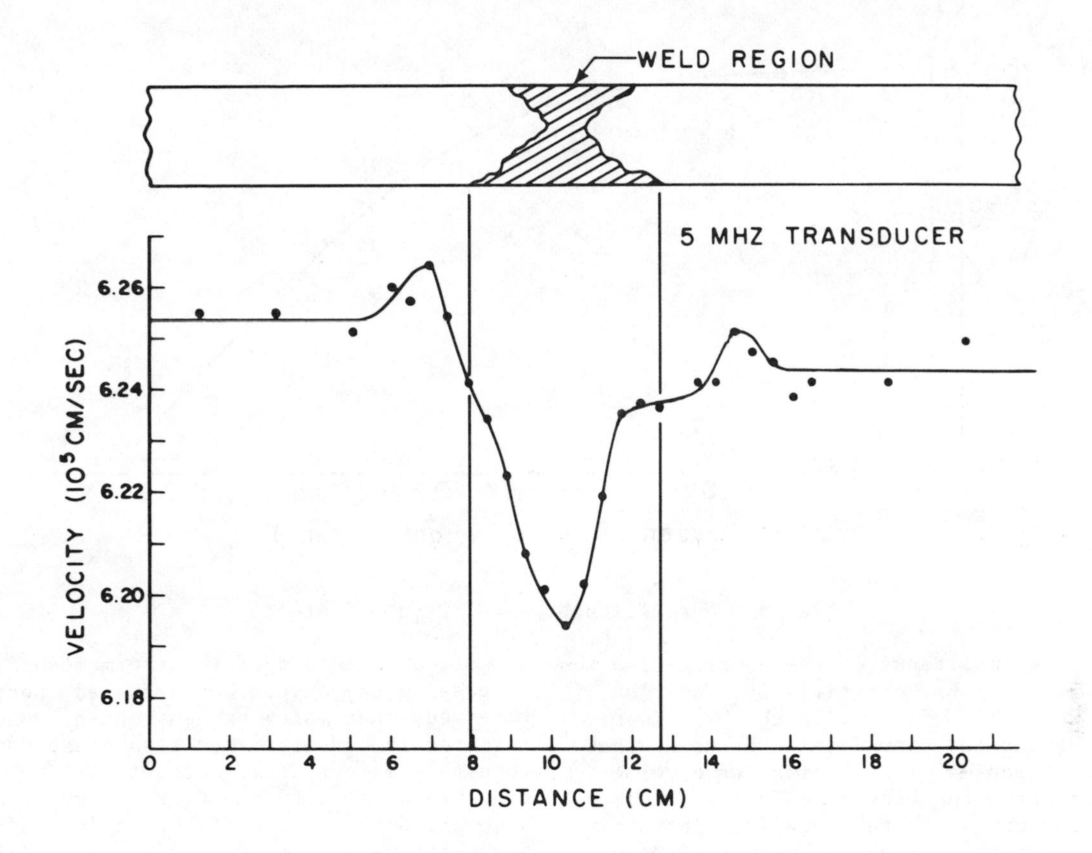

Fig. 6 - Longitudinal Wave Velocity vs. Transducer Location Along a Welded Ti-6211 Specimen

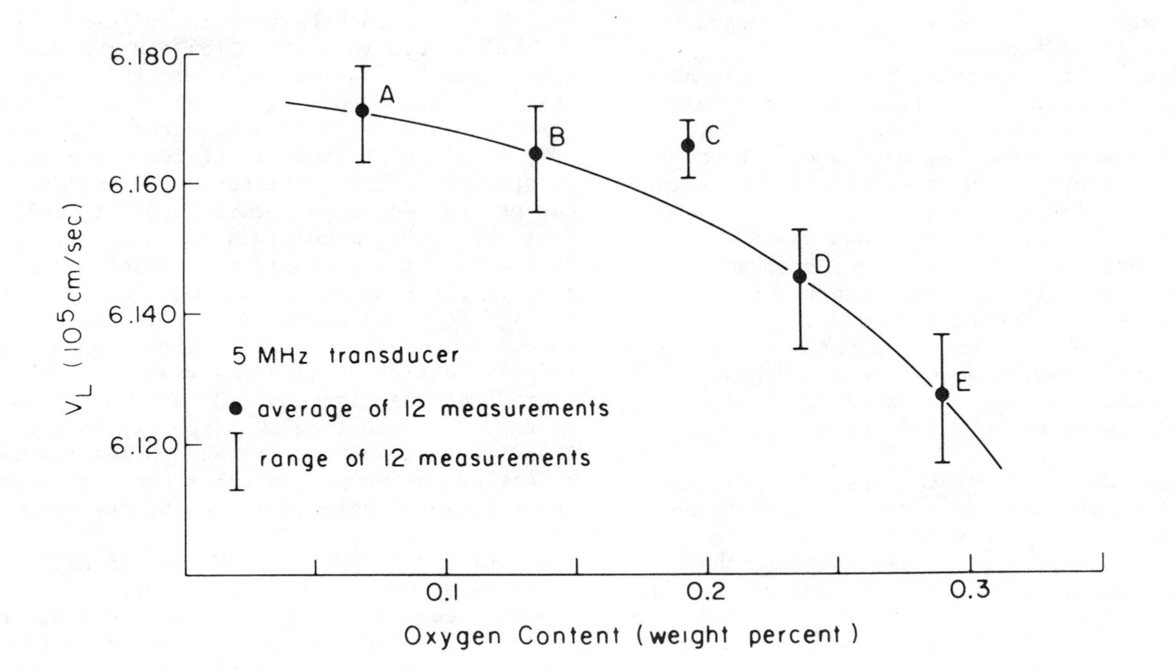

Fig. 7 - Longitudinal Wave Velocity vs. Oxygen Content

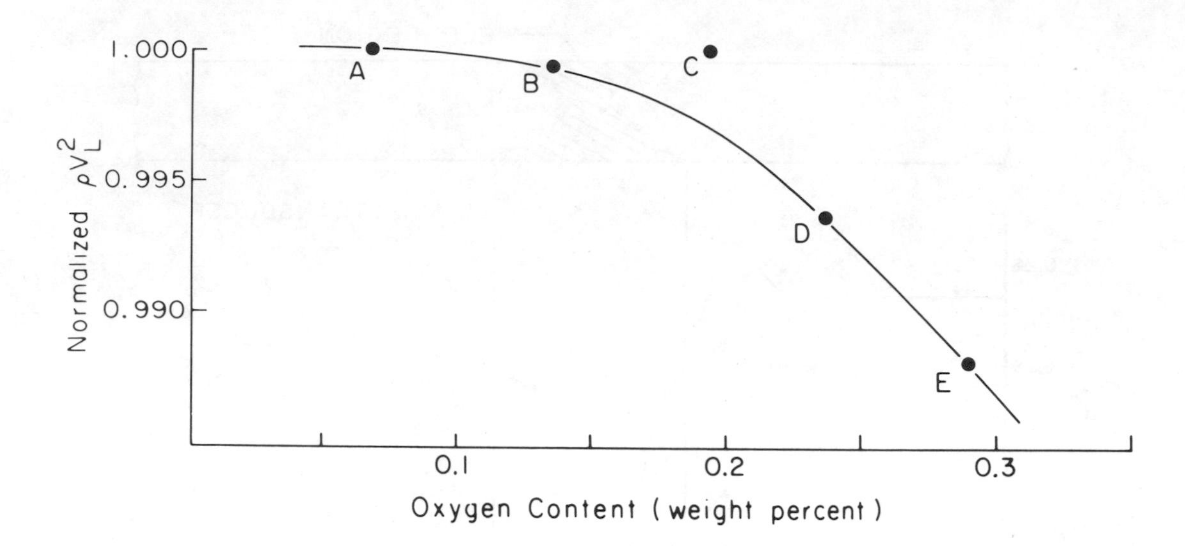

Fig. 8 - Normalized ρV_L^2 vs. Osygen Content

not consider the importance of the complex Ti-O electronic interaction. Additionally, Zener's conclusions were in conflict with the experimental results of Hsu and Conrad (6) who showed that in 100% alpha titanium, longitudinal wave velocity and, therefore, the elastic moduli, increased with increasing oxygen content. Hence, it is doubtful that the theoretical arguments of Zener (3) account for the results obtained here. In fact, if the only parameter changing was the amount of oxygen in the alpha-phase lattice of the Ti-6211 alloy, following Hsu and Conrad (6) we would expect V_L to increase with oxygen content rather than decrease.

Changes in the relative amounts of the alpha and beta phase could also affect the wave velocity in both the welded specimens and the oxygen contaminated specimens. As mentioned before, the Ti-6211 alloy is a near alpha alloy containing primarily alpha phase (HCP) metal with some beta phase (BCC) metal. The sound wave velocity should be lower in the beta phase, because it has a more open crystal structure than the alpha phase. Hence, the decreased sound velocity observed in the weld region could possibly be caused by an increased amount of beta phase metal present in the weld region. X-ray diffractometer measurements were performed to test this hypothesis.

Initial diffractometer scans indicated that the BCC (110) and the HCP (101) peaks could be used to qualitatively assess changes in the relative amounts of alpha phase and beta phase material. Subsequent scans demonstrated that the weld region of the specimen tested contained more beta phase metal than the base plate. These results, along with consideration of the thermal history of the weld zone, support the hypothesis presented above. The rapid cooling rates present during welding should result in more retained, high temperature beta phase metal and, hence, lower sound wave velocities.

If as a result of the thermomechanical processing the oxygen contaminated specimens received, increasing oxygen content somehow resulted in more retained beta phase material, then this effect could account for the longitudinal wave velocity behavior observed in Fig. 7. However, both x-ray diffraction analysis and quantitative metallographic analysis failed to support this hypothesis and in fact indicated that, as would be expected, the amount of beta phase actually tended to decrease with increasing oxygen content.

Other factors that could influence V_L would be differences in texture (8) and residual stress introduced by rolling the plate specimens. X-ray texture measurements performed at the Naval Research Laboratory in Washington, D.C. showed very little difference in texture from specimen to specimen. The contribution of residual stress can be studied by performing a stress relieving heat treatment (anneal) on the specimens and then repeating the ultrasonic wave velocity measurements. The results of this work are presented in Fig. 9; after annealing the longitudinal wave velocity increased slightly with oxygen content. This is consistent with the results of Hsu and Conrad (6) and Mignogna (12) which are also shown in Fig. 9. Annealing the five oxygen contaminated Ti-6211 specimens enables one to see the real effect of the oxygen on the elastic properties of the titanium lattice unmasked by any residual stress effects.

ULTRASONIC ATTENUATION MEASUREMENTS - Attenuation measurements were performed on the as-received oxygen contaminated specimens to evaluate the usefulness of attenuation for quantitatively detecting the presence of oxygen. The elastomer coating discussed above provided a fairly reproducible transducer-specimen coupling and helped decrease the data scatter typically encountered in attenuation measurements with contact transducers. Between 24 and 32

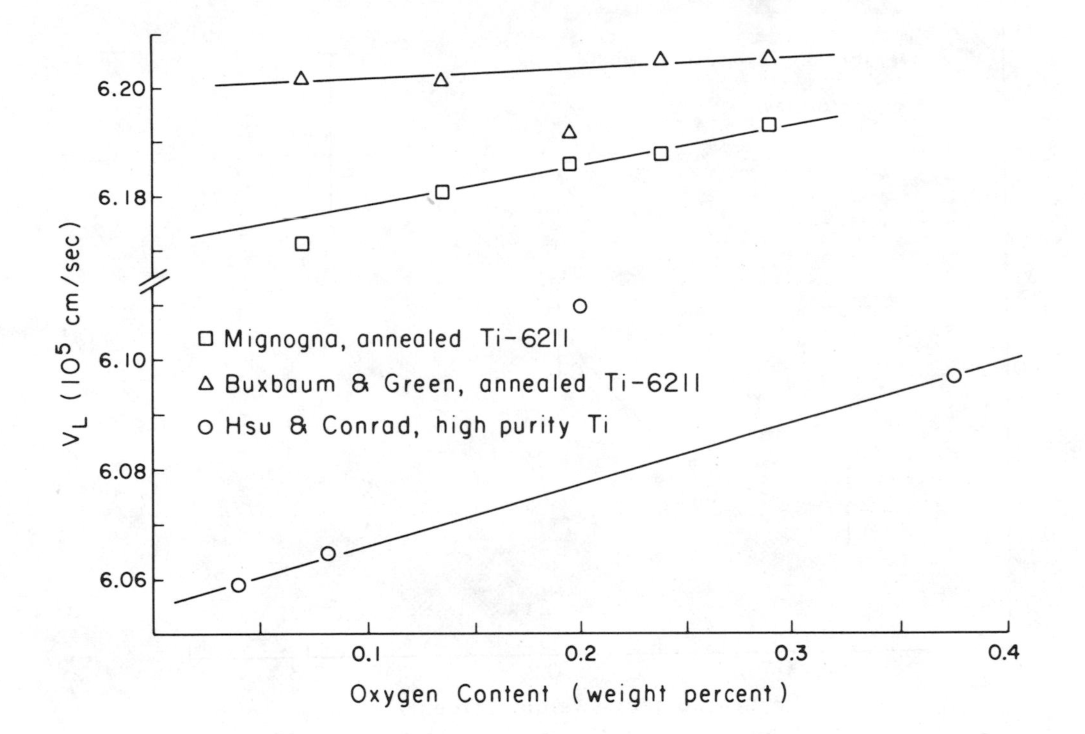

Fig. - 9 - Longitudinal Wave Velocity vs. Oxygen Content for Annealed Ti-6211 and High Purity Ti

attenuation measurements were performed on each specimen; the average and standard deviation of these measurements are presented in Fig. 10. The results show considerable scatter due to the sensitivity of the attenuation measurement to slight inhomogeneities within a given Ti-6211 plate. However, the data also indicate that ultrasonic attenuation may have potential for quantitatively detecting the presence of oxygen in Ti-6211. It is not known whether the decrease in attenuation with increasing oxygen content is due to the effect of the oxygen on the alpha titanium lattice, the change in the amount of retained beta phase or the microstructural variations induced by oxygen and thermal history.

HARDNESS MEASUREMENTS - Rockwell "C" hardness, a measure of material response to plastic deformation, was found to increase with

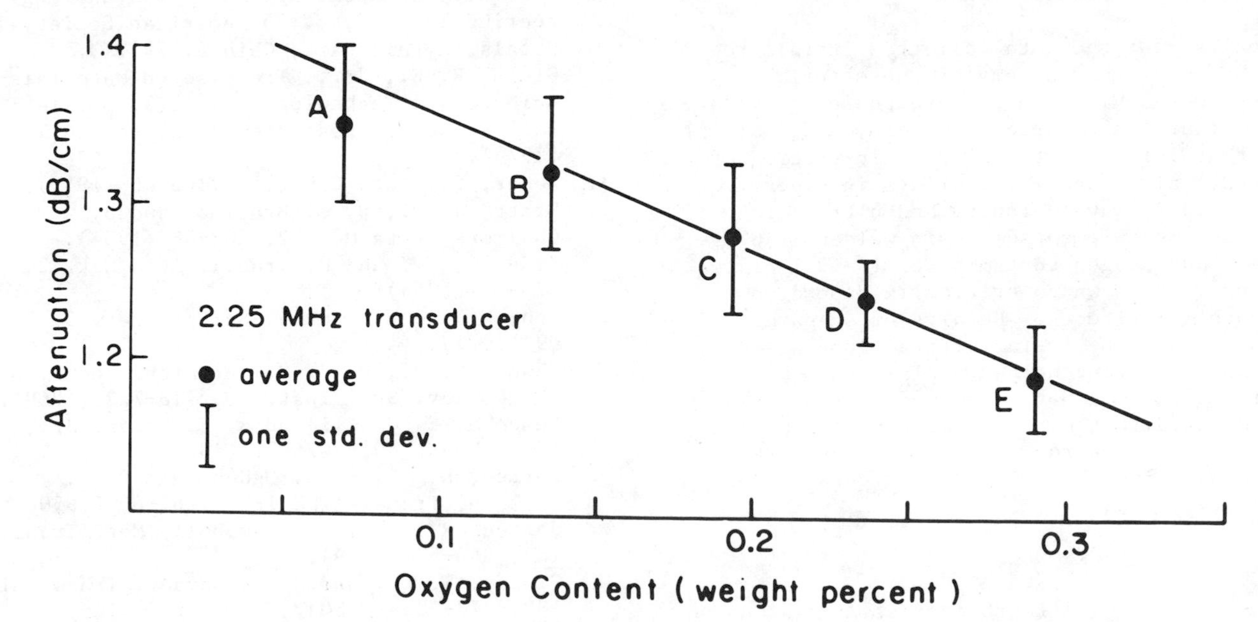

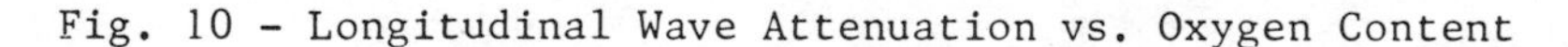

Fig. 10 - Longitudinal Wave Attenuation vs. Oxygen Content

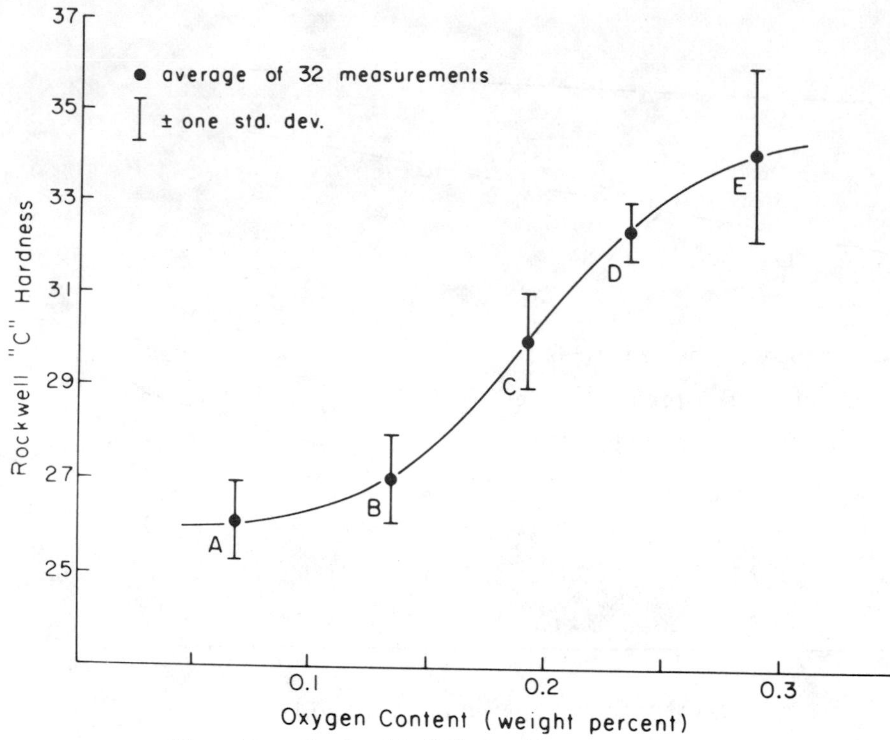

Fig. 11 - Rockwell "C" Hardness vs. Oxygen Content

increasing oxygen content (see Fig. 11). The increase was probably due to the solid solution strengthening provided by the interstitial oxygen, which partitions preferentially into the alpha phase (9-11). This increase is consistent with results reported in the literature. An alternative explanation could involve the changing distribution of alpha phase and beta phase material as the oxygen content changes.

CONCLUSIONS

Analysis of the data collected in this research has underscored some of the material effects such as high temperature phase retention, microstructural alteration, texture, and residual stress that influence the ultrasonic testing of the Ti-6211 alloy and metal alloys in general. Both ultrasonic wave velocity and ultrasonic attenuation measurements made in welded Ti-6211 specimens and oxygen contaminated Ti-6211 plate specimens were shown to critically depend on these material effects. Results of scanning electron microscopy, x-ray diffraction analysis, and hardness testing helped explain the variations in ultrasonic data observed. This work also indicated that ultrasonic attenuation shows promise as a nondestructive tool for quantitatively evaluating the amount of oxygen present in a Ti-6211 sample.

ACKNOWLEDGMENTS

The authors wish to thank Miss Angela Guarda for her technical assistance; Dr. B.B. Djordjevic of Martin Marietta Laboratories in Catonsville, Maryland, for providing the elastomer coating which served as a couplant for the attenuation measurements; Mr. Henry Chaskelis of the Naval Research Laboratory, Dr. Bruce MacDonald of the Office of Naval Research and Dr. Hans Vanderveldt of the Naval Sea Systems Command for supporting this work.

REFERENCES

1. "Glossary of Metallurgical Terms and Engineering Tables", p. 35, American Society for Metals, Metals Park, Ohio (1979).
2. Green, R. E., Jr., "Treatise on Materials Science and Technology", Vol. 3, pp. 145-151, Academic Press, New York, New York (1973).
3. Zener, C., Acta Cryst. 2, 163-66 (1949).
4. Pratt, J. N., W. J. Bratina, and B. Chalmers, Acta Met. 2, 203-08 (1954).
5. Ying, C. F., and R. Truell, Acta Met. 2, 374-79 (1954).
6. Hsu, N. and H. Conrad, Scripta Met. 5, 905-09 (1971).
7. Chung, D. H., D. J. Silversmith, and B. B. Chick, Rev. Sci. Inst. 40, 718-722 (1969).
8. Henneke, E. G., II and R. E. Green, Jr., J. Appl. Phys. 40, 3626-31 (1969).
9. Jaffee, R. I., H. R. Ogden, and D. J. Maykuth, Trans. AIME 188, 1261-66 (1950).
10. Jaffee, R. I., I. E. Campbell, Met. Trans. 185, 646-54 (1949).
11. Finlay, W. L., and J. A. Snyder, Trans. AIME 188, 277-286 (1950).
12. Mignogna, R. B., private communication, March, 1983.

ULTRASONIC WAVE INTERACTION IN ANISOTROPIC SYSTEMS

Laszlo Adler
Department of Welding Engineering
The Ohio State University
Columbus, OH 43210

Ken Bolland
Department of Welding Engineering
The Ohio State University
Columbus, OH 43210

Antal Csakany
Department of Welding Engineering
The Ohio State University
Columbus, OH 43210

Alain Jungman†
Department of Welding Engineering
The Ohio State University
Columbus, OH 43210

Ben Oliver
Department of Metallurgy
University of Tennessee
Knoxville, TN 37916

†Permanent address: Groupe de Physique des Solides, Universite Paris VII.

ABSTRACT

Ultrasonic waves have been used as a valuable tool to determine material properties in homogeneous isotropic materials. Real materials, such as welds and castings, however, due to heat treatment, may become anisotropic as well as inhomogeneous. Such additional complexities will alter the wave propagation resulting in losses due to single and multiple scattering, mode conversion, etc. Earlier work by Kupperman et al. (1) and Simpson et al. (2) is of particular interest in addressing the effects of elastic anisotropy on wave propagation. Here, in order to approach these complex problems, theoretical models and experimental developments will be presented, which will include reflection and transmission through bicrystal interfaces (grain boundaries), as in (2), but for samples having large anisotropy.

DESCRIPTION OF THE SAMPLES

The sample is pure nickel, with an approximately cylindrical shape (height = 1.506 cm, diameter = 2 cm). It has two flat sided walls, to permit contact with a transducer (Fig. 1). The numbered areas 1-6, delineated on the top view, correspond to nickel cyrstals with different orientations. The coordinate system, as determined by x-ray diffraction for the two largest crystals, are represented in Figs. 2 and 3.

THEORETICAL CONSIDERATIONS

ENERGY-TRANSPORT AND THE WAVE NORMAL - The direction of wave propagation (normal to the wavefront) does not, in general, coincide with the direction of energy transport for anisotropic materials. It is possible to compute the angular deviation between the energy-transport and wave propagation vectors (compare Ref. (1)). Theoret-

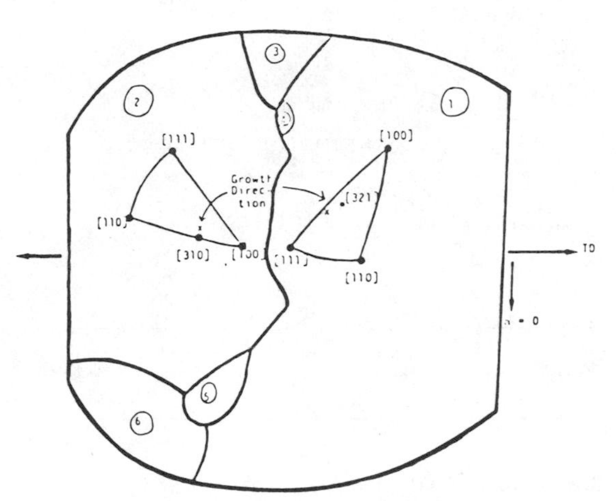

Fig. 1. Top view of the nickel sample, showing two major crystals.

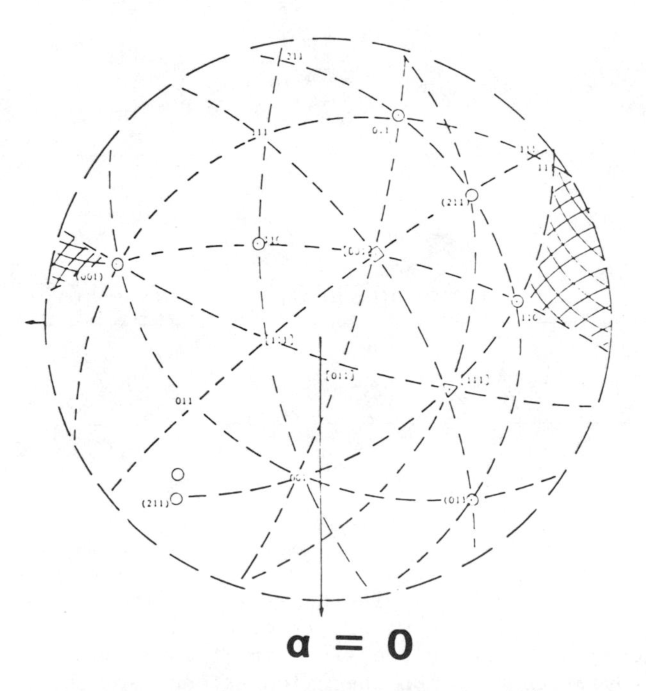

Fig. 2. Coordinate System for Crystal 1.

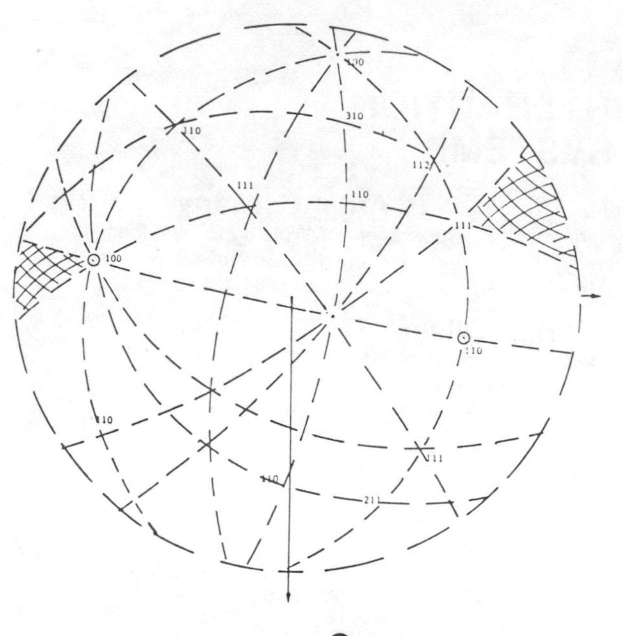

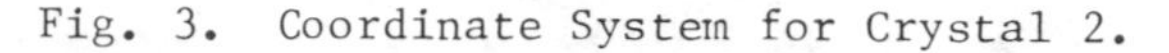

Fig. 3. Coordinate System for Crystal 2.

Table I

Orientation of the Axes of Nickel Crystals 1 and 2.

	REFERENCE ORIENTATION	GROWTH DIRECTION	
	REFERENCE ORIENTATION	MEASURED	THEORY
	[110]	20°	19.10°
X - al 1	[111]	24°	22.20°
2° off [321]	[100]	34°	36.70°
	[110]	28°	26.57°
X - al 2	[111]	40°	43.10°
2° - 4° off [321]	[100]	18°	18.43°

(a) Growth Direction

	REFERENCE ORIENTATION	TRANSVERSE DIRECTION
X - al 1	[100]	16°
	[110]	25°
X - al 2	[100]	20°
	[110]	30°

(b) Transverse Direction

ical plots of the deviation as a function of incident angle for waves directed toward a bicrystal boundary are given in Figs. 4 and 5. (Properties used for the calculations were those of the nickel sample).

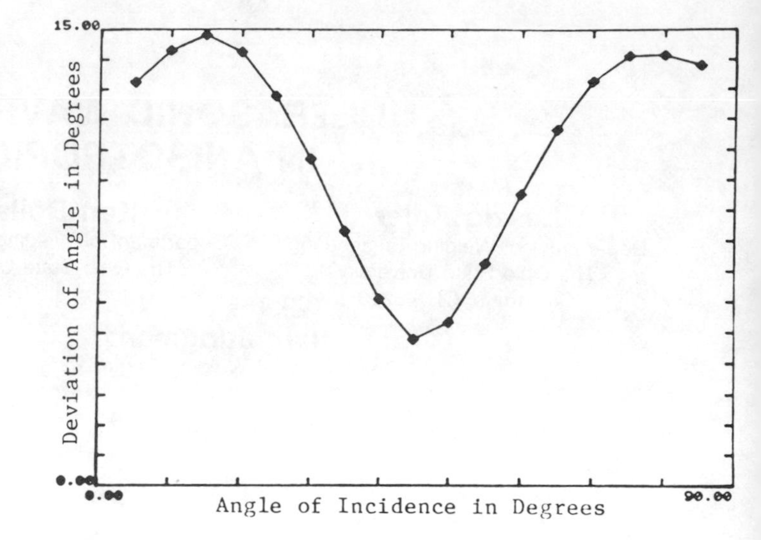

Fig. 4. Angular Deviation Between the Energy-Transport Vector and the Direction of Wave Propagation as a Function of the Angle of Incidence on a Bicrystal Boundary. (Incident quasilongitudinal wave, directed from crystal #2, onto the interface 2-1).

Fig. 4 shows the angular deviation between the incident wave direction in crystal 2 and the energy transport vector. Fig. 5 illustrates the variation in angular deviation for a wave reflected from the boundary between crystals #2 and #1 (wave is in crystal #2). The angular difference in energy transport and wave direction for the wave transmitted from crystal #2 into crystal #1 can also be calculated. From first principles, the velocities of the different waves have been derived.

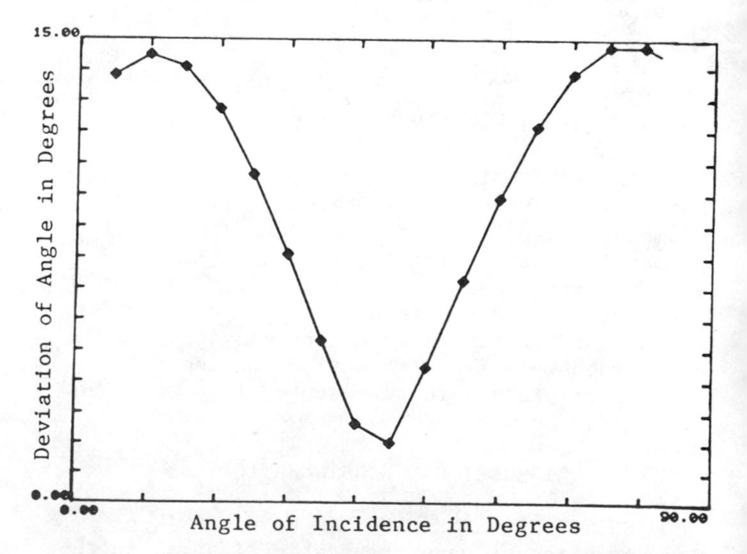

Fig. 5. Angular Deviation Between the Energy-Transport Vector and the Direction of Wave Propagation as a Function of the Angle of Incidence on a Bicrystal Boundary. (Reflected quasilongitudinal wave from the bicrystal, 2-1, boundary).

REFLECTION AND TRANSMISSION AT THE BICRYSTAL INTERFACE - The acoustic impedance discontinuity at the bicrystal boundary can be attributed to the difference in the crystallographic orientation. It is desired to experimentally study the transmission and reflection of ultrasound at the interface and make comparisons with theoretical predictions. However, problems arise due to the geometry of the interface; the surface is neither smooth, nor parallel to the side walls of the specimen. Hence, the reflection and the transmission coefficients contain information including the relative orientation of the crystallographic axes, the local roughness of the surface and its local orientation with respect to the incident beam. These parameters contribute to a scattering effect from the boundary, which is superimposed on the reflection effect due to the discontinuity of the symmetry directions. Also, amplitude measurements of the reflected wave from such an interface include information about shape, orientation and roughness of the surface. In order to address these problems a new computer-controlled ultrasonic system has been developed which is capable of obtaining point-by-point measurements of the ultra-sonic parameters.

EXPERIMENTAL SYSTEM

COMPUTER-CONTROLLED ULTRASONIC MEASUREMENT - Fig. 6 gives the general scheme of the computer-controlled, ultrasonic experiment. The turntable holding the sample may be rotated and the transducer may be scanned linearly. The transducers both transmit ultrasonic pulses toward the specimen and receive the backscattered signals. The computer controls the stepping motors, measures the backscattered signal amplitude and stores the results. Post-processing to emphasize certain details is also done by the microcomputer.

The operating parameters of the system are itemized below:

- Angle resolution: 1/40 degree
- Linear resolution: 0.08 mm
- Stepping rate: 40 impulses/sec
- A-D conversion: 10 bit
- Number of stored measurement points: 400

THE MICROCOMPUTER AND ITS PERIPHERAL DEVICES - An AIM-65/ROCKWELL microcomputer is used as the central part of the ultrasonic backscattering measurement system. The 6502 microprocessor-based system can easily be developed due to the very convenient software and hardware interface elements.

On Fig. 7 the overall scheme of the measuring set-up can be found. Included are:

--AIM-65 microcomputer with on-board printer, keyboard, 20 character display unit.
--Memory extension/RAM: 32 k byte
--General purpose input/output and timer unit.

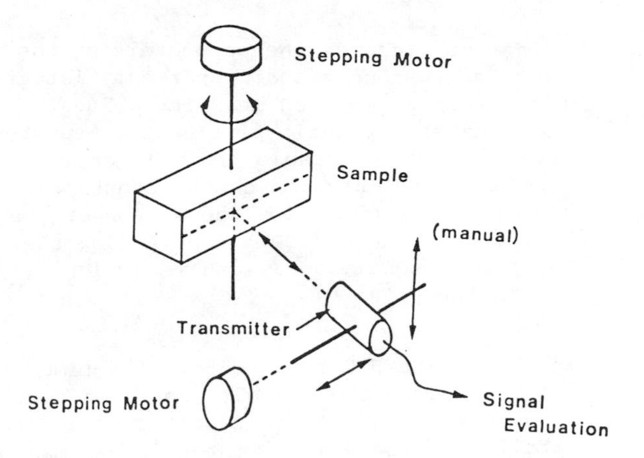

Fig. 6. Computerized Acquisition of Ultrasonic Backscattered Data.

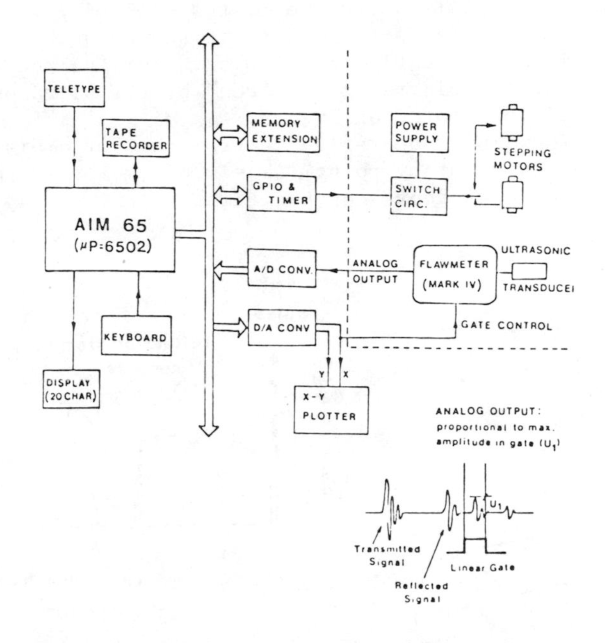

Fig. 7. Hardware Components of the Computer-Controlled Ultrasonic Data Acquisition System.

--Analog-to-digital converter
--Digital-to-analog converter
--Teletype
--Magnetic tape recorder

The programs developed for backscattering measurements now contain the following options.

1. "Manual control" makes it possible to move the turntable by means of specific keys on the keyboard.
2. "Search for maximum" determines the position in a given angular range where the backscattered signal amplitude is a maximum. The turntable then returns the sample to this position.
3. "Measurement" acquires ultrasonic backscattering data versus angle. The signal amplitude is also displayed for the preselected range. The data acquisition may be halted at preselected positions (i.e. to permit modifying the amplification of the flaw detector).

4. "Print-out" records the parameters of the measurement process and prepares a histogram of the self-normalized measured values.
5. "Plot" makes a graphic plot of the measured values in two selectable froms (rectangular or polar representation can be chosen).
6. "Plot the screen" copies the screen of the flawmeter to the plotter by advancing the linear gate continuously and measuring amplitude in the gate.

MAPPING OF THE BICRYSTAL - Both the time-of-flight (which gives the velocity) and the amplitudes (which give the attenuation) along a direction parallel to the axis of the cylinder may be measured; giving information about the profile of the interface as well as the properties of single crystals which are listed on Table 2. The geometrical arrangement for scanning the sample is shown on Fig. 8. The structure on the face of the bicrystal as "seen" by the ultrasonic beam is presented on Fig. 9.

The amplitude reflected from the bottom is recorded point by point ($\Delta x = 0.07$ mm, $\Delta z = 1$ mm) for ten different scan positions.

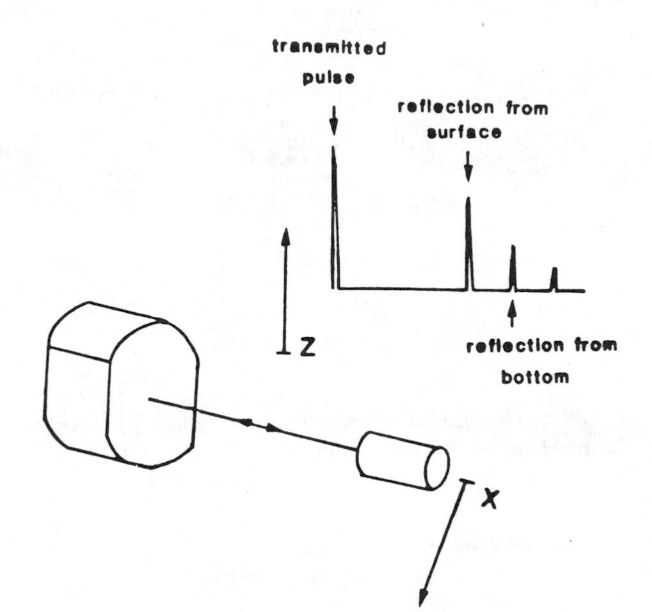

Fig. 8. Arrangement of the Sample and Transducer for Measuring Amplitude Reflected from the Bottom of the Bicrystal Sample as a Function of Position. Measured value: reflected amplitude from the bottom of the sample. Measuring frequency: 5 MHz.

These scans are presented isometrically in Fig. 10. The two zones with different reflected amplitudes on Fig. 10, which are separated by a minimum, correspond to the two main crystals. Each nickel crystal has a different longitudinal velocity. Fig. 11 shows the positioning of the gate for measurements on the two crystals. The left section of the scans in Fig. 10 corresponds to a higher attenuation and a slower velocity

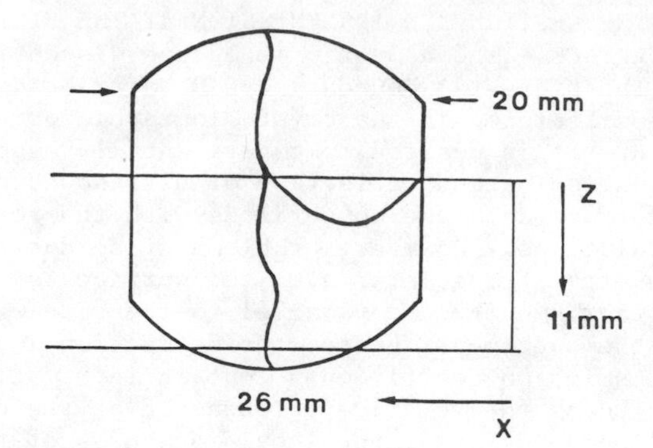

Fig. 9. View of the Bicrystal Surface Scanned by the Ultrasonic System.

Table II

Experimental and Theoretical Values of the Quasi-Longitudinal Wave Speed in the Nickel Bicrystal.

	Directon of The Wave	Crystal	Experiment	Theory*	Error %
V (ms^{-1})	Parallel to the Brenystal Interface	(1)	6008	6200	3
		(2)	5686	5510	3
V (ms^{-1})	Transverse to the Brenystal Interface	(1)	5271	6210	16
		(2)	5760	5660	2

* The elastic constants

$C_{11} = 2.5 \times 10^{11}$ Pa

$C_{12} = 1.5 \times 10^{11}$ Pa

$C_{44} = 1.25 \times 10^{11}$ Pa

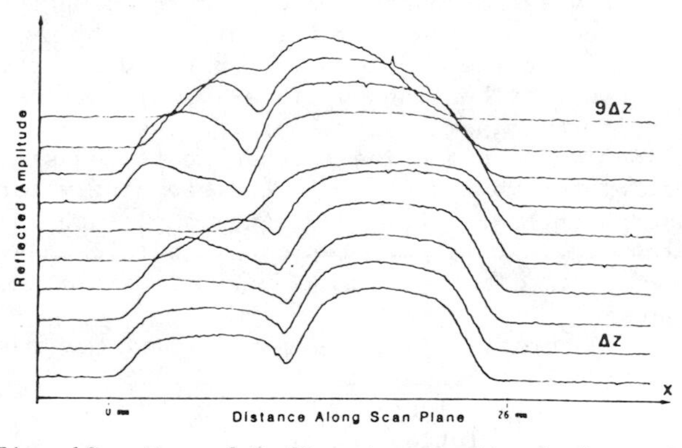

Fig. 10. Map of Reflection Amplitude from the Back Surface of the Nickel Sample. The two separable parts of the curves are due to reflections with different time parameters. The time difference between the two signals is approximately 0.6 μs.

(refer to Table 2). From such rear surface reflection plots it is possible to obtain a mapping which shows the bicrystal boundary. A program to acquire both time-of-flight and amplitude of the reflected wave, has also been written.

The bicrystal sample was next oriented so the ultrasonic beam was normal to the plane surfaces on the sides of the sample. The bicrystal was scanned and reflected amplitudes from three depth planes were acquired (surface, plane 1 and plane 2). Fig. 12 shows a perspective view of the sample and isoamplitude lines of reflection from

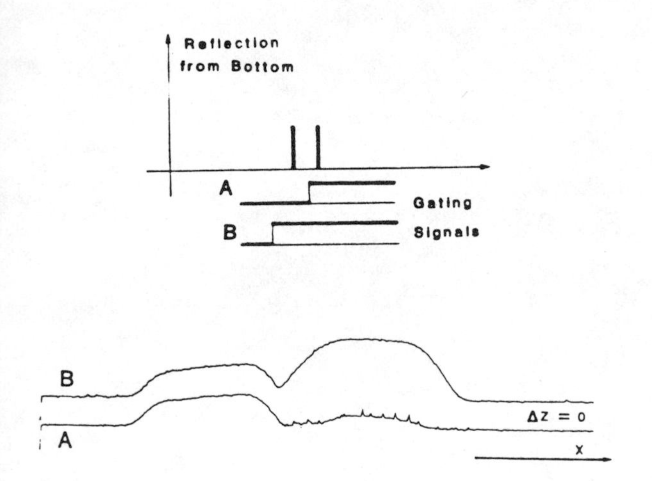

Fig. 11. Effects of Gate Position upon Reflection Scan Profile.

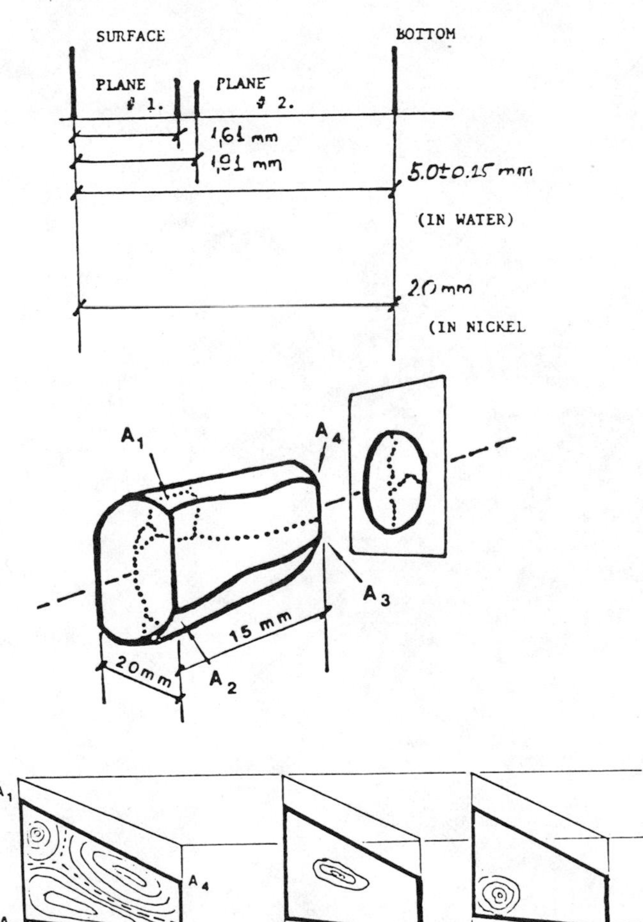

Fig. 12. Mappings of the Reflected Amplitude from Planes Within the Bicrystal Specimen.

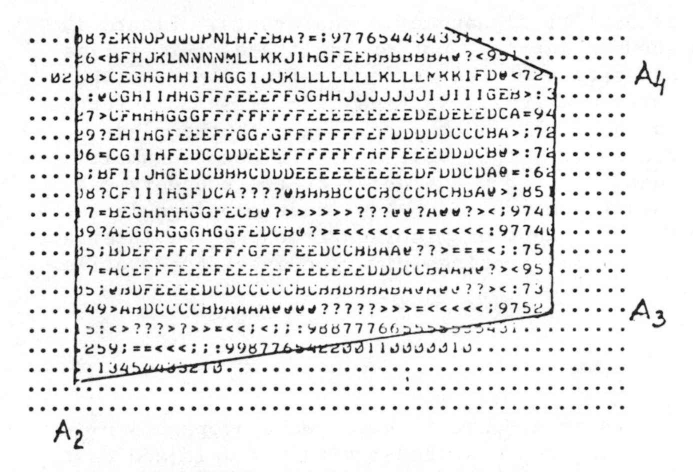

Fig. 13. Mapping of the Reflected Amplitude from the Surface of the Bicrystal Specimen (Amplitudes are encoded as symbols).

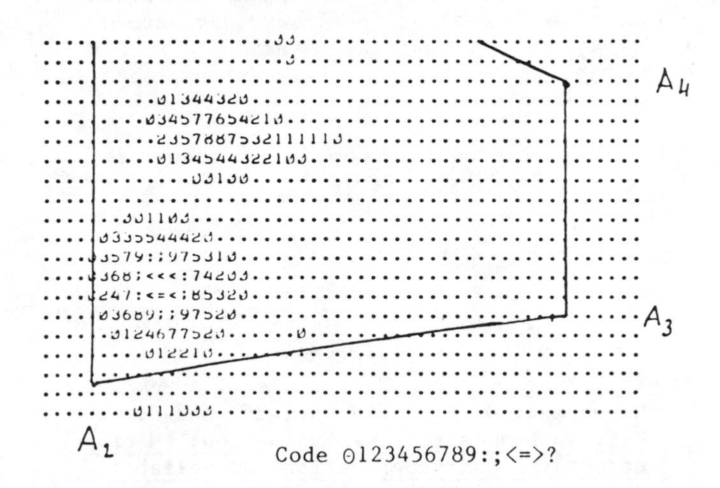

Fig. 14. Mapping of the Reflected Amplitude from the Interface Plane in the Bicrystal Specimen. (Amplitudes are encoded as symbols).

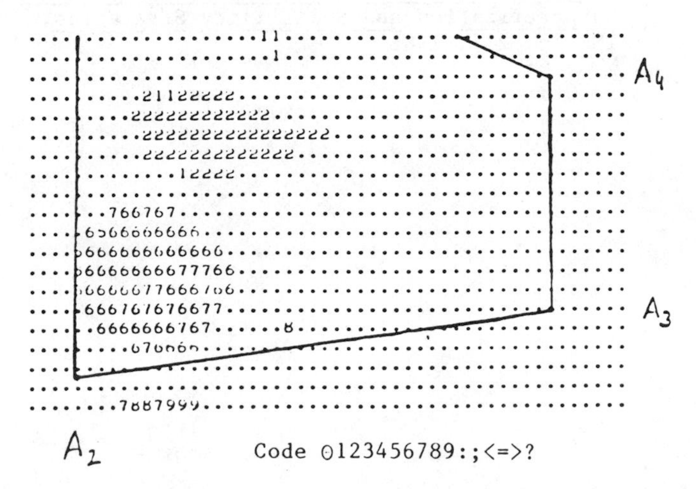

Fig. 15. Mapping of the Time Difference Between the Bicrystal Interface Reflection and the Front-Surface Reflection. (Time difference is encoded as symbols).

each of the three planes. Mappings of the amplitude from the surface and interface planes are shown on Figs. 13 and 14. The amplitudes are encoded bi-alphanumeric characters. Figure 15 shows a mapping, not of amplitude, but of time separation of the surface reflection from the interface echo. In Figs. 13-15, symbols further to the right on the scale beneath the plots represent larger values of the encoded amplitudes. The 3-D character of the boundary can thus be determined. Amplitude mappings indicate the roughness or misorientation of the interface, while time mappings give the interface profile.

SUMMARY

In an attempt to study wave propagation in an anisotropic system-typical of weld and cast materials, work has been described in a nickel bicrystal. Theoretical results to describe wave propagation in each of the crystals and through their interfaces have been presented. A new computer-controlled ultrasonic system has been described which is capable to obtain point by point evaluation of wave parameters (velocity and attenuation). Various displays of mapping of the boundary between the two crystals have been presented.

† Permanent address: Groupe de Physique des Solides, Universite Paris 7.

REFERENCES

1. Kupperman, D. S., K. J. Reiman, and D. I. Kim, in Nondestructive Evaluation: Microstructural Characterization and Reliability Stragegies, AIME, p. 199, 1981.

2. Simpson, W. A., L. Adler, and T. K. Bolland, in Nondestructive Evaluation: Microstructural Characterization and Reliability Strategies, AIME, p. 217, 1981.

II. Powder Metallurgy and Ceramics

APPLICATION OF NDE METHODS TO GREEN CERAMICS: INITIAL RESULTS

David S. Kupperman
Argonne National Laboratory
Argonne, Illinois 60439

Henry B. Karplus
Argonne National Laboratory
Argonne, Illinois 60439

Roger B. Poeppel
Argonne National Laboratory
Argonne, Illinois 60439

William A. Ellingson
Argonne National Laboratory
Argonne, Illinois 60439

Harry Berger
Industrial Quality, Inc.
Gaithersburg, Maryland 20879

C. Robbins
National Bureau of Standards
Washington, DC 20234

E. Fuller
National Bureau of Standards
Washington, DC 20234

ABSTRACT

This paper describes a preliminary investigation to assess the effectiveness of microradiography, ultrasonic methods, nuclear magnetic resonance, and neutron radiography for the nondestructive evaluation of green (unfired) ceramics. The objective is to obtain useful information on defects, cracking, delaminations, agglomerates, inclusions, regions of high porosity, and anisotropy.

The application of microradiography to ceramics is reviewed, and preliminary experiments with a commercial microradiography unit are described.

Conventional ultrasonic techniques are difficult to apply to flaw detection in green ceramics because of the high attenuation, fragility, and couplant-absorbing properties of these materials. However, velocity, attenuation, and spectral data were obtained with pressure-coupled transducers and provided useful information related to density variations and the presence of agglomerates.

Nuclear magnetic resonance (NMR) imaging techniques and neutron radiography were considered for detection of anomalies in the distribution of porosity. With NMR, areas of high porosity might be detected after the samples are doped with water. In the case of neutron radiography, regions of high binder concentration (and thus high porosity) should be detectable, although, in general, imaging the binder distribution throughout the sample may not be feasible because of the low overall concentration of the binder.

INTRODUCTION

The strength of ceramic materials is controlled by the stress necessary to enlarge a flaw of given size and orientation to a critical size, that is, a size sufficient to cause failure.[1,2] This critical size may be as small as 10-100 μm,[3] which is orders of magnitude smaller than for metals. Most ceramic materials exhibit a slow rate of crack growth which varies with temperature, stress, and environment. If these conditions are known and the material is adequately characterized as to the type and size of existing flaws, the probability of failure can be predicted; moreover, if it can be established that flaws are smaller than a predetermined critical size and the material properties meet appropriate specifications, adequate component life can be assured. Improved nondestructive evaluation (NDE) methods can facilitate detection of these small flaws. Furthermore, development of NDE methods for the unfired (green) state can result in considerable cost savings in fabricating structural components, as the final firing stage would be eliminated for defective components.

The research discussed below is part of an ongoing effort to study and develop acoustic, radiographic, and other techniques to characterize structural ceramics with regard to the porosity, cracks, inclusions, and amounts of secondary phases. Since little information was found in the open literature concerning NDE for green ceramics, the approach followed in this effort included broad scoping studies to establish potentially useful NDE techniques and the major parameters that will affect the detection of defects in green ceramics.

SPECIMENS

Green ceramic specimens were provided by the Ceramics Group of the Materials Science and Technology Division at Argonne National Laboratory (ANL) and the Materials Chemistry Division of the National Bureau of Standards (NBS). The specimens were generally 50-60% of theoretical density and very fragile. The main experimental effort was carried out with silicon nitride disks 3.3 cm in diameter and 0.6 cm thick, silicon carbide disks of various sizes, and magnesium aluminate spinel disks 3.7 cm in diameter and 0.6 cm thick. Some other materials were also included. The specimens were cold-pressed with various loading pressures and additions of polyvinyl alcohol (PVA) or Carbowax (CW) binder, and with or without agglomerates. Table I provides additional information about the specimens.

Table I. Green Ceramic Specimens Used in the Present Investigation

No.	Diam x Thickness (cm)	Powder Grain Size (μm)	Cold Pressing Load (MPa)	Sp. Gravity
Si_3N_4 + <2 wt.% PVA (ANL)				
601[a]	3.3 x 0.6	<5	110	1.61
602[a]	↓	↓	110	1.61
603[a]			220	1.73
604[a]			220	1.75
605[a]			165	1.69
606[b]			110	1.39
607[b]			110	1.41
608[b]			165	1.45
609[b]			165	1.47
610[b]			220	1.57
611[b]			220	1.58
612[c]			137	1.66
613[c]			210	1.71
614[c]			210	1.73
615[c]			210	1.78
616[c]			210	1.80
SiC + 10 wt.% CW[d] (ANL)				
SILC	3.3 x 1.9	<1	138	ND[e]
SI2	3.3 x 0.6	↓	69	↓
SC-3	3.3 x 0.3		104	
SC-4	3.3 x 0.3		69	
SC-5	3.3 x 0.3		69	
$YCrO_3$ + <1 wt.% PVA (NBS)				
YC-1	irregular	ND	104	ND

No.	Diam x Thickness (cm)	Powder Grain Size (μm)	Cold Pressing Load (MPa)	Sp. Gravity
$MgAl_2O_4$ (Spinel) + <1 wt.% PVA (NBS)				
NB1[f]	3.7 x 0.6	<5	210	ND
NB2[f]	↓	↓	↓	1.84
NB3[f]				ND
NB4[f]				↓
NB5[g]				
NB6[g]				
NB7[g]				1.82
NB8[g]				ND
NB9[g]				↓
NB10[h]				
NB11[h]				
NB12[h]				
NB13[h]				1.85
NB14[h]				ND
MgO + 20 wt.% CW (ANL)				
A10K	3.3 x 0.8	0.3	ND	ND
B10K	↓	↓	ND	ND
C20K			ND	ND
1			69	1.52
2			69	1.48
4			138	1.63
5			138	1.66
6			138	ND

[a] 2% agglomerates (10-30 μm) and deflocculant added.
[b] No agglomerates or deflocculant.
[c] With deflocculant only.
[d] Except SC-5, which contained 10 wt.% PVA.
[e] ND = not determined.
[f] No agglomerates.
[g] 2% agglomerates (75-100 μm) added.
[h] 20% agglomerates (75-100 μm) added.

MICRORADIOGRAPHY

APPLICATION TO NDE OF CERAMICS - Radiography[4-10] is a well-recognized and widely used NDE method. Several types of radiation can be used for radiography; these include x-rays, gamma rays, neutrons,[11,12] and charged particles.[13,14] Industrial radiographic inspection largely involves x-rays, and this report is primarily concerned with x-radiography. However, many of the following comments apply to radiography in general.

With conventional radiographic methods, the difference in attenuation produced by a change in object thickness on the order of 1-2% can be visualized. Thickness or density changes of as little as 0.5% and high-contrast features as small as several micrometers in size can be detected under optimum circumstances (good geometry, thin object, low x-ray energy, fine-grain photographic film, etc.). As the contrast decreases to a few percent, the resolution capability is degraded.

Microradiography[15-21] refers to the production of radiographic images that display improved spatial resolution. Generally, the term is applied to techniques that provide magnified images (usually in the range of 2 to 100X). Microradiography offers the potential capability to detect and display small inhomogeneities so they can be located, sized, and evaluated as to type and severity. This capability is important with all materials, but it is particularly important with ceramics and other brittle materials because, as discussed earlier, defects that can lead to failure tend to be relatively small (10 to 100 μm). Readily available microradiographic methods, which may be classified as either contact or projection techniques, now permit one to detect some inhomogeneities in that size range.

Contact microradiography offers two advantages: (1) it can be accomplished with conventional x-ray equipment and (2) large areas (the size of the film) can be inspected at one time. Disadvantages of the contact method include the time required for the extra step of enlarging the image and the fact that scatter from the object is readily detected, thereby contributing to reduced image contrast.

Projection microradiography provides better contrast because the object is well removed from the detector and scatter is detected to a lesser degree. Also, the enlargement of the x-ray image offers the possibility of performing radiography in real time with fast detectors, such as image intensifier systems,[22] which have inherently poorer resolution capabilities. Image intensifiers usually show a spatial resolution of 0.125 mm (4 line pairs/mm) to 0.50 mm (1 line pair/mm). For a typical 20X enlargement system, object details 10 times or more smaller than the above values should be detectable at reasonable contrast. Real-time methods offer the advantage of object manipulation to obtain the optimum radiographic view (to align a crack with the radiation beam, for example). This is important because if the alignment is off by as much as 6 degrees in a 1.25-cm (0.5-in.)-thick sample, detectability is greatly reduced.[8] A disadvantage of the projection magnification method is that only a small area of the object can be inspected at one time. For example, at 20X magnification, only about 1 cm of the object would be in a given 22.5-cm image intensifier field.

In ceramics, heavy inclusions such as tungsten carbide or iron usually present good x-ray contrast to the low-atomic-number host material. Detection of such high-atomic-number inclusions in the size range of 25 μm (0.001 in.) has been reported in silicon nitride and silicon carbide ceramics.[23] An example of a defect that is similar in attenuation to the host material, and thus difficult to detect even if relatively large, is a silicon inclusion in silicon nitride. Voids should be detectable if the size range fits the resolution and contrast capability of the inspection system. For example, in a 5-mm (0.2-in.)-thick specimen, a 100-μm (0.004-in.) void would represent a contrast sensitivity of 2%; this should be detectable in a ceramic sample if an x-ray energy appropriate to the material is used and the geometric unsharpness U_G is less than the size of the magnified image. In a typical case, where an x-ray tube with a 10-μm focal spot is used with a source-to-object distance of 25 mm (1 in.) and an object-to-detector distance of 500 mm (20 in.), the magnification would be 20X and U_G would be 0.2 mm. The image contrast will clearly be reduced as the image size approaches and goes below the unsharpness value. In this example, one would predict that the projected image of a 10-μm spherical void would be slightly larger than U_G, and could therefore be detected. However, to maintain at least the 2% contrast, this size void should be in a specimen no thicker than 0.5 mm.

Volatile binders designed to be removed by firing are relatively low-atomic-number materials that should have little influence on x-ray attenuation. Therefore, radiographic differences between green and fired ceramics are expected to be relatively small when x-rays are used.

EXPERIMENTAL RESULTS - The limited experimental results obtained thus far for green ceramic specimens compare prints of projection microradiographs with prints of conventional radiographs that were photographically enlarged to about the same value (4X). The projection microradiographs were taken at NASA-Lewis (Cleveland, Ohio) with a microfocus unit made by the Nicolet Instrument Corporation. The unit was operated at 35 kV and had a 10-μm focal spot; the source-to-film distance was 30 cm and the source-to-object distance was 7 cm. The conventional radiographs were taken at ANL; the x-ray unit was operated at 65 kV and had a 0.5-mm focal spot. The source-to-film distance was 75 cm; the object was essentially in contact with the detector. The U_G values that would result from these geometries is 33 μm for the microfocus system and about 15 μm for the conventional radiograph; both values are sufficiently low that their influence on total unsharpness should be small.

Figure 1 shows microcracking in ceramic sample SILC, which is about 1.9 cm (0.75 in.) thick. When radiographic prints are compared, as is done here, much more cracking is evident in the microfocus radiograph. The unprinted conventional radiograph showed almost all the microcracking, but with less overall clarity than the unprinted microfocus radiograph.

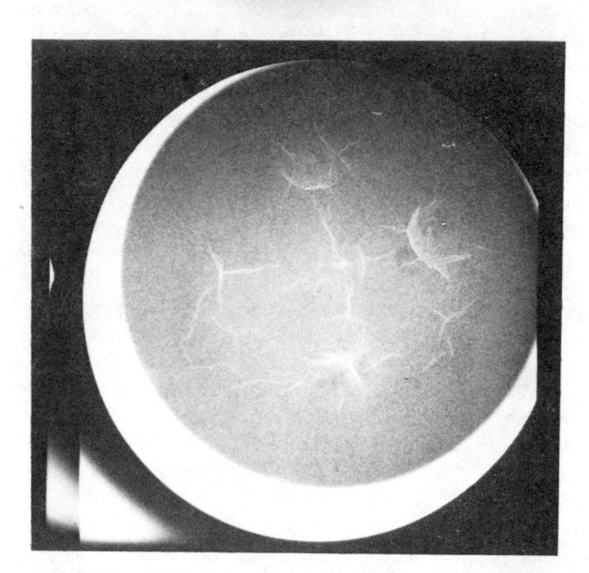

Fig. 1 - Prints of a conventional radiograph (top) and a microfocus radiograph (bottom) of green ceramic sample SILC, showing microcracking

Figure 2 shows prints of radiographs of ceramic sample NB10 which is 0.63 cm (0.25 in.) thick. The radiographic conditions were similar to those described above. The radiographs show small inclusions and porosity. Although the print comparison again shows a marked superiority for the microfocus radiograph, the unprinted conventional radiograph showed most of the same detail.

Fig. 2 - Prints of a microfocus radiograph (top) and a conventional radiograph (bottom) of green ceramic sample NB10, showing small inclusions (black dots)

CONCLUSIONS - Analyses and literature results show that microradiography of ceramics can provide useful inspection information. The small defects that can lead to failure of ceramics can be detected by radiography under some circumstances. The experimental results obtained in this program thus far are limited. The prints comparing microfocus radiographs with enlarged conventional radiographs tend to be misleading in that the original negatives of the conventional radiographs showed much of the same detail. Nevertheless, greater clarity was noted on the microfocus radiographs.

ULTRASONIC TECHNIQUES

The difficulties associated with ultrasonic examination of green ceramics are formidable. For example, low-frequency sound waves are not scattered sufficiently to allow the detection of flaws much smaller than 1 mm in size; on the other hand, sound waves with frequencies $\gtrsim 3$ MHz generally undergo excessive attenuation in samples >3 mm thick. Also, ordinary couplants (e.g., water or glycerol) are often absorbed by green ceramics; this not only mitigates their coupling function, but may also affect the subsequent fabrication process. This problem can be avoided by using pressure alone to couple the transducer to the specimen, but great care is required because the applied pressure can affect the data or even result in sample breakage.

In spite of these difficulties, studies of ultrasonic attenuation and acoustic velocity (including dispersion and frequency spectra) in green ceramics may provide useful information related to density variations, porosity content, presence of agglomerates and delaminations, elastic anisotropy, and material quality in general; some examples of promising approaches are presented below. Further development of ultrasonic testing techniques, coupled with signal enhancement techniques, may lead to improved flaw detection sensitivity.

ELASTIC ANISOTROPY - The elastic anisotropy of a green ceramic specimen can be determined from the change in sound velocity that occurs when a shear-wave transducer is rotated with respect to the specimen, thus varying the polarization of the shear waves propagating in a particular direction. Sound velocity data were acquired for SiC, MgO, and $YCrO_3$ specimens with a Panametric 5052UAX ultrasonic transducer analyzer and a Tektronix 7904 oscilloscope with 7B85 and 7B80 time bases. Panametric 2.25-MHz normal-incidence shear-wave transducers (13 mm in diameter) and Aerotech 2.25-MHz alpha transducers (6 mm in diameter) were employed for the measurements. The velocity was measured by overlapping successive echoes in the pulse-echo mode and determining the time delay from the oscilloscope. The time base was calibrated by checking the oscilloscope readings against a sequence of precisely timed pulses from a Tektronix type 184 time mark generator. With the SiC and MgO specimens, which have a high content of Carbowax binder, Panametric shear-wave couplant was successfully used for both longitudinal and shear waves. With the chalk-like $YCrO_3$ specimens (<1% PVA binder), no couplant was needed for either type of wave.

Figure 3 shows the echo pattern for 2.25-MHz longitudinal and shear waves in a 3.8-mm-thick $YCrO_3$ specimen. (About 6 μs separate the first and second longitudinal echoes; about 10 μs separate the shear echoes.) Apparent attenuation is clearly rather high, on the order of 10-20 dB/cm. Furthermore, the effect of polarization on shear-wave velocity shows that the material is elastically anisotropic: A maximum velocity variation of about 3% was reproducibly observed as the shear-wave transducer was rotated about its axis while remaining over the same point on the sample. (The pressure of the transducer on the sample was controlled so that the velocity variations could not be attributed to variations in transducer loading.) Table II summarizes the sound velocity data for the $YCrO_3$ sample. The low value of Poisson's ratio implies that under stress, the volume of the specimen is significantly reduced; this is consistent with the porous, low-density nature of the sample. Since the modulus of elasticity is related to the velocity of sound, variations in shear velocity with polarization can indicate variations of modulus with direction. This variation in modulus could affect the performance of a component made from such a sample.

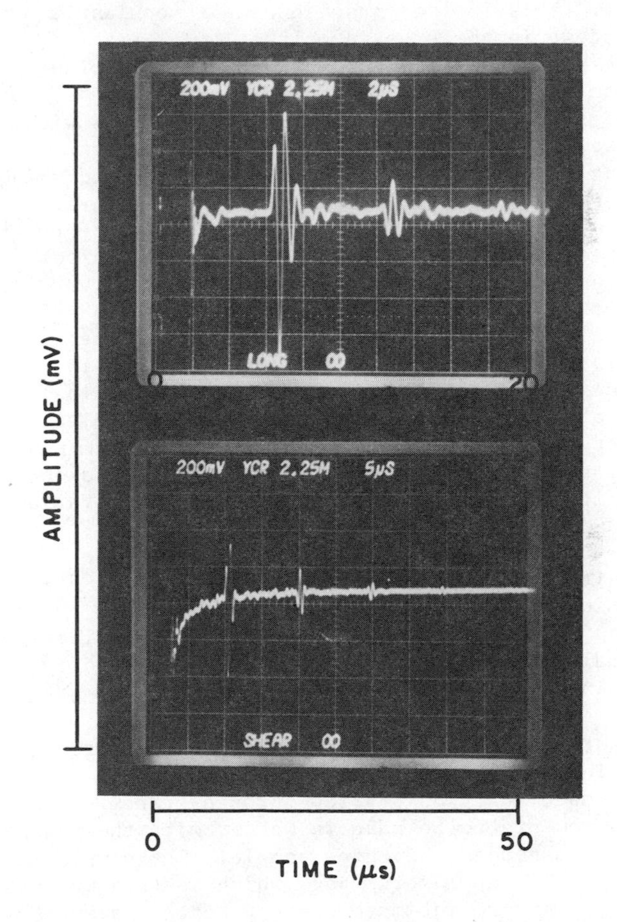

Fig. 3 - Echoes from 3.8-mm-thick $YCrO_3$ sample insonified with (top) longitudinal waves and (bottom) shear waves at 2.25 MHz. Axis dimensions are as follows: (left) 200 mV/div, (top) 2 μs/div, and (bottom) 5 μs/div

Table II. Sound Velocity in $YCrO_3$ Sample with PVA Binder

Longitudinal Velocity (10^5 cm/s)	Shear Velocity (10^5 cm/s)	Poisson's Ratio, σ
1.43	0.947 max	0.11
1.43	0.919 min	0.15

Another example of variation with polarization was observed in a 3-mm-thick section of a spinel disk, insonified with shear waves propagating in the plane of the disk. With polarization parallel to the disk axis, the velocity (error <0.5%) was about 3.5% higher than with polarization perpendicular to the disk axis. No variation of velocity with polarization was found for wave propagation parallel to the disk axis. This implies that the shear modulus is greatest in the pressing direction.

VARIATION OF VELOCITY WITH DENSITY - The velocity of longitudinal waves has been measured as a function of density for MgO specimens with 20 wt.% Carbowax binder. Density variations among these specimens result from differences in loading pressure, i.e., in porosity. Figure 4 shows a plot of density vs sound velocity for four samples. One of the samples is anomalous (it has visible delamination). Based on the other three samples, a linear relationship between velocity and density is apparent; the variation in density is 1/3 the variation in velocity. Also shown in Fig. 4 is the nominal variation in density for three of the samples (rectangles around the data points), determined from variations in velocity measurements taken at five different points on each sample. Variations in density of up to 2% were observed. These results suggest that not only can the density (and thus porosity) of a sample be determined from sound velocity measurements, but if sufficiently small transducers were available, anomalous microstructures could be detected by mapping local variations in density.

As discussed in a later section, the measured sound velocity can vary with transducer loading in some instances. A technique has been developed that would allow through-transmission measurements to be made in water, with the transducer suspended above the sample. The sample is enclosed in a plastic "bag" which is then evacuated to assure intimate contact between sample and plastic. Tests have shown that sound can be transmitted through the immersed bag and sample, allowing velocity scans of the sample to be acquired with minimal variation in acoustic coupling. We plan to optimize this "immersion" testing technique so that sound velocity variations in green ceramic samples can be reproducibly mapped.

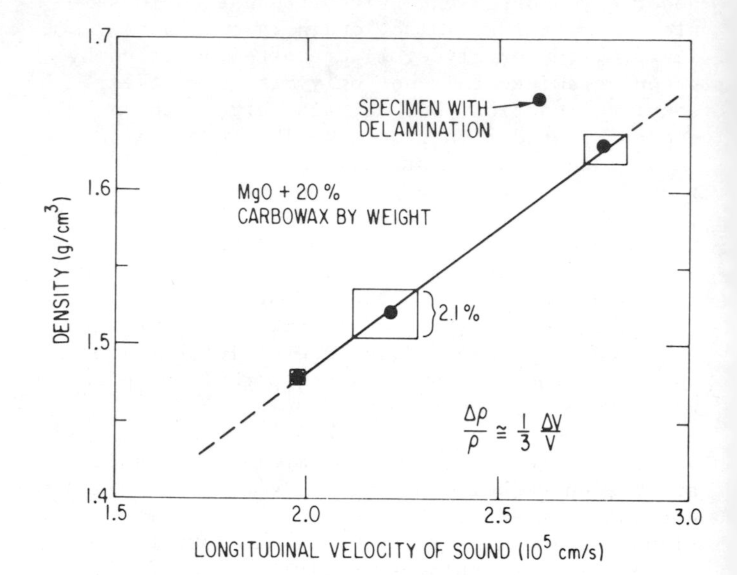

Fig. 4 - Sample density vs longitudinal velocity of sound for MgO + 20% Carbowax. Rectangles show variation within each sample

The maximum density (0% porosity) of MgO-20 wt.% Carbowax, calculated from the densities of the two constituents, is 2.7 g/cm^3. The velocity obtained at this density by extrapolating the curve of Fig. 4 is 8.1 x 10^5 cm/s. This velocity can be compared to theoretical models that predict upper and lower bounds for sound velocity in a homogeneous composite material with a soft matrix and hard filler, as a function of the volume fraction of one of the constituents. Theoretical values for the longitudinal velocity were calculated as a function of volume fraction using both the Voigt model[24,25] (which assumes constant strain and gives an upper bound for velocity) and the Reuss model[24,25] (which assumes constant stress and gives a lower bound). The velocities were calculated from the modulus according to the following relationships:

$$\rho v_{\ell_V}^2 = f_1 \rho_1 v_{\ell_1}^2 + f_2 \rho_2 v_{\ell_2}^2 \quad \text{(Voigt velocity)}, \quad (1)$$

$$\frac{1}{\rho v_{\ell R}^2} = \frac{f_1}{\rho_1 v_{\ell_1}^2} + \frac{f_2}{\rho_2 v_{\ell_2}^2} \quad \text{(Reuss velocity)}, \quad (2)$$

where f_1, f_2 are fractional volumes, ρ_1, ρ_2 are densities, and v_{ℓ_1}, v_{ℓ_2} are longitudinal velocities for constituents 1 and 2, respectively, and $\rho = f_1\rho_1 + f_2\rho_2$. Figure 5 shows the resultant upper (Voigt) and lower (Reuss) velocity limits for MgO/Carbowax composites vs the volume fraction of Carbowax. The extrapolated velocity for a volume fraction of 37.5% Carbowax (the amount used in the ANL pellets) is indicated. This value is within the theoretical bounds and is close to the Voigt limit.

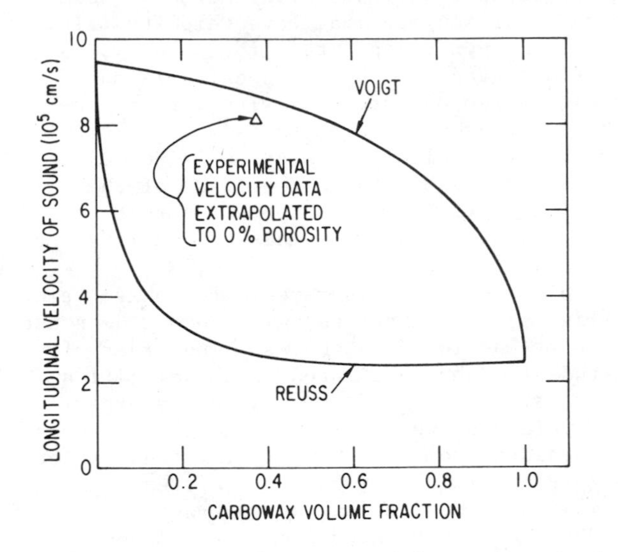

Fig. 5 - Theoretical upper and lower bounds for longitudinal velocity vs Carbowax volume fraction. Experimental value determined from extrapolation of data in Fig. 4 is also shown

These results indicate that sound velocity measurements can provide quantitative information on porosity and local fluctuation in density, and can perhaps indicate anomalous microstructures. The analysis also indicates that data acquired in this manner may be predicted by models for composites, further suggesting that sound velocity data and relationships to anomalous microstructures may be understood from a fundamental point of view.

EFFECT OF AGGLOMERATES ON ULTRASONIC PROPERTIES - The presence of agglomerates in the green state can have deleterious effects on the properties of the ceramic specimen after firing.[26] The use of ultrasonic techniques to obtain information on the presence of agglomerates was investigated, on the assumption that ultrasonic scattering from groups of agglomerates might alter the spectral characteristics or attenuation of ultrasonic waves even if individual agglomerates cannot be detected. Since ordinary attenuation is difficult to measure accurately, spectral data were examined. The wavelength of longitudinal waves is about 1.3 mm at 1 MHz. Agglomerates in the 100-μm (0.1-mm) size range should scatter in the Rayleigh region, where scattering is frequency dependent. Figure 6 shows frequency spectra (0-2 MHz) for longitudinal waves propagating in spinel disks with 0, 2, and 20 wt.% agglomerates (75-100 μm in size) in a matrix with particle size <5 μm. The spectrum for Plexiglass is also shown to permit correction for the transducer characteristics. Rayleigh scattering dominates the frequency dependence of the attenuation. Larger particles contribute more to the higher frequency scattering and thus the introduction of agglomerates is expected to accentuate the high-frequency attenuation. This appears to be the case in Fig. 6.

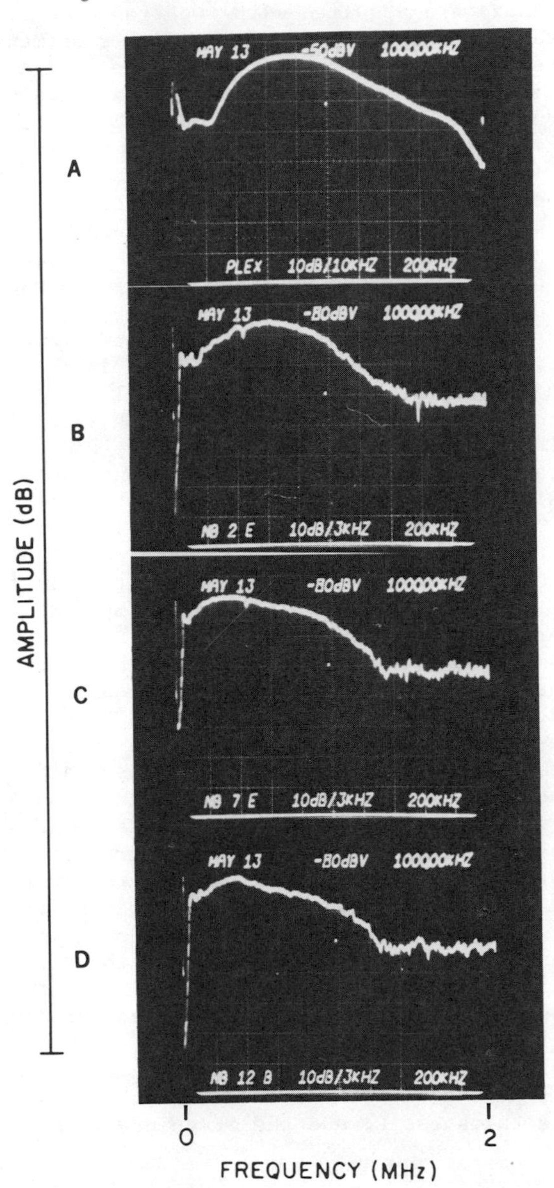

Fig. 6 - Frequency spectra for longitudinal waves propagating in (A) Plexiglass and (B-D) spinel disks with agglomerate contents of (B) 0%, (C) 2%, and (D) 20%. Horizontal axis shows 200 kHz/div; vertical axis shows 10 dB/div

These results suggest that frequency data may provide qualitative information on the presence of agglomerates in green ceramics. Further efforts are obviously needed to establish the validity of this concept. The effect of agglomerates on the velocity of longitudinal and shear waves was also examined. No statistically significant difference was noted between samples with and without agglomerates.

DISPERSION AND EFFECTS OF PRESSURE ON VELOCITY AND ATTENUATION - The high attenuation of ultrasonic waves in green ceramics makes it difficult to obtain reliable ultrasonic data with standard single-transducer, pulse-echo techniques. To obtain more accurate measurements of the group and phase components of velocity, data were acquired with two transducers, placed on opposite sides of the disk specimens. The receiver could then be coupled to a high-gain, low-noise amplifier and a signal obtained even under highly attenuating conditions, without contamination by residual transducer ringing.

In order to keep transmitting and receiving transducers aligned while changing and repositioning specimens, one transducer was held in the chuck of a sturdy drill press. Weights were hung on the drill press handle to apply steady pressure. The force on the transducers was calibrated using a suitable force gauge in the position of the specimens. Friction in the press limited accuracy and repeatability; this problem was ameliorated with thorough lubrication, vigorous tapping with a mallet, and separate calibration of increasing and decreasing weight ranges.

To measure group velocity, impulses were sent from the transmitting transducer to the receiving transducer with and without the sample in place, and the arrival times were compared. To measure phase velocity, the phase of the through-transmission signal was compared with the phase of the transmission signal with the sample removed, and the frequency that gave 0° and 180° phase shifts between the transmitted and received wave was determined. The phase shift between the transmitting and receiving transducers in the absence of a sample was determined at that frequency to calculate a correction factor. The frequency was then progressively reduced until a very small number of waves (1/2, 1, or 1-1/2) could be shown unambiguously to be present in the sample. By counting the number of 0° and 180° phase shifts and adding ϕ, the exact number of wavelengths can be calculated at each frequency. The phase velocity c is then calculated from the whole number of 180° phase shifts N, the transducer phase shift ϕ, the sample thickness L, and the frequency f, where

$$c = f\lambda \quad (3)$$

and

$$\lambda = (L/(N + \phi/360). \quad (4)$$

Phase velocity is then plotted as a function of frequency and the slope dc/df is determined graphically. The group velocity, u, can then be determined from the formula

$$u = c/[1 - (f/c)(dc/df)]. \quad (5)$$

This method of determining phase velocity works only for highly attenuating samples in which the standing waves are very weak. Attenuation was estimated from the spectrum of the received pulse. The spectra were obtained using a Tektronix 7903 with a 3L5 plug-in module. Group velocity was found to vary with applied transducer pressure and sample density. Shear-wave velocity (Fig. 7) shows a monotonic rise with applied pressure and with increasing sample density. For longitudinal-wave velocity (Fig. 8) the rise is even steeper than for shear waves and the dependence on sample density is more complex. A few measurements with larger low-frequency transducers tend to indicate a significantly lower velocity. Caution is needed in interpreting these data in the light of the rather long-duration (low-frequency) received pulse. (Measurements were made to the first zero crossing.)

The dispersion measurements for phase velocity (Fig. 9, dashed curves) also show a decrease in velocity at low frequencies. The order of magnitude for the calculated group velocities (Fig. 9, solid curves) agrees with the experimental values shown in Figs. 7 and 8. The spectrum obtained from a received pulse transmitted through Plexiglass is shown for reference in Fig. 10. The longitudinal waves transmitted with 2.25-MHz transducers attained higher frequencies (Fig. 11) than the shear waves (Fig. 12). Also, the spectrum and amplitude were much more strongly pressure dependent for the longitudinal waves than for the shear waves. Shear-wave transmission for the spinel samples was also very limited in frequency (Fig. 13).

CONCLUSIONS - The acoustical properties of green ceramics are obviously more complex than the properties of other familiar materials. Although the propagation of ultrasonic waves in green ceramics is subject to very high attenuation, the sensitivity of the present techniques permitted studies on samples 6-10 mm thick with longitudinal- wave frequencies as high as 3 MHz and shear-wave frequencies as high as 1.7 MHz. Strong changes in attenuation and velocity were observed, particularly for the higher frequency longitudinal waves, with applied pressure. The variation of sound velocity with applied pressure may yield important information on the structure of green ceramics. Investigations of a wider range of materials, as well as the relation of the observed parameters to the properties of fired ceramics, are needed in the future.

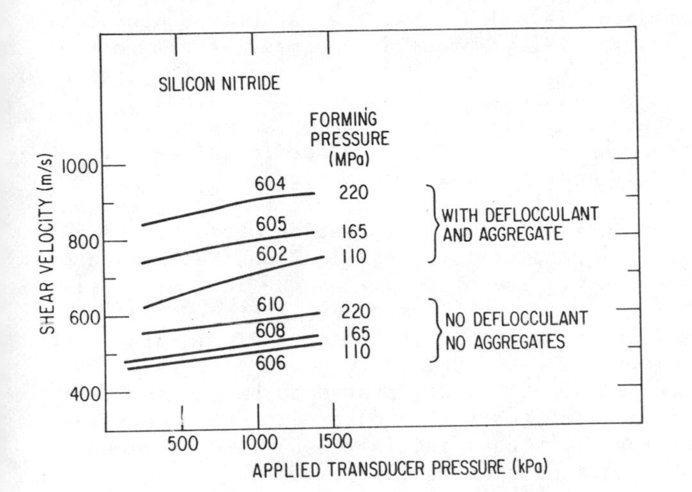

Fig. 7 - Shear-wave velocity vs applied transducer pressure for silicon nitride greenware. Sample numbers (see Table I) are shown on the curves

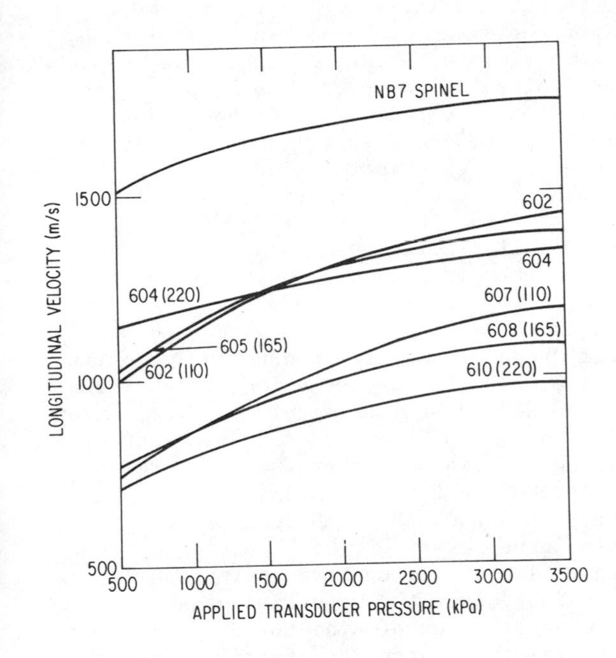

Fig. 8 - Longitudinal-wave velocity vs transducer pressure for spinel (upper curve) and silicon nitride greenware. Sample numbers (see Table I) and forming pressures (in MPa) are shown on the silicon nitride curves

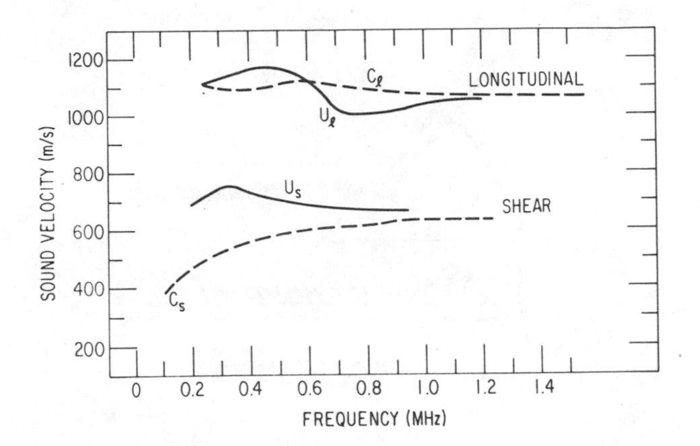

Fig. 9 - Phase velocities (dashed curves) and group velocities (solid curves) of longitudinal and shear waves as a function of frequency

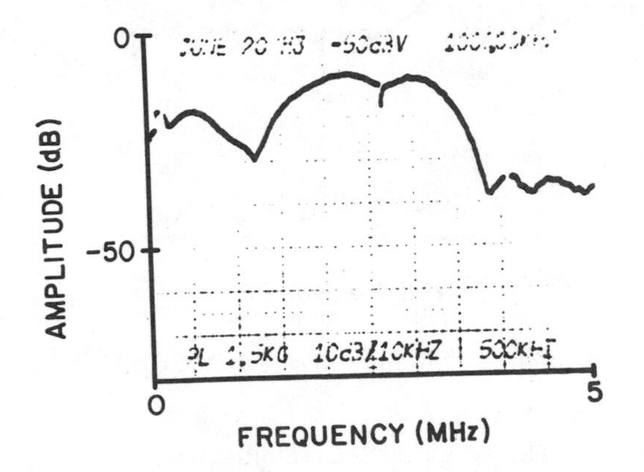

Fig. 10 - Reference transmission through Plexiglass. Horizontal axis shows 500 kHz/div; vertical axis shows 10 dB/div

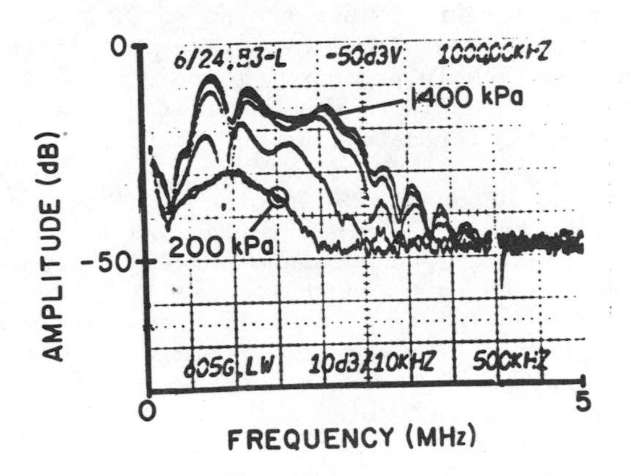

Fig. 11 - Variation of received longitudinal-wave spectrum with transducer pressure. The attenuation decreases with increasing pressure in the range from 200 to 1400 kPa

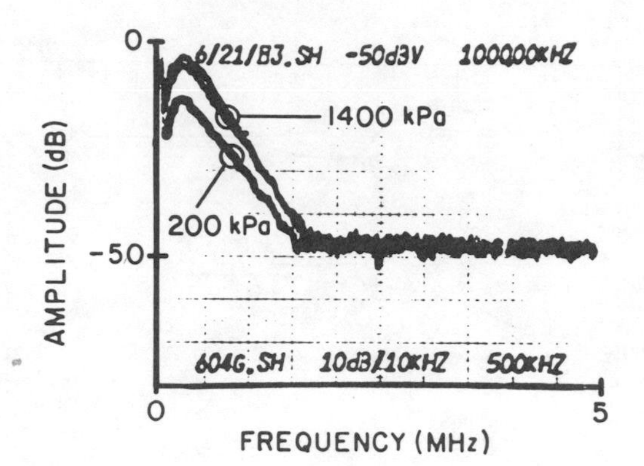

Fig. 12 - Comparison of received shear-wave spectra obtained with transducer pressures of 200 and 1400 kPa

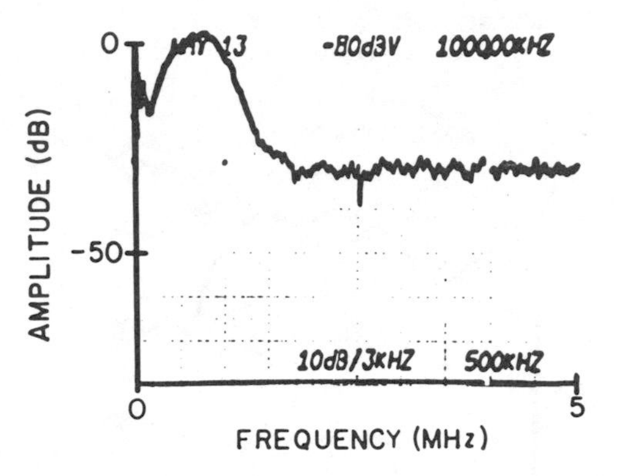

Fig. 13 - Shear-wave spectrum for spinel sample NB2

UNCONVENTIONAL TECHNIQUES

NUCLEAR MAGNETIC RESONANCE - An important objective of NDE for green ceramics is to establish the distribution of porosity and, in particular, the presence of pores on the order of 100 μm or larger. Since the pores of green ceramics are filled with a hydrogen-rich binder, they will be rich in protons. Therefore, nuclear magnetic resonance (NMR) was considered as a possible method for detecting and imaging the pores.

The technology of NMR imaging is discussed in detail in Refs. 27-29. Briefly, NMR is a quantum mechanical phenomenon observed in atomic nuclei that have spin and thus a small magnetic moment. The nuclei tend to align themselves with an externally applied magnetic field. The resonant oscillation (precession) of the nuclei during the alignment process is the NMR phenomenon. The frequency of precession (Larmor frequency) is proportional to the applied magnetic field and, in practice, occurs in the radio frequency (RF) band. To observe an NMR signal, a second time-dependent magnetic field (at the Larmor frequency) must be added to the static field. An induced signal following an RF pulse is detected by a tuned RF coil.[29] If a magnetic gradient is added, then the Larmor frequency will vary spatially across a specimen. After Fourier transformation, the NMR signal takes the form of a wave related to the shape of the sample. Computer techniques similar to those used in x-ray tomography are employed to obtain three-dimensional spatial information.

A preliminary evaluation of NMR imaging has been carried out in cooperation with the General Electric Medical Systems Division, New Berlin, Wisconsin. The objective of the initial feasibility tests was to establish whether commercially available NMR tomographic imaging technology could be applied to green ceramics. Detection of PVA binder proved to be extremely difficult, but tests with water-doped SiC were encouraging. These initial tests were conducted with equipment designed for medical rather than structural-ceramics applications, but the system could be modified to enhance the sensitivity for application to ceramic materials.

Silicon carbide disk SILC was doped with water (~3% by weight) and then cold pressed. Figure 14 shows axial views of two volumetric slices of the disk, each 8 mm in axial length, which were visible because of their proton-rich regions (water-filled pores); their positions are indicated by circles. An orthogonal view is seen in Fig. 15. The two views are consistent with respect to the distribution of proton-rich areas. Quantitative information can also be obtained from the intensity of the image. The use of a smaller coil would enhance the sensitivity and make it possible to resolve volumes as small as 100 μm.

Since the introduction of water (possibly as steam) to fill the interconnecting porosity network may have an adverse effect on the final fired component, the most likely application of NMR techniques would be in the development of processing techniques rather than in screening components before firing.

NEUTRON RADIOGRAPHY - Hydrogen-rich binders will also have relatively high neutron cross sections. Therefore, neutron radiography[30] was also considered for the detection of areas with a high concentration of binder.

The relative intensity of radiation that will be transmitted through an object under inspection is given by

$$I = I_o e^{-\mu x}, \tag{6}$$

where I is the radiation transmitted through an object of thickness x (in cm) and linear absorption coefficient μ (in cm^{-1}).

The absorption coefficients[31] for glass (0.22 cm^{-1}) and polyethylene (3.5 cm^{-1}) were used to estimate the likelihood of detecting the PVA binder in a 0.6-cm-thick green ceramic (spinel) sample as follows:

$$I/I_o \text{ (ceramic)} = e^{-\mu x} = 0.87;$$

$$I/I_o \text{ (ceramic + 1\% binder)} = e^{-(\mu_1 x_1 + \mu_2 x_2)} = 0.81.$$

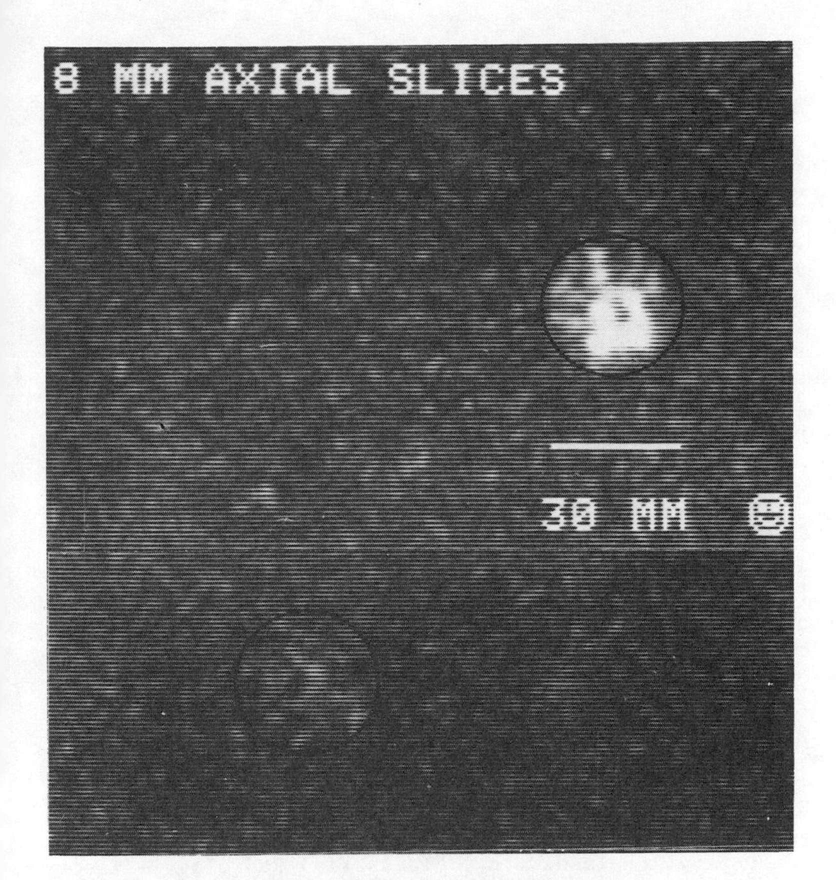

Fig. 14 - NMR image of slices through a water-doped SiC disk (axial view). Two images of the same sample are evident. The upper right shows a center portion and the lower left a section near the sample edge. The white areas indicate the presence of water

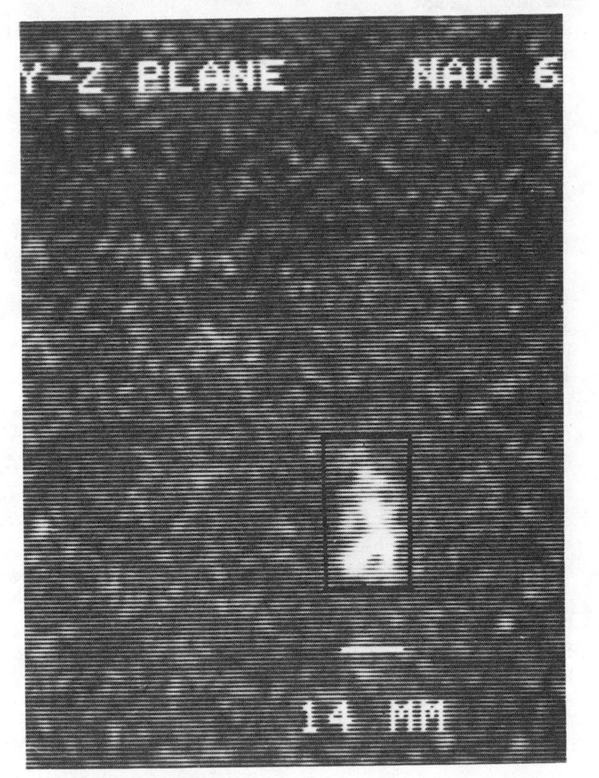

Fig. 15 - NMR image of same sample as Fig. 14 (side view)

Thus, the binder reduces transmission by ~7%. From the characteristic curve for Kodak "R" film (automatic processing) at a density of 2.0, one can calculate that a 7% change in exposure will lead to a film variation of ±0.04 density units, about twice the minimum detectable variation. Thus, it appears that a 1% concentration of binder should be observable.

Thermal neutron radiographs of sample NB1 (with PVA binder) and NB14 (from which the binder had been burned out) were prepared by Aerotest Operations, San Ramon, California. The samples were placed side by side, and successive radiographs were made with varying average film densities. Unfortunately, no difference between the samples could be detected, partly because neither sample had a uniform radiographic density. Future efforts will concentrate on preparing standards with varying amounts of PVA to establish the sensitivity of this technique, and on increasing the sensitivity by use of cold neutrons.

SUMMARY

A preliminary investigation was carried out to assess the effectiveness of microradiography, ultrasonic methods, nuclear magnetic resonance, and neutron radiography for the nondestructive evaluation of green (unfired) ceramics. The objective was to obtain useful information on defects, cracking, delaminations, agglomerates, inclusions, regions of high porosity, and anisotropy.

Microradiographs are sharper than ordinary radiographs, particularly if samples are reasonably thick (>1 cm) or of odd geometry. Another major advantage of microradiography is the capability to observe enlargements (20-40X) of an object in real time while manipulating the object. Preliminary experiments were carried out with a commercial microradiography unit (courtesy of NASA-Lewis). Very small (<50-μm) defects can be detected by microradiography if they are sufficiently different in density from the surrounding material. Voids are much more difficult to detect than high-density inclusions.

Conventional ultrasonic techniques are difficult to apply to flaw detection in green ceramics because of the high attenuation, fragility, and couplant-absorbing properties of these materials. However, it proved possible to couple longitudinal- and shear-wave transducers to green specimens by pressure alone. The amount of pressure strongly affected the measured velocity and attenuation, particularly for high-frequency longitudinal waves, and this was taken into account. Velocity, attenuation, and spectral data were shown to provide useful information on density variations and the presence of agglomerates. The elastic anisotropy of the green state (an important parameter related to the fabrication process) and variations in dispersion (which may also be related to microstructural variations) were also successfully measured. Future efforts will establish the correlation of these parameters with the ultimate performance of the fired material.

A nuclear magnetic resonance imaging technique was considered for detection of porosity in the green state. When the binder was burned out and replaced with a proton-rich dopant such as water, areas of high porosity could be detected. Neutron radiography was considered for detection of anomalies in the distribution of binders, which have a relatively high neutron cross section. Although imaging the binder distribution throughout the sample may not be feasible because of the low overall concentration of binder, regions of high binder concentration should be detectable.

ACKNOWLEDGMENTS

The authors wish to thank Aerotest Operations for providing the neutron radiographs, G. Glover and L. Edelheit of General Electric Medical Systems for providing access to NMR imaging equipment, S. Klima for providing access to the NASA microradiography facility, and R. Lanham and J. Picciolo of ANL for technical assistance on this project. The authors also wish to thank E. Stefanski for editing this manuscript.

REFERENCES

1. Evans, A.G., G.S. Kino, P.T. Khuri-Yakub and B.R. Tittman, Mater. Eval. 35, 85-96 (1977)
2. Kino, G.S., Science 206, 173-180 (1979)
3. Goebbels, K. and H. Reiter, in "Progress in Nitrogen Ceramics," F.L. Riley, ed., pp. 627-634, Martinus Nijhoff Publ., Boston and The Hague (1983)
4. Anon., "Nondestructive Examination," ASME Boiler and Pressure Vessel Code, Section V, ANSI/ASME PVB-V, ASME, New York (1983)
5. Anon., "Radiography in Modern Industry," Eastman Kodak Co., Rochester, New York, Fourth Edition (1980)
6. Anon., "Nondestructive Inspection and Quality Control," Metals Handbook, Vol. 11, pp. 105-156, ASM, Metals Park, Ohio (1976)
7. Bryant, L., ed., "Radiography and Radiation Methods," Nondestructive Testing Handbook, Vol. 3, ASNT, Columbus, Ohio (1983)
8. Halmshaw, R., ed., "Physics of Industrial Radiology," Heywood Books, London (1966)
9. Halmshaw, R., "Industrial Radiology - Theory and Practice," Applied Science Publishers, London (1982)
10. McMaster, R.C., ed., "Nondestructive Testing Handbook," two volumes, Ronald Press, New York (1959)
11. Berger, H., "Neutron Radiography," Elsevier, Amsterdam (1965)
12. Berger, H., ed., "Practical Applications of Neutron Radiography and Gaging," ASTM STP 586, ASTM, Philadelphia (1976)
13. Koehler, A.M. and H. Berger, in "Research Techniques in Nondestructive Testing," Vol. 2, R.S. Sharpe, ed., pp. 1-30, Academic Press, London and New York (1973)
14. West, D. and A.C. Sherwood, Nondestr. Test. (Chicago), pp. 249-258 (October 1973)
15. Cosslett, V.E. and W.C. Nixon, "X-Ray Microscopy," Cambridge Monographs in Physics, Cambridge University Press, Cambridge (1960)
16. Rodgers, E.H., "A Report Guide to Autoradiographic and Microradiographic Literature," Report No. MS-64-10, Army Materials and Mechanics Research Center, Watertown, Massachusetts (August 1964)
17. Freitag, V., W. Stetter and P. Charbit, Electromedica (Germany) 2, 45-49 (1974)
18. Braunbeck, J., Laser Elektro-Opt. (Germany) 8, 28-30 (1976)
19. Spiller, E., R. Feder and J. Topalian, Phys. Technol. 8, 22-28 (1977)
20. Sharpe, R.S., J. Microsc. (Oxford) 117, Pt. 1, 123-43 (1979)

21. Schmahl, G., D. Rudolph and B. Niemann, Recherche 12, 1136-37 (1981); McClung, R.W., Mater. Res. and Stds. 4(2), 66-69 (February 1964); Foster, B.E., E.V. Davis, and R.W. McClung, "High Resolution Boreside Radiography of Small Diameter Tube-to-Tubesheet Welds," ORNL-5474 (February 1979).
22. Parish, R.W., Brit. J. Nondestr. Test. 24, 210-213 (1982)
23. Khuri-Yakub, B.T., in "Encyclopedia of Materials Science and Engineering," H. Berger, NDE Subject Editor, Pergamon Press, Oxford, in press
24. Anderson, O.L., in "Physical Acoustics," Vol. 111b, W.P. Mason, ed., pp. 43-95, Academic Press, New York (1965)
25. Lees, S. and C.L. Davidson, IEEE Trans. Sonics Ultrason. SU-24(3), 222 (May 1977).
26. Lange, F.F., Panel Discussion on New Directions for Fabrication Reliability, Energy Materials Coordinating Contractors Meeting on Problems and Opportunities in Structural Ceramics, September 29-30, 1982, Germantown, MD
27. Pykett, I.L., Sci. Am., pp. 78-88 (May 1982)
28. Manfield, P. and P.G. Morris, "NMR Imaging in Biomedicine," Academic Press, New York (1982)
29. Battomley, P.A., IEEE Spectrum, p. 32 (February 1983)
30. Berger, H., "Neutron Radiography," Elsevier Publishing Co., New York (1965)
31. Anon., ASTM Standard E 748-80, in "1983 Annual Book of ASTM Standards," Section 3, pp. 669-685, ASTM, Philadelphia (1983)

POROSITY STUDY OF SINTERED AND GREEN COMPACTS OF $YCrO_3$ USING SMALL ANGLE NEUTRON SCATTERING TECHNIQUES

K. Hardman-Rhyne
Center for Materials Science
National Bureau of Standards
Washington, DC 20234

N. F. Berk
Center for Materials Science
National Bureau of Standards
Washington, DC 20234

E. D. Case
Lawrence Berkeley Lab. and Dept. of Matls. Science
Univ. of Cal.
Berkeley, CA 94720

ABSTRACT

Sintered and "green" compact samples of $YCrO_3$ are studied to determine the void sizes and density using small angle neutron scattering techniques which have been extended to the beam broadening regime to detect sizes larger than 0.15μm. This approach can be used with other on-line processing NDE techniques such as ultrasonics to standardize their results. Although the density ratio of the voids in the "green" compact and sintered material of $YCrO_3$ are very different (0.42 and 0.03 respectively), the average void radius is very similar (0.17 and 0.18 μm respectively).

INTRODUCTION

Porosity in ceramic materials is a critical aspect of the sintering process. To elucidate its extent, a quantitative study has been conducted at the National Bureau of Standards with small angle neutron scattering (SANS) to determine average pore size. Rather than restricting the SANS measurements to the typical 1 to 100 nm size regime of SANS diffraction, the neutron beam broadening regime is explored by extending the SANS characterization into the tens of micrometer size regime. This extension of SANS technique to larger sizes is an important result because it allows a greater overlap of SANS characterization with other NDE methods, such as ultrasonics. In general, thermal neutrons are an excellent nondestructive probe of microstructure because of their neutral charge, low flux and energy; and they are not absorbed by most materials. Also, since neutrons interact primarily with the atomic nuclei, the neutron beam is highly penetrating without disturbing the sample. In contrast, x-ray and electron beams tend to be more damaging and more sensitive to surface phenomena. While techniques such as mercury porosimetry and optical and electron microscopy can detect defects, voids, and second phases, SANS can quantify these effects nondestructively throughout the bulk of a material. Unfired ceramics, in particular, are often difficult to probe by other methods because the compressed powder may not maintain its structural integrity during the measurement. Thus SANS provides the only effective means of studying some "green" compacts and a powerful tool for determining void sizes in both compacts and sintered powders.

THEORY

The nature of small angle scattering from a monodisperse population of spherical particles or voids is determined by the phase shift ρ that a plane wave suffers in traversing a single particle;

$$\rho = (4\pi/\lambda)\Delta nR, \qquad (1)$$

where λ is the neutron wavelength, $\underline{R}$ is the particle radius, and

$$\Delta n = \Delta b\lambda^2/(2\pi), \qquad (2)$$

is the index of refraction of the particle or void relative to the matrix medium, which is assumed to be homogeneous and thus, acting alone, to produce no angular divergence in the neutron beam. In Eq.(2) $\Delta\underline{b}$ is the relative scattering length density or contrast of the particle (or void);

$$\Delta b = \sum_{cell} b_i/V_{cell} - b_{matrix}, \qquad (3)$$

where the sum is over the coherent scattering lengths $\underline{b}_i$ of the material formula unit for a crystalline unit cell, $\underline{V}_{cell}$ is the unit cell volume, and where $\underline{b}_{matrix}$ is defined by the analogous average for the matrix material. Thus combining Eqs.(1)-(3) the phase shift parameter ρ can be written in the useful form

$$\rho = 2\Delta bR\lambda, \qquad (4)$$

which shows the three independent factors on which it depends: 1) material contrast, $\Delta\underline{b}$; 2) particle size, R; 3); neutron wavelength, λ.

In the limit of small phase shift, $\rho<<1$, single-particle scattering is described by the Born approximation or equivalently the Rayleigh-Gans model which in SANS is identified with small angle diffraction. In this limit, the neutron differential cross-section--i.e. the relative probability for scattering into angle ε, which is equal to 2Θ--is expressed exactly as the Fourier transform of the single-particle density self-correlation function, which is sensitive to the details of particle shape and size. The theory and measurements in the diffraction regime are usually described in terms of the scattering wavevector

$$Q = 2\pi\varepsilon/\lambda. \qquad (5)$$

The angular deviations of scattered neutrons are proportional to λ and inversely proportional to particle radius. Generally, the diffractive regime applies to particles or voids of radius less than 0.1 μm. Several reviews[1-3] treat SANS diffraction in detail.

In the opposite limit, $\rho>>1$, the scattering from a single particle is well described by ray optics with each particle refracting neutrons as a lens. In this regime the neutron angular deviations produced by a particle are determined by the contrast index of refraction, Eq. (2), and are therefore proportional to λ^2. Moreover, in probabilistic terms every particle scatters, which has two important consequences. First, the observed intensity is not separable into an unscattered "incident beam" and a weak, scattered part (as in the diffraction regime); rather the incident beam appears to be broadened beyond instrumental resolution without "residual" scattering. Second, because multiple scattering produces this "beam broadening", particle size influences the measurement indirectly through the macroscopic configuration of the scatterers (for example, the mean particle spacing for fixed volume fraction). Usually, the refractive regime is reached by particles larger than 10 μm.

For intermediate values of the phase shift, $\rho\approx1$, the scattering is not well described in terms of either limiting case or as a simple combination of diffractive and refractive effects. In work described elsewhere[4] a general formal expression for the scattering cross-section for a uniform sphere--valid for all ρ values--as derived by Weiss [5], has been incorporated into the multiple scattering formalism of Snyder and Scott[6], modified for the relevant "pin-hole" geometry of the typical SANS instrument. The synthesis of these classical formalisms provides a theory of multiple scattering that is applicable over a wide range of single particle phase shifts, encompassing the transition from diffractive to refractive behavior, and a useful tool for extracting particle size from the wavelength dependence of beam broadening data. For this intermediate regime, the predicted neutron intensity as a function of scattering angle is approximately Gaussian with width roughly proportional to λ^2. However, every particle does not scatter and the mean number of scattering events is not constant with respect to wavelength as is the case in the multiple refraction regime. Although both regimes have approximately λ^2 behavior, this behavior results from different "mechanisms" and depends on the particle radius in very different ways. This intermediate regime is particularly relevant for voids in ceramic materials.

EXPERIMENT

Two samples of $YCrO_3$ were fabricated from pure powders by isostatic pressing at 207 MPa (30,000 psi); one sample was then sintered at 1750°C[7]. The sintered $YCrO_3$ sample was 7.8 mm thick and the "green" compact of $YCrO_3$ was 12.2 mm thick. The average equivalent diameter value of the powder was 1.3 μm determined by sedimentation methods[8]. The average grain size of the sintered material was approximately 6.0 μm [9], as determined from the linear intercept on scanning electron micrographs of fracture surfaces. The density of the "green" compact was

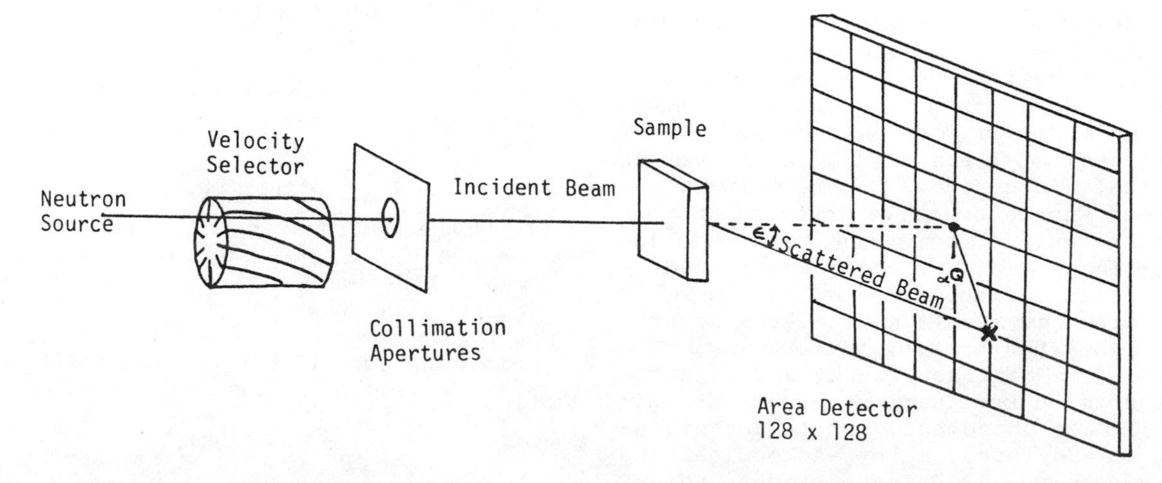

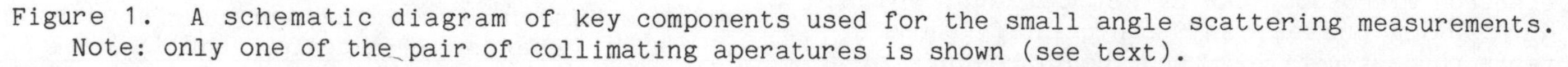
Figure 1. A schematic diagram of key components used for the small angle scattering measurements. Note: only one of the pair of collimating aperatures is shown (see text).

approximately 57% of the theoretical density (i.e., 0.43 void density ratio) and that of the sintered material was approximately 94% (0.06 void density ratio). The density values were obtained by measuring the mass and volume of the samples.

SANS measurements and analyses were performed at the National Bureau of Standards. A schematic of the major components of the SANS facility is shown in Figure 1. The mean wavelength, λ, can be varied from 0.4 to 1.0 nm by selecting the appropriate speed of a rotating helical-channel velocity selector. This is important in these beam broadening experiments because the wavelength dependence of the neutron scattering is a necessary part of the analysis and the resulting neutron beam widths are most sensitive to λ at λ greater than 0.7 μm. Collimating apertures define the beam direction and divergence and consist of a pair of cadmium pinhole irises, one after the velocity selector and another before the multiple sample chamber. The scattered neutrons are detected on a 64 cm x 64 cm position-sensitive proportional counter and are divided into 128 columns and 128 rows. Further details of the SANS facility can be found in [10].

The angle between the incident beam and the scattered beam is the scattering angle, ε, (see Fig. 1). The magnitude of the scattering vector Q is $(4\pi/\lambda)\sin \varepsilon/2$ which is $2\pi\varepsilon/\lambda$ in the small angle limit. The distance on the detector between the incident and scattered beam is proportional to Q; thus the intensity is often expressed as a function of Q in reciprocal angstroms ($1 \text{Å}^{-1} = 10 \text{nm}^{-1}$). For detailed analysis, the beam broadening data is better expressed as a function of scattering angle ε because the full width at half maximum of the incident beam, ε_b, is a constant for all wavelengths. These experiments required only 3 minutes to 2 hours depending on the wavelength, void size, density, and thickness of the sample, and were similar in technique to transmission measurements. In these measurements the beamstop which normally attenuates the transmitted direct beam was removed and the exact center of the scattering patterns was determined from the constant intensity contours. The method of analysis of the SANS data will be described in the next section.

RESULTS AND DISCUSSION

SANS measurements were obtained at six or seven wavelengths of the following: 0.485, 0.545, 0.625, 0.70, 0.80, 0.90, 0.95, and 1.0 nm. The neutron scattering spectra are circularly symmetric around the center of the incident beam and are approximately Gaussian in shape at the low Q values ($\leq 0.015 \text{Å}^{-1}$). A linear slice (one of the columns on the two dimensional detector) through the center of the scattering plane can be plotted as intensity versus the row number to qualitatively determine the beam broadening effect. The results reveal a striking difference between the sintered $YCrO_3$ sample

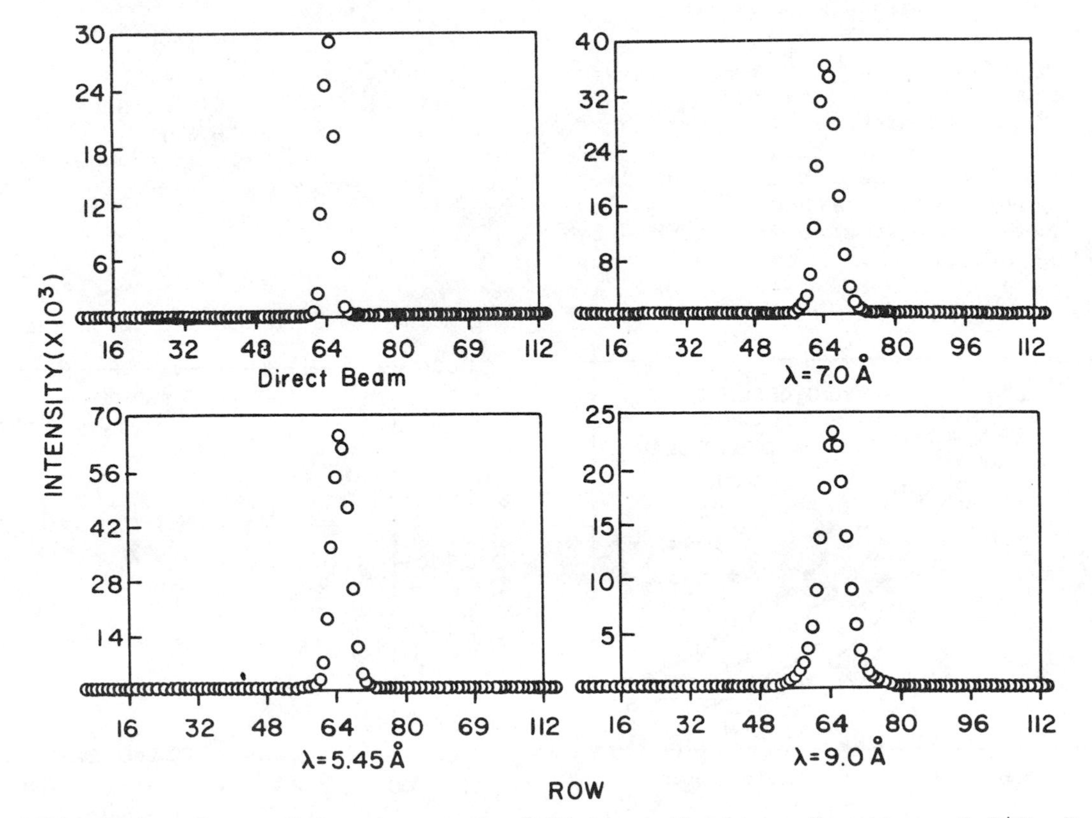

Figure 2. SANS spectra for a sintered compact of $YCrO_3$ at three wavelengths: 0.545, 0.7, 0.9 nm. Plotted is the scattering intensity versus a linear column slice through the center of the neutron scattering plane (as indicated by the row number).

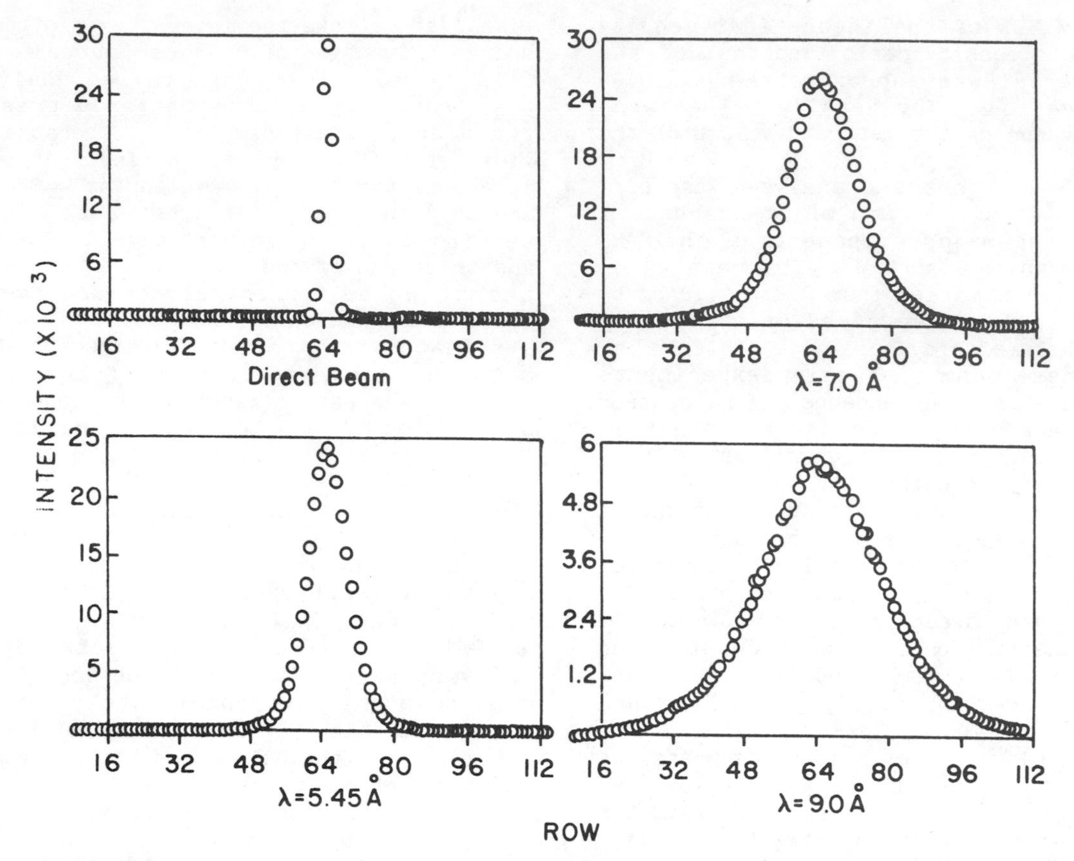

Figure 3. SANS spectra for a "green" compact of $YCrO_3$ at three wavelengths: 0.545, 0.7, and 0.9 nm. Plotted is the scattering intensity versus a linear column slice through the center of the neutron scattering plane (as indicated by the row number).

and the "green" compact $YCrO_3$ sample as illustrated in Figures 2 and 3 respectively. The data from the sintered material (Fig. 2) shows slight wavelength dependence, but the "green" compact reveals dramatic beam broadening which is strongly wavelength dependent (see Fig. 3). This dependence is illustrated in Figure 4 by plotting the normalized intensity versus scattering vector Q for five wavelengths. In Figure 5 the wavelength dependence for the sintered $YCrO_3$ sample is shown.

At low Q (Q < 0.016Å) these beam broadening data can be represented by a Gaussian distribution where the full width at half maximum, $\Delta\varepsilon$, can be determined from the Gaussian standard deviation σ_G as shown below:

$$\Delta\varepsilon = 0.3748\lambda\sigma_G. \qquad (6)$$

The measured width contains both the beam broad-

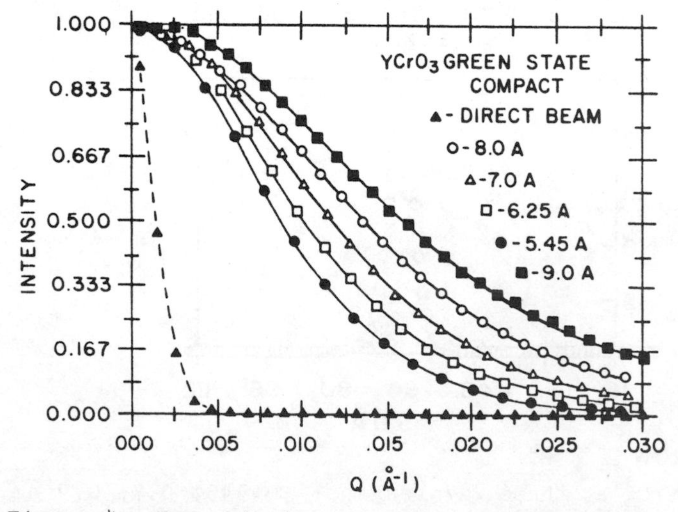

Figure 4. Normalized neutron scattering intensity versus scattering vector, Q, for the "green" compact of $YCrO_3$ at five wavelengths.

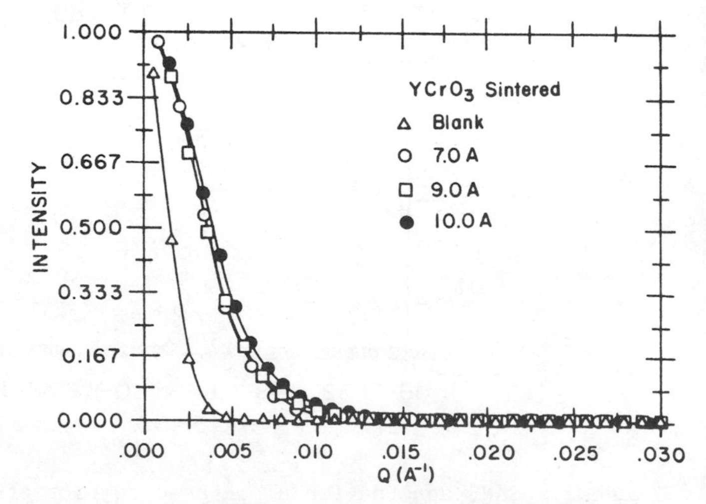

Figure 5. Normalized neutron scattering intensity versus scattering vector Q for the sintered $YCrO_3$ sample at four wavelengths.

ening scattering and that due to the wavelength independent broadening of the incident beam resulting from finite collimation, which is determined for the fine pin-hole geometric configuration (12 and 8 mm) to be $4.62 \times 10^{-3} \pm 10^{-5}$ radians. This instrumental contribution, ε_b, must be subtracted from the experimentally determined width, ε, so that

$$\Delta\varepsilon = [\varepsilon^2 - \varepsilon_b^2]^{1/2}. \qquad (7)$$

Although the qualitative aspects of the data clearly demonstrate a strong effect of ceramic processing on the population of neutron scatterers in these materials, quantitative measures of the particle or void size, shape and size distribution are less straightforward. Moreover, the Δb, λ, and R values correspond to phase shifts (Eq. (4)) well within the intermediate range of values for which the neutron scattering is not expected to be analyzable by multiple refractive behavior alone. Therefore, as mentioned above, a generalized beam broadening theory[4] relevant for this region has been developed at NBS to quantitatively analyze the SANS data for densified ceramics and other distributed defects in this size regime. The radius and void (or particle) density ratio can be obtained from this theory, which can be extended in principle to consider particle packing, polydispersity, and particle or void shapes other than spheres. Specifically we extract the void radius from the experiment by computing theoretical scattering curves and adjusting the radius and void fraction parameters until a pair of values are found that reproduce the wavelength dependence of the measured beam broadening widths over the whole wavelength range. Sample thickness and the measurement wavelength are

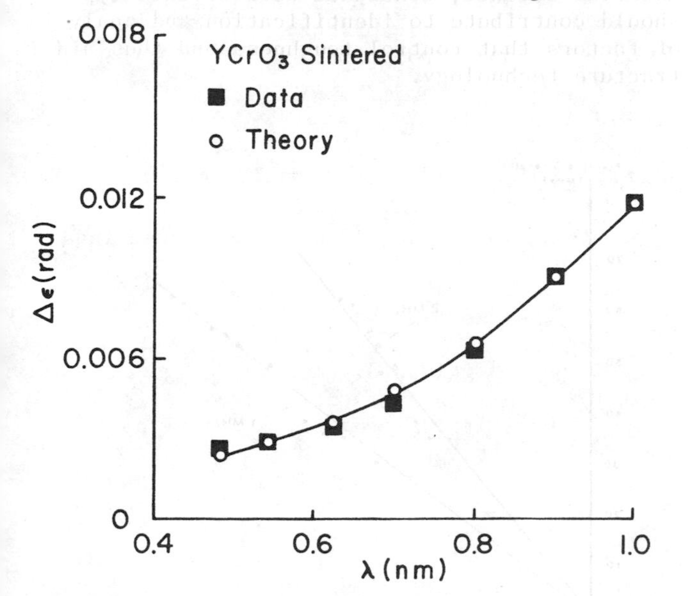

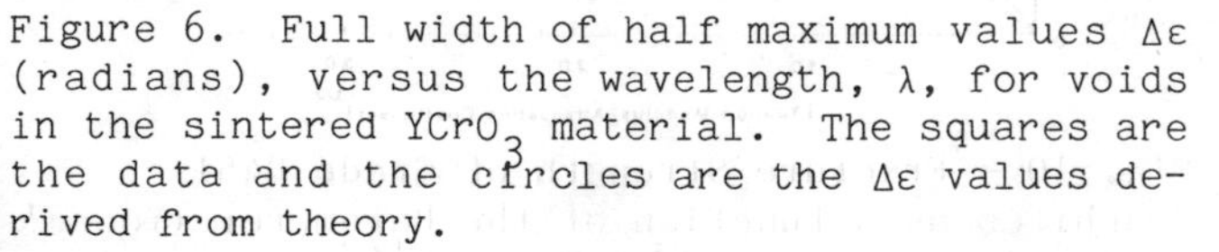
Figure 6. Full width of half maximum values $\Delta\varepsilon$ (radians), versus the wavelength, λ, for voids in the sintered $YCrO_3$ material. The squares are the data and the circles are the $\Delta\varepsilon$ values derived from theory.

Table 1

$\Delta\varepsilon$ Values for $YCrO_3$-sintered, where $\Delta b = 5.277 \times 10^{-4}$ nm^{-2}, void density ratio = 0.03, thickness = 7.8 mm and R = 0.18 μm.

λ(nm)	Δε(radians)-Data	Δε(radians)-Theory
0.485	0.002770	0.002299
0.545	0.002903	0.002910
0.625	0.003517	0.003684
0.70	0.004197	0.004831
0.80	0.006221	0.006682
0.90	0.009092	0.009143
1.00	0.011782	0.011670

treated as fixed parameters in the fitting procedure. In fact since in this work both the experimental and computed shapes are nearly Gaussian over about two standard deviations the theoretical width can be accurately estimated numerically without having to generate the scattering curve, although the computations are still time consuming.

Excellent agreement of the data and theoretical values for $\Delta\varepsilon$ can be obtained for both samples as shown in Figure 6 and Table 1 for

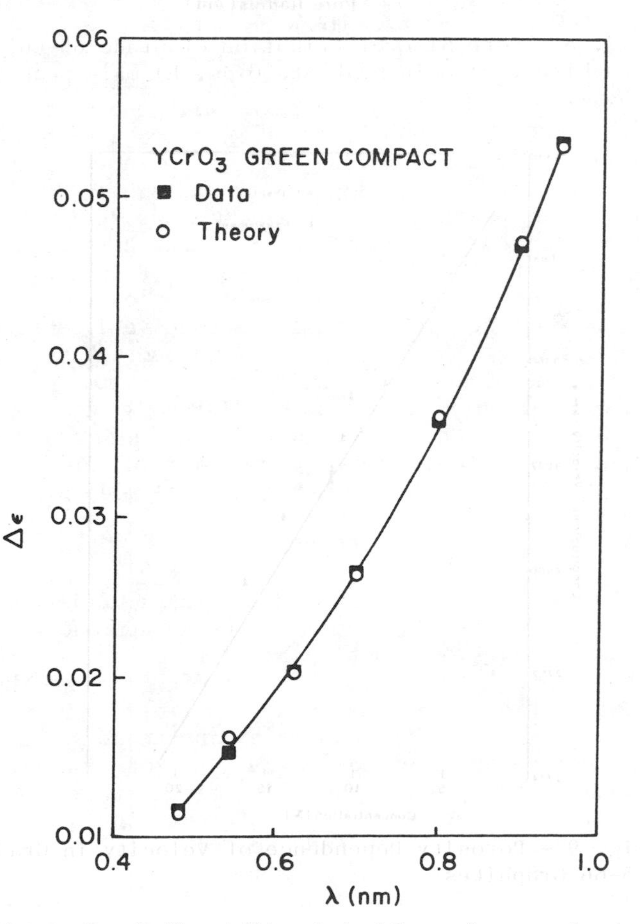

Figure 7. Full width at half maximum values, $\Delta\varepsilon$ (radians), versus the wavelength, λ, for voids in the "green" compact of $YCrO_3$. The squares are the data and the circles are the $\Delta\varepsilon$ values derived from theory.

Table 2

$\Delta\varepsilon$ values for $YCrO_3$-green compact where Δb = $5.277X10^{-4}$ nm^{-2}, void density ratio = 0.42, thickness = 12.2 mm and R = 0.17 μm

λ(nm)	Δε(radians)-Data	Δε(radians)-Theory
0.485	0.01156	0.01137
0.525	0.01525	0.01611
0.625	0.02038	0.02028
0.70	0.02660	0.02645
0.80	0.03598	0.03617
0.90	0.04700	0.04711
0.95	0.05338	0.05341

for sintered $YCrO_3$ and in Figure 7 and Table 2 for "green" compact $YCrO_3$. The experimental error bars are smaller than the symbol size in the figures, and the theoretical points shown are sensitive to variations of radius within 2%, especially at the higher wavelengths. The best-fit void radius in the sintered $YCrO_3$ material is 0.18 μm with a void density ratio of 0.03. The difference between this fit-determined void density ratio and the measured value (0.06) probably results from void sizes outside the intermediate regime; void sizes greater than 30 μm would not have been detected in these experiments. The best-fit void radius in the $YCrO_3$ "green" compact material is 0.17 μm with a void density ratio of 0.42 compared with the measured value, 0.43. We note that these data (Table 2 and Fig. 7) could not be fit to the known loose powder size[8] for any density ratio.

The theory used here assumes the voids are monodispersed spheres in an otherwise homogeneous material. This simple model for the observed neutron scattering seems appropriate for the densified $YCrO_3$ with a measured void density ratio of 0.06. For the "green" compact sample with a measured void density ratio of 0.43 the scattering model may be consistent with a pore model based on a network of interconnecting spheres. However, we do not attempt here to relate the fitted values to material models.

CONCLUSION

The neutron scattering beam broadening technique discussed here offers an approach for quantitatively evaluating pore size and volume changes in the material occurring during the sintering process in a nondestructive manner. Ways are being explored to expand these studies to include nonspherical powder or void shapes, agglomeration defects and polydispersivity; and to evaluate the sensitivity of the technique to these effects. While SANS techniques have limited applications for on-line process control under manufacturing conditions, they do provide a powerful laboratory tool for understanding the densification process; and they constitute a unique laboratory standard with which alternate NDE methods can be compared.

ACKNOWLEDGEMENTS

The authors thank Carl Robbins, Taki Negas and Lou Domingues for their preparation and characterization of the "green" compact and sintered $YCrO_3$ materials.

REFERENCES

1. Weertman, J. R., Nondestructive Evaluation: Microstructural Characterization and Reliability Strategies, edited by O. Buck and S. M. Wolf, American Institute of Mining, Metallurgical, and Petroleum Engineers, New York, pp. 147-168 (1981).

2. Kostorz, G., Treatise on Materials Science and Technology, Vol. 15, edited by G. Kostorz, Academic Press, New York, pp. 227-289 (1979).

3. Herman, H. "Nondestructive Evaluation of Materials with Cold Neutron Beams", report to Naval Air Systems Command, Contract no. N00019-77-M-0418, Washington, DC, Dec. 1977.

4. Berk, N. F., and K. Hardman-Rhyne, to be published.

5. Weiss, R. J., Phys. Rev. 83, 379 (1951).

6. Snyder, H. S. and W. T. Scott, Phys. Rev. 76, 220 (1949).

7. Negas, T., L. P. Domingues (private communications, 1983).

8. Robbins, Carl (private communications, 1983).

9. Case, E. D., and C. J. Glinka, to be published in the J. of Matl. Sci.

10. Glinka, C. J., AIP Conference Proceedings 89, 395 (1982).

NONDESTRUCTIVE EVALUATION OF POWDER METALLURGY MATERIALS

B. R. Patterson
University of Alabama in Birmingham
Birmingham, Alabama 35294

K. L. Miljus
B, E&K Engineering Co.
Birmingham, Alabama 35201

W. V. Knopp
P/M Engineering and Consulting
Wyckoff, New Jersey 07481

ABSTRACT

Ultrasonic velocity and resonant frequency measurements have proven to be valuable tools for nondestructively evaluating the processing and properties of powder metallurgy (P/M) materials. Measured values of ultrasonic velocity and resonant frequency increase with increasing P/M part density, time and temperature of sintering, and part strength. Although these nondestructive methods are sensitive to the pore structure of P/M materials, which strongly affects properties, they are less sensitive to the matrix microstructure and composition which also affects the strength. Different quality control curves relating part strength to nondestructive measurement values must be developed for different P/M alloys. The direction of ultrasonic velocity measurement must be standardized for a particular part or test bar since anisotropy of the pore structure in high density P/M materials can also produce anisotropic velocity measurements. Resonant frequency measurements must also be correlated with strength for a particular part type or test bar since the measured frequency will vary with part dimensions. Quality control curves of velocity versus strength may also need to be standardized for particular blends of metal powders, since addition of porous iron powder to dense atomized powders produces differences in velocity measurements for similar density and sintering conditions. In tests employing P/M stainless steel production parts, ultrasonic velocity measurements have proven capable of discriminating well sintered parts with satisfactory microstructures and high crushing strength from identically appearing parts sintered in a contaminated atmosphere with resulting unsatisfactory microstructures and low crushing strength.

INTRODUCTION

Nondestructive characterization techniques, such as ultrasonic velocity and frequency measurements, hold considerable promise for evaluating the quality of powder metallurgy (P/M) parts and materials (1-4). A principal benefit of these techniques is that they provide nondestructive assessment of the extent of interparticle bonding, pore size and shape in addition to detecting gross flaws such as cracks. This is an especially valuable capability with P/M materials, which often exhibit wide ranges of strength arising from variations in processing variables and powder characteristics, even when cracks are not present.

Variations in sintering time, temperature, atmosphere, powder type, contaminants, and other variables can produce parts which appear sound but have undesirably low mechanical properties. Techniques for rapid assessment of the extent of sintering in individual parts would be extremely useful for monitoring product quality. This paper reviews the results of several prior studies relating ultrasonic velocity and resonant frequency measurements to the strength of selected P/M materials and illustrates the application of these methods for examining production P/M parts.

The ultrasonic velocity through a material is the speed of transmission of a high frequency (MHz range) sound wave. The velocity is dependent on the interatomic forces in the metal, but microstructural features also have an important effect. In recent years it has been found that the microstructural characteristics which control the mechanical strength of some materials, such as P/M and cast iron (5), also control their ultrasonic velocity.

The microstructures of these two types of material are somewhat similar in that both contain a minor phase, i.e., graphite in cast iron and porosity in P/M, which reduces the strength compared to the pure fully dense matrix material. The morphology of the weaker phase strongly affects the mechanical strength, even more so than the volume fraction of the phase. Irregular, interconnected porosity in P/M parts and flake graphite in cast iron reduces the strength much more than when rounded and disconnected. The presence of an irregular second phase also reduces the elastic modulus and it is this property coupled with the density that basically controls the speed of sound transmission. The interparticle bonding in P/M materials also affects the ability for sound transmission and the strength of the material.

The resonant frequency of a part is the frequency at which the part "rings" when struck or vibrated by other means. Since this frequency is dependent on the velocity of sound within the part and the part length, it is also closely related to ultrasonic velocity and is sensitive to the same microstructural and mechanical properties.

Although both resonant frequency and ultrasonic velocity measurements have proven useful for inspection of P/M parts, each technique has particular capabilities and limitations. The resonant frequency of a part can be measured with precision but varies with part size and cannot easily be made on very small parts. The ultrasonic velocity through a material is usually not as sensitive to specimen size but can be affected by other variables such as the frequency of the ultrasonic signal, direction of measurement within the specimen if the microstructure is anisotropic, and other microstructural factors which affect the ultrasonic wave. For the above reasons, one nondestructive technique or the other may be better suited for evaluating the strength of a particular type of P/M part or test bar.

It is the goal of this review to examine the effects of the above mentioned variables of microstructure, processing parameters, and measurement parameters on the relationships between velocity and resonant frequency measurements and mechanical properties of P/M materials.

EXPERIMENTAL

ULTRASONIC MEASUREMENTS-

In the following studies, ultrasonic velocity and resonant frequency measurements were performed on standard flat powder metallurgy tensile bars (ASTM E8/MPIF Std. 10), rectangular transverse rupture specimens (MPIF Std. 41), and some production parts for comparison with mechanical properties. Longitudinal ultrasonic velocity was measured across the center of the gage sections of the tensile bars, in the pressing direction, using the apparatus schematically illustrated in Figure 1. Velocities were also measured in the pressing direction of the production parts and in several directions on the transverse rupture bars. The signal frequency for all velocity measurements was 1 MHz.

The apparatus used for measuring resonant frequency is illustrated in Figure 2. The test parts were resonated by vibrating an exciter plate adjacent to the part. At the resonant frequency, the part vibrated at an increased amplitude which was detected by a pick-up plate. The frequency of vibration depends on the part shape as well as sintering conditions, but comparison of similar parts eliminates the shape variable and allows evaluation of the extent of sintering.

Resonant frequency was determined only for the tensile bars and some production parts. For the tensile bars, flat ends were ground on the grips and the specimens were vibrated longitudinally.

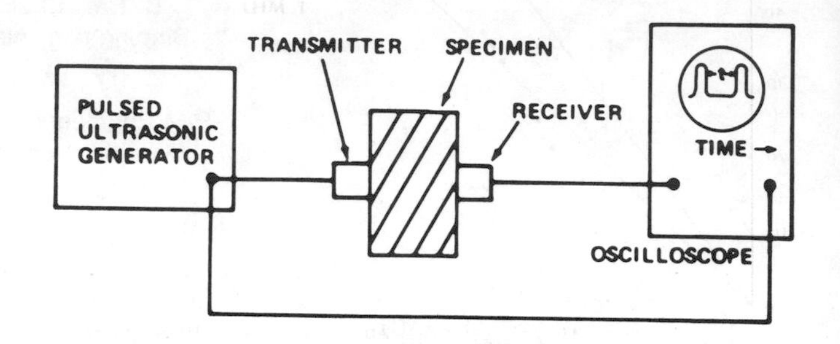

$$\text{VELOCITY} = \frac{\text{PART LENGTH}}{\text{TIME FOR SOUND TRANSMITTAL}} = \frac{1 \text{ IN.}}{5.71 \times 10^{-6} \text{ SEC}} = 0.1750 \times 10^{6} \text{ IN./SEC.}$$

Figure 1. Schematic of ultrasonic velocity measurement apparatus.

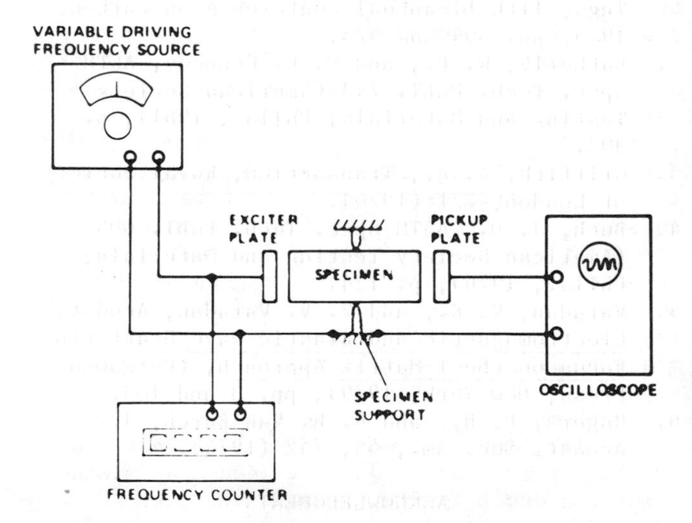

Figure 2. Schematic of resonant frequency measurement apparatus.

EFFECT OF SINTERING TIME AND TEMPERATURE -

The velocity and resonant frequency measurements were made on tensile bars which were companion specimens to bars which had previously been mechanically tested. The bars were pressed from atomized iron powder with 1 percent graphite and 3/4 percent lubricant to a green density of 6.70-6.80 gm/cm^3, and sintered for a variety of times and temperatures (Table I) to

Table I. P/M Tensile Bar Production Parameters

MPIF STANDARD NO. 10 TENSILE BARS:

Powder Blend:	99% Ancorsteel 1000 1% Graphite 3/4% Acrawax C
Green Density:	6.70 - 6.80 gm/cm^3

SINTERING CONDITIONS:

Temperature:	1832, 1922, and 2050°F(1000, 1050, and 1121°C)
Time:	10, 20, and 30 Min.
Atmosphere:	Endothermic, +30°F Dew Point

Table II. Mechanical Property and Nondestructive Test Data for P/M Tensile Bars

	SINTERING TIME (min)	TENSILE STRENGTH (ksi)	YIELD STRENGTH (ksi)	ULTRASONIC VELOCITY (in./μsec)	RESONANT FREQUENCY (Hz)
1832°F (1000°C)					
	10	28.6	26.1	0.1586	24,635
		24.9	24.9	0.1448	23,897
	20	32.1	27.1	0.1653	24,927
		29.7	27.8	0.1542	24,926
	30	33.3	27.0	0.1705	25,162
		33.3	31.4	0.1681	24,981
1922°F (1050°C)					
	10	32.0	29.5	0.1639	24,924
		29.2	28.5	0.1536	24,511
	20	34.9	29.9	0.1732	25,250
		35.9	33.4	0.1685	25,225
	30	39.0	29.6	0.1755	25,251
		42.6	36.0	0.1771	25,634
2050°F (1121°C)					
	10	41.7	33.2	0.1716	25,212
		38.5	34.3	0.1709	25,677
	20	43.6	34.1	0.1804	25,712
		47.2	37.6	0.1858	26,434
	30	45.0	33.3	0.1831	25,978
		48.9	37.2	0.1874	26,602

produce a range of strength values. The sintering temperatures were 1000, 1050, and 1121°C (1832, 1922, and 2050°F). The sintering times were 10, 20, and 30 minutes at each temperature. The yield and tensile strengths, ultrasonic velocity, and resonant frequency data for the bars are presented in Table II.

The sintering cycles produced specimens with yield strengths ranging from 25 to 38 ksi and ultimate tensile strengths ranging from 25 to 49 ksi. Ultrasonic velocity and resonant frequency values also showed large variations, from 0.145 to 0.187 in/μsec and 23.9 to 26.6 KHz, respectively.

The relationship between ultimate tensile strength and the ultrasonic velocity and resonant frequency values are illustrated in Figures 3 and 4, respectively. The velocity and resonant frequency increased with the strength of the material irrespective of whether the

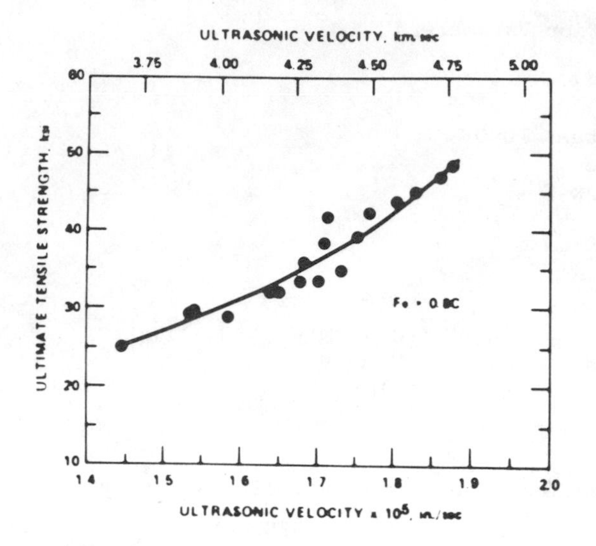

Figure 3. Relationship between ultimate tensile strength and ultrasonic velocity for P/M steel tensile bars.

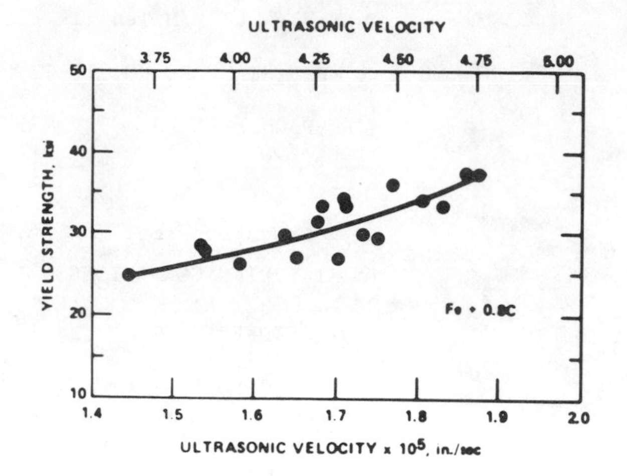

Figure 5. Relationship between yield strength and ultrasonic velocity for P/M steel tensile bars.

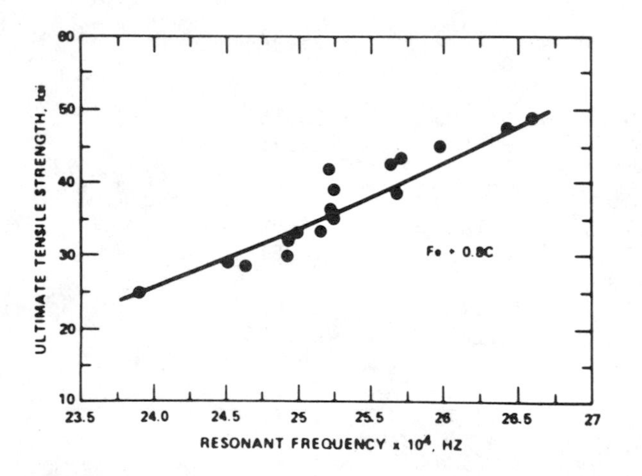

Figure 4. Relationship between ultimate tensile strength and resonant frequency for P/M steel tensile bars.

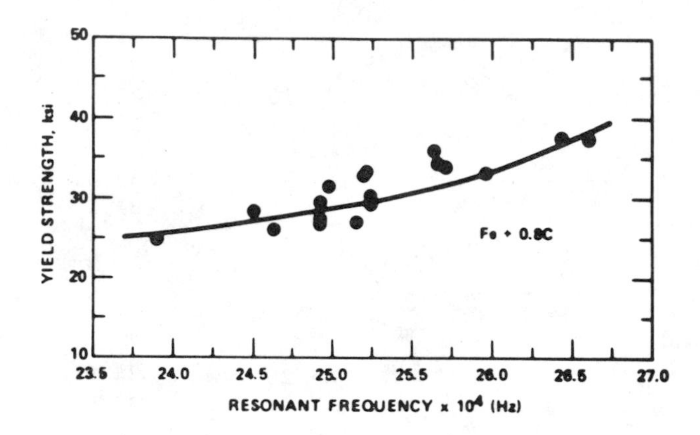

Figure 6. Relationship between resonant frequency and yield strength for P/M steel tensile bars.

strength change was a result of a longer sintering time or a higher sintering temperature. Similar relationships between yield strength and velocity and frequency are illustrated in Figures 5 and 6 respectively.

The correlation of velocity and resonant frequency with both yield and tensile strength suggests that these methods may provide a nondestructive means for assessing the quality of P/M parts. Plots similar to those in Figures 3 and 4 were made using 12 tensile bars, those sintered for 10 and 30 minutes, and best-fit curves were drawn through the data points. The strengths of the specimens sintered for 20 minutes were predicted from the measured ultrasonic velocity and resonant frequency values.

Figure 7 compares the mechanically measured tensile strengths with the nondestructively predicted strengths for specimens sintered for 20 minutes. Very little deviation existed between the predicted and the actual strength values. The measured and predicted values and the difference between the two values are presented in Table III. Most predicted values differed no more than one or two ksi from the measured values and were less than 10% of the average strength (except in one case). The deviation between the predicted strengths from nondestructive measurements and the actual strength might have been even less if measurements had been performed on the same bar rather than similarly processed companion specimens.

Table III. Measured Strengths of P/M Tensile Bars and Values Predicted From Ultrasonic Velocity and Resonant Frequency Curves.

SPECIMEN NO.	MEASURED TENSILE STRENGTH (ksi)	PREDICTED TENSILE STRENGTH (ksi)		[MEASURED-PREDICTED] (ksi)	
		VEL.	FREQ.	VEL.	FREQ.
2125	32	33	34	-1	-2
2224	35	38	37	-3	-2
2324	44	44	41	0	3
5124	30	28	34	2	-4
5224	36	35	37	1	1
5324	47	48	48	-1	-1

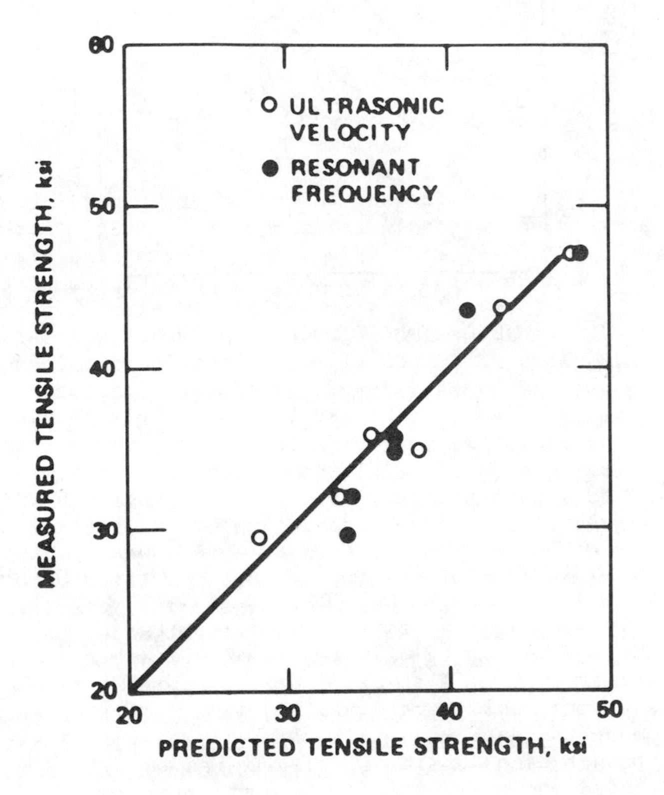

Figure 7. Comparison between measured strength and strength predicted from ultrasonic velocity and resonant frequency.

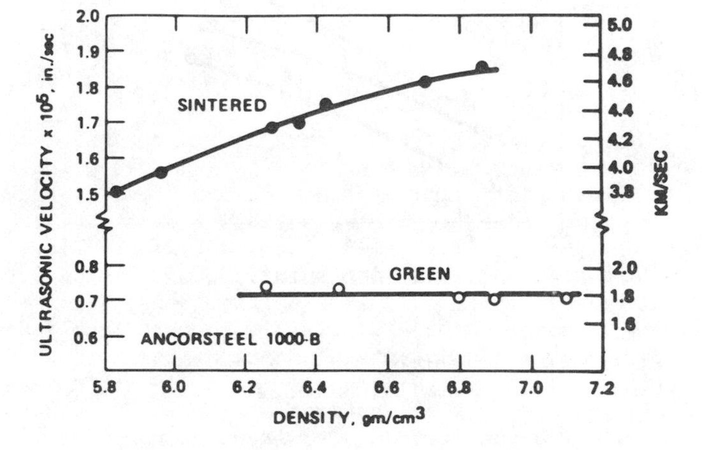

Figure 8. Relationship between ultrasonic velocity and density of green and sintered cyclindrical specimens.

EFFECT OF DENSITY - The effect of density on the ultrasonic velocity through cylinders 1.275 in. long x 0.305 in. diameter, pressed from atomized iron powder (Ancorsteel-1000 B) with 1 percent lubricant is illustrated in Figure 8. The extent of compaction had essentially no effect on velocity in the unsintered condition. The lack of bonding between particles overrides any dependence on pore volume fraction. Specimens sintered identically for 1/2 hour at 1038°C (1900°F), however, had a very close correlation between density and velocity. Figure 9 illustrates an equally good correlation between resonant frequency and the density of the sintered cylindrical pieces. The un-

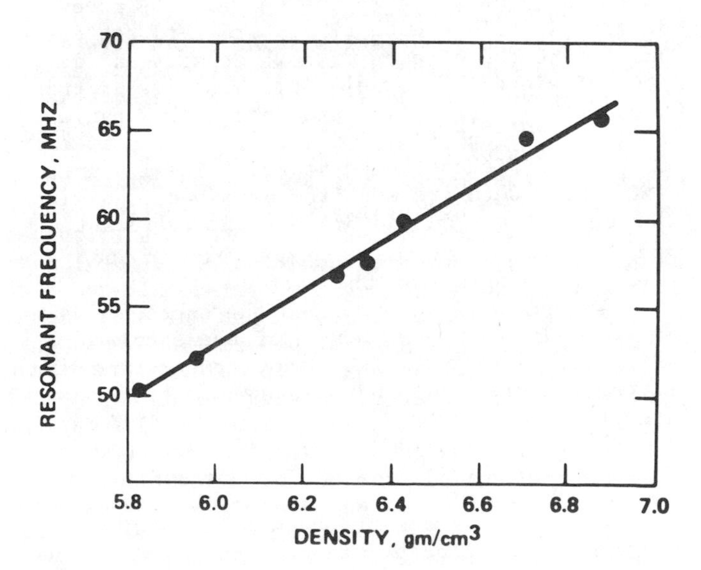

Figure 9. Relationship between resonant frequency and density of sintered cylindrical specimens.

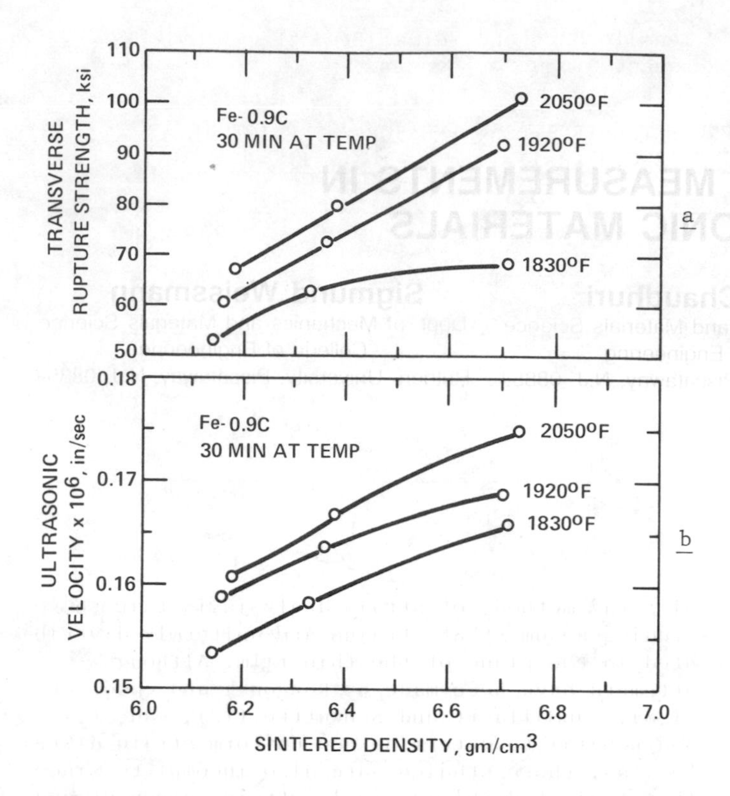

Figure 10. Effect of density and sintering temperature on rupture strength and ultrasonic velocity of Fe-0.9C.

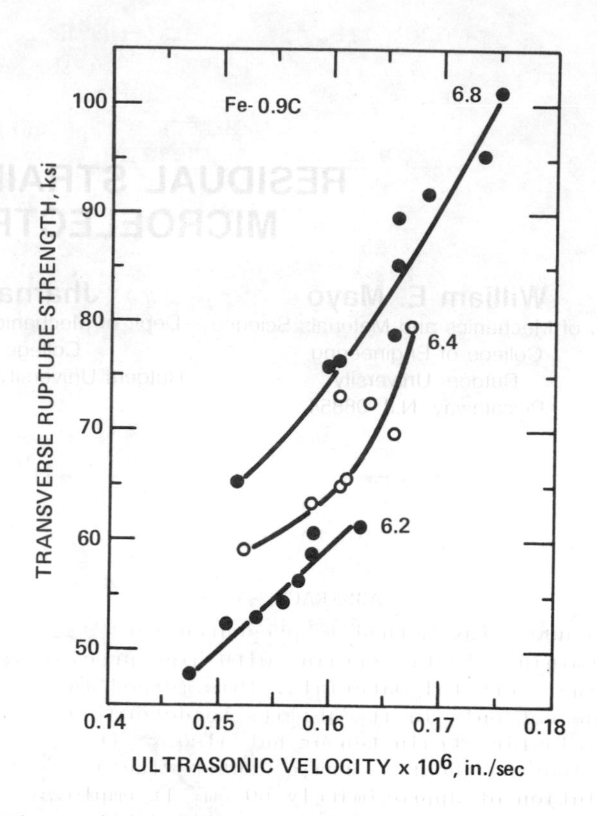

Figure 11. Relationship between transverse rupture strength and velocity for Fe-0.9C bars of different densities.

sintered parts could not be resonated in the test device.

The above results demonstrate that for constant sintering time and temperature, the velocity through atomized iron increases with part density, and that velocity increases with sintering time and temperature when the pressed density is held constant. In order to better understand the effects of both density and degree of sintering on velocity, a series of transverse rupture bars were produced with a range of densities and sintered for 30 minutes at each of three different temperatures. The 1/4 in. thick bars were pressed from atomized iron with 0.9 percent carbon and 3/4 percent lubricant added. All velocities were measured in the pressing direction of the bars.

Figure 10(a) illustrates the expected increase in rupture strength of the specimens with both density and sintering temperature, with strength increasing with temperature at all densities. Figure 10(b) shows a similar increase in velocity with density for these specimens, again with velocity increasing with sintering temperature at all densities. These results indicate that while density cannot be obtained directly from velocity measurements, the degree of sintering, and most importantly, the strength of a part can be determined.

Figure 11 illustrates the relationship between transverse rupture strength and velocity for these and other bars of atomized iron with 0.9 percent carbon addition, pressed to 6.2, 6.4, and 6.8 gm/cm^3 and sintered from 10 to 30 minutes at 1000, 1050, and 1121°C.

Within each density group velocity increased uniformly with degree of sintering and strength. A single curve of velocity versus strength would have exhibited more variability among the data than individual curves drawn through each density group. It can be concluded, however, that for a particular density, a very close relationship exists between strength and degree of sintering of P/M parts to be evaluated nondestructively.

EFFECT OF CARBON AND ALLOYING ADDITIONS - Although velocity is sensitive to interparticle bonding and density, it has not previously been known what effect the matrix structure of P/M materials exerts on the previously shown relationship between strength and velocity. Figure 12 illustrates the relationship between ultrasonic velocity and strength for atomized iron with 0, 0.4, 0.7, and 0.9 percent carbon additions. These bars were pressed to green densities from 6.2 to 7.0 gm/cm^3 and sintered for times of 10 to 30 minutes at 1121°C. The bars of each carbon level produced a separate curve of strength versus velocity, with lower carbon levels having lower strength at comparable velocities. This result is as might be expected from prior velocity versus strength studies on cast irons (5) where the percentage of pearlite in the matrix had little effect on

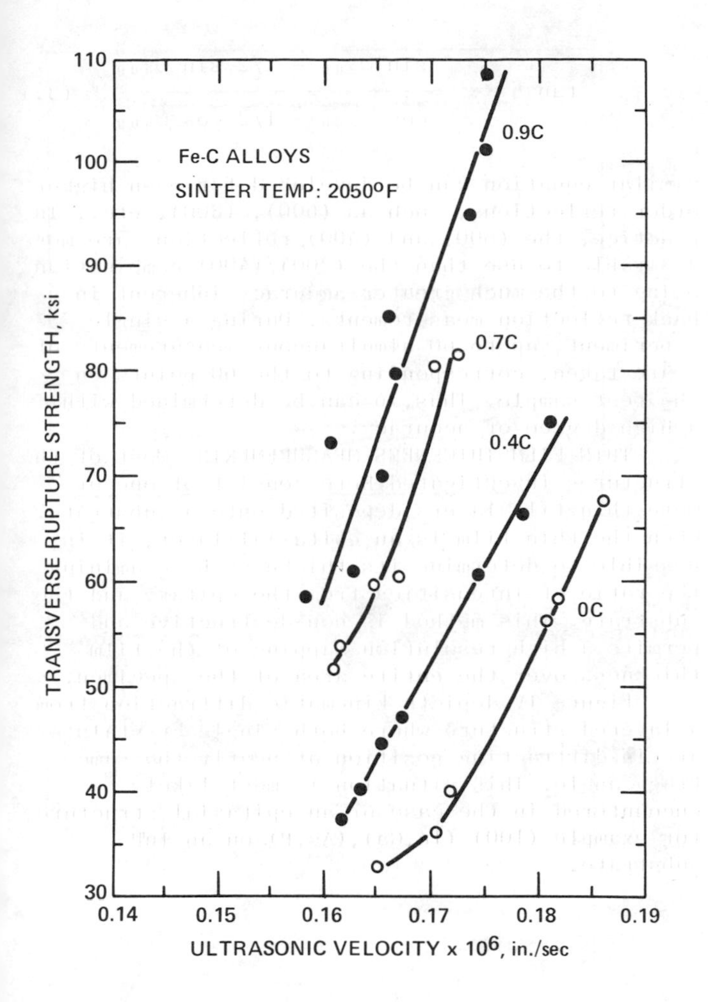

Figure 12. Relationship between transverse rupture strength and ultrasonic velocity for atomized iron with different levels of carbon addition.

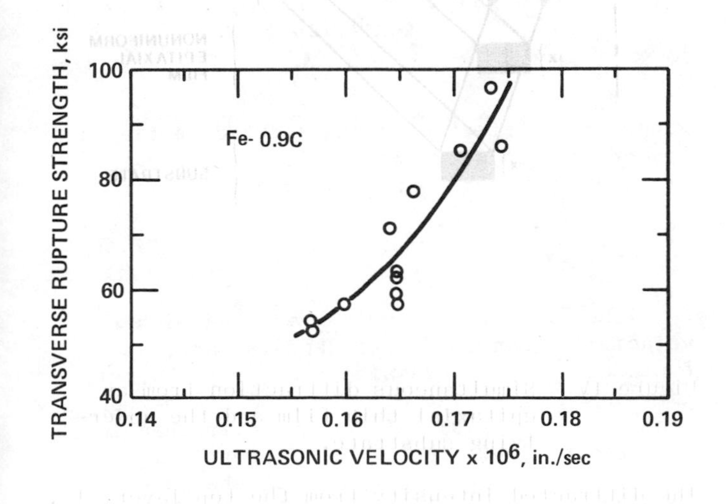

Figure 13. Strength versus velocity curve for Fe-0.9C bars of different densities, sintering times and temperatures.

velocity, although it increased strength.

In both P/M materials and cast irons the sound wave is more sensitive to the shape and amount of the weak, low density phase, i.e., porosity in P/M materials and graphite in cast iron, than to the matrix structure. The implication of these results for quality control of P/M production parts is that a separate standard curve of strength versus velocity must be used for parts of different carbon levels.

The need for different strength versus velocity quality control curves for different alloys is also evident when comparing curves for P/M alloys Fe-0.9C, Fe-2Ni-1Cu-0.9C and Fe-2Cu-0.9C in Figures 13, 14, and 15. Each

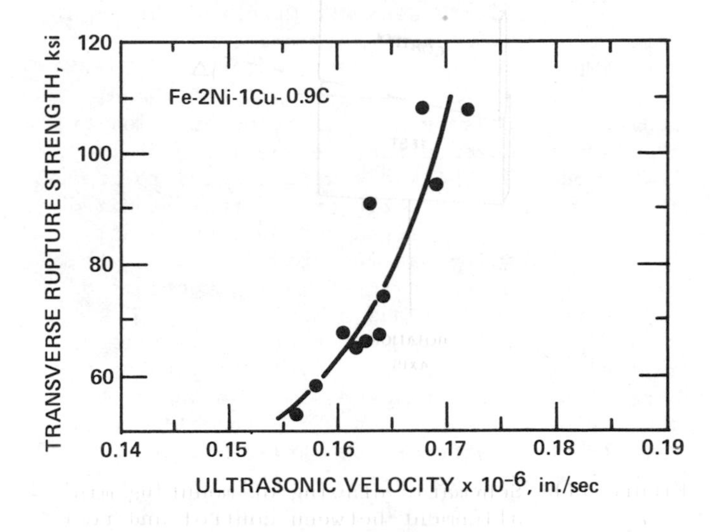

Figure 14. Strength versus velocity curve for Fe-2Ni-1Cu-0.9C bars of different densities, sintering times and temperatures.

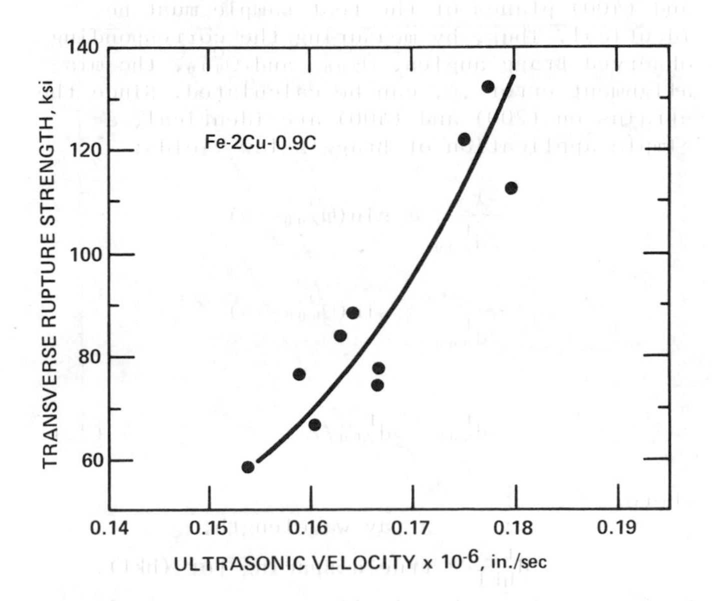

Figure 15. Strength versus velocity curve for Fe-2Cu-0.9C bars of different densities, sintering times and temperatures.

curve represents bars with green densities of 6.2, 6.4, and 6.8 gm/cm^3 sintered for 10 and 30 minutes at 1121°C. In each case strength and velocity increased with density, sintering time and temperature, to produce curves of strength versus velocity which could be used for quality control. Comparison of the three curves together in Figure 16, however, shows each alloy to have a unique curve, as was evident for the bars with different carbon levels in Figure 12. As with varying carbon levels, the copper and nickel additions strengthened the matrix of the materials, improving the overall part strength, without similar effect on velocity which is controlled primarily be pore shape. Thus, it would be necessary to use a separate QC curve for each alloy from which parts are produced.

EFFECTS OF MEASUREMENT VARIABLES - To examine the effects of part wall thickness, length, and measurement direction on velocity, standard transverse rupture specimens of 1/8", 1/4", and 1/2" thickness were pressed to green densities of 6.2 and 7.0 gm/cm^3 from atomized iron powder and sintered for 30 minutes at 1121°C. Velocity was measured in the length, width, and thickness (pressing) directions of each bar, using a 1 MHz signal.

Figure 17 shows that there was little or no consistent difference in velocity among the bars of different thickness. This indicates that velocity measurements in production parts may also be unaffected by variations in wall thickness, as least as thin as 1/8". Measurements in parts with thinner walls may possibly show an effect of decreased velocity value due to interaction of the sound wave with the part surface.

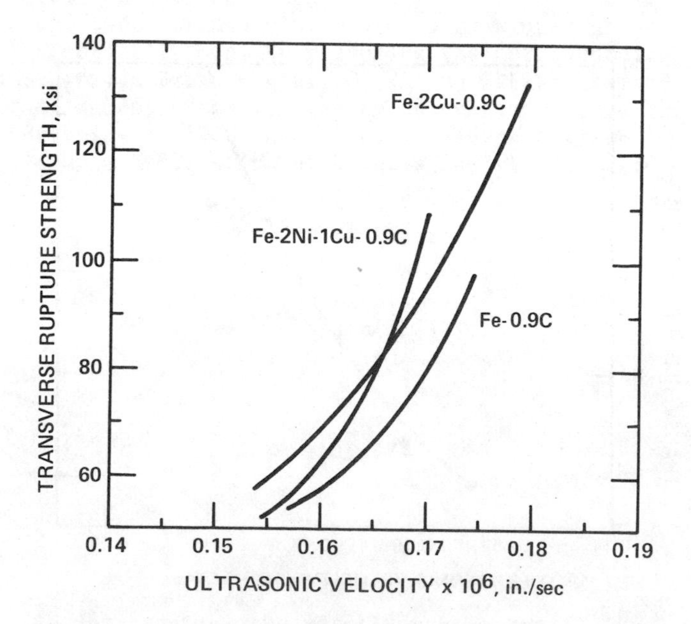

Figure 16. Comparison of strength versus velocity curves for atomized iron with different alloying additions.

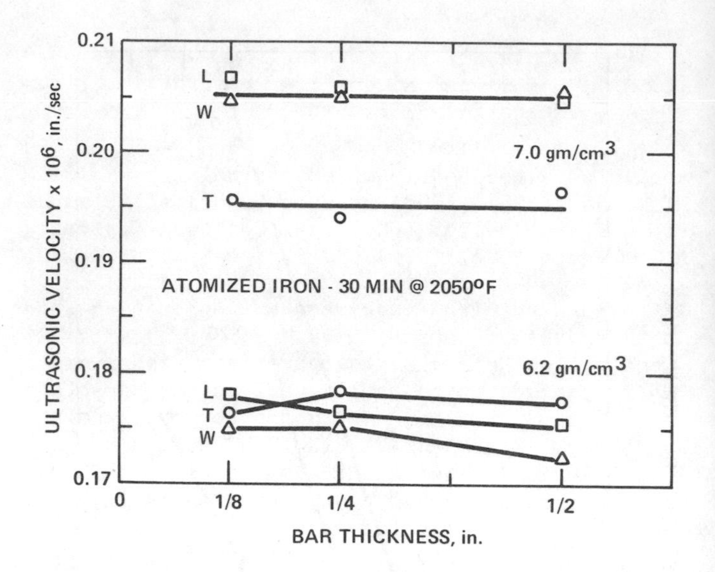

Figure 17. Effect of bar thickness, density, and measurement direction on ultrasonic velocity.

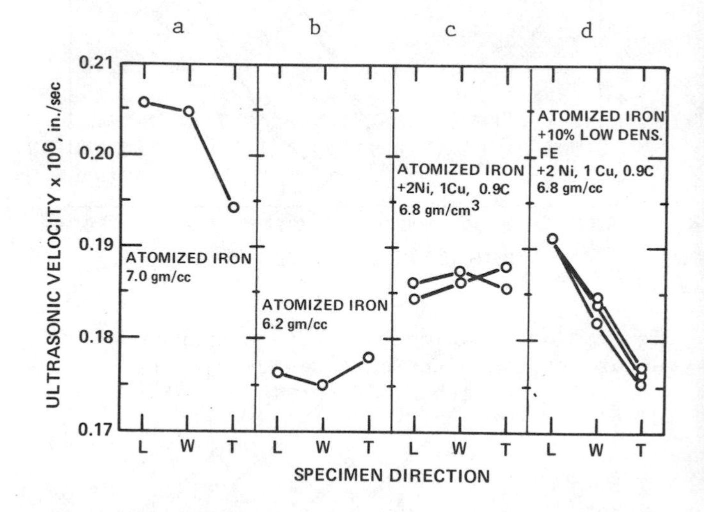

Figure 18. Effect of measurement direction and powder type on ultrasonic velocity through 1/4" transverse rupture bars.

The 7.0 gm/cm^3 density bars had higher velocities than the lower density bars, as expected. These data also show that the velocity in the pressing direction of the higher density parts is significantly lower than in directions within the plane of pressing. This is probably due to flattening of the pores, which reduces the elastic modulus and velocity of sound in that direction. The 6.2 gm/cm^3 specimens did not show this velocity reduction in the pressing direction, probably due to less pore antisotropy.

The above results indicate that the direction of measurement within a part should be standardized when using velocity as a production quality control tool. Measurement in the pressing direction is generally most practical since flat and parallel surfaces on opposite sides of the specimen are desirable for accurate velocity measurements.

Figures 18(a) and (b) illustrate the difference in velocity in the length, width, and thickness directions of the 1/4" bar pressed from atomized iron, from Figure 17. Figure 18(c) shows the velocities in the three directions of two identically prepared 1/4" thick atomized iron bars with 2 percent nickel, 1 percent copper and 0.9 percent carbon additions. The velocities of these 6.8 gm/cm^3 bars are intermediate between the 7.0 and 6.2 gm/cm^3 bars and show little consistent velocity difference with direction. In Figure 18(d), however, the same alloy blend containing 10 percent of a low density (porous) iron powder addition shows a marked decrease in velocity from the length to width to thickness directions. The reason for this difference between directions, and length and width in particular, is not known, but is expected to be due to an interaction between the ultrasonic sound wave and the very fine porosity in the low density powder. It is important to note that within the plane of pressing the measurements showed a difference with the length of path traveled by the sound waves. Measurement in a consistent direction, such as thickness, would eliminate such variability in production quality control.

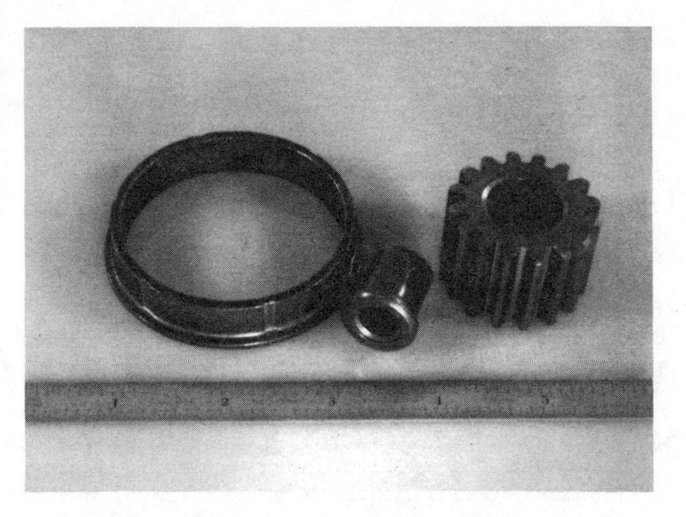

Figure 19. P/M parts evaluated by ultrasonic velocity and resonant frequency measurement. Stainless steel insert is at center.

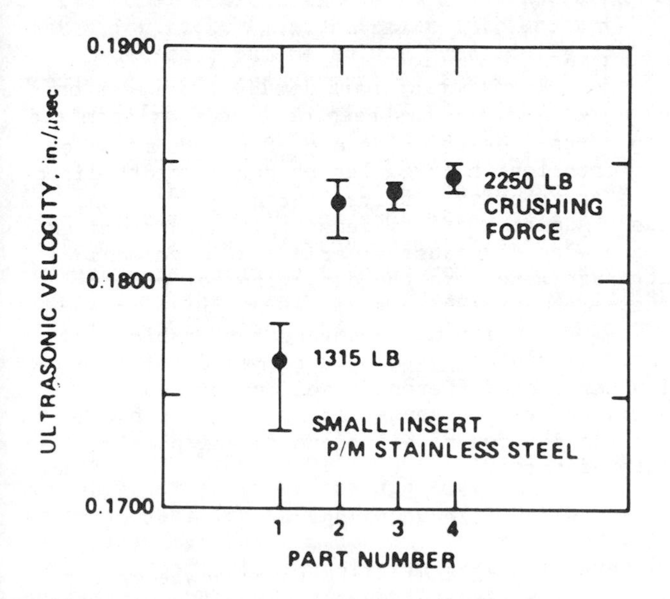

Figure 20. Crushing force and ultrasonic velocity of small inserts.

EVALUATION OF PRODUCTION PARTS - In order to demonstrate the value of the above methods for evaluating the quality of production P/M parts, nondestructive methods were applied to several types of parts supplied by a commercial P/M parts producer. One of these parts, illustrated in Figure 19, was a 3/4" long stainless steel insert with a density of 6.8-6.9 gm/cm^3. Several inserts were supplied from lots known to be good and lots with unsatisfactory properties. One of the small inserts was from a lot with a crushing strength of about 1315 lb. while the crush strength of the other inserts averaged 2250 lb.

The ability of velocity measurements to discriminate between good and bad parts is illustrated in Figure 20. Three tubular inserts from the lot with an average crushing strength of 2250 lb. exhibited average velocities of approximately 0.1840 in/µsec. These values are well above the velocity of 0.1765 for the part with the low crushing strength. The error bars show the variability of the velocity measurement at locations around the circumference of each part.

Differences in the microstructure in the good and bad stainless steel inserts are readily apparent in Figures 21 and 22. The higher quality parts showed a clean, well sintered structure while the prior particle boundaries of the weaker part were lined with intermetallic particles.

The above examples illustrate good relationships between velocity and strength of test specimens and production parts. The development of good correlations between strength or microstructure of parts and NDE signals might result in minimum acceptable velocity or resonant frequency values to be set for individual parts. Periodic inspection of parts could allow statistical control of part quality. The ease and rapidity of these methods could allow more frequent inspection and perhaps provide early indications of processing deviations. Good and bad parts might also be sorted by these means.

In addition to inspecting individual parts after sintering, these methods could also be used to examine standard test bars sintered with the parts. Correlations between the strength and velocity or resonant frequency of standard bars and the strengths of the parts might provide a criteria for acceptance of parts. The test

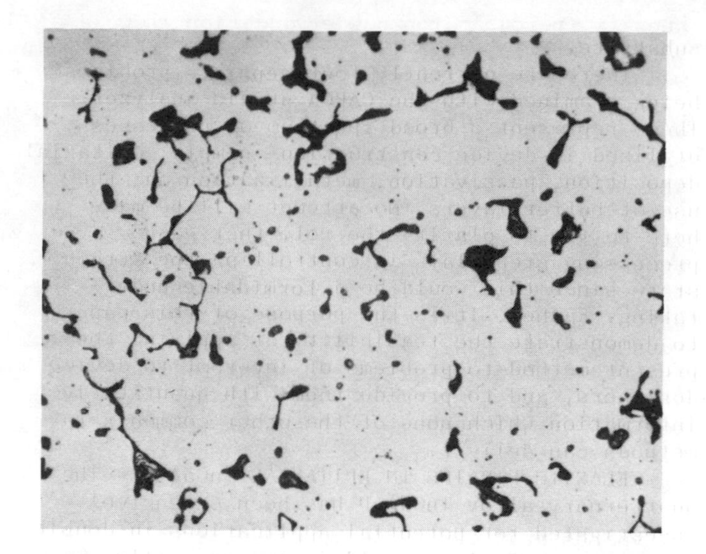

Figure 21. Microstructure of strong P/M stainless steel insert. Unetched, 500X.

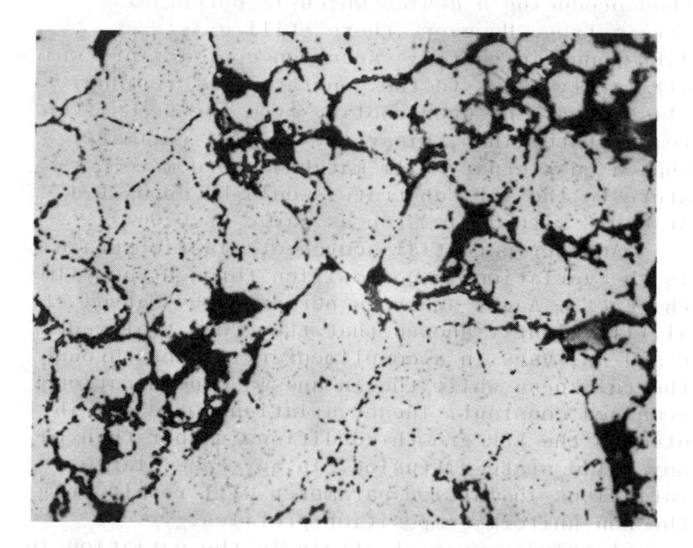

Figure 22. Microstructure of low strength P/M stainless steel insert. Unetched, 500X.

bar could be inspected rapidly and would provide a record of the sintering run. A standard test specimen would also allow standards to be set for different grades of materials, based on frequency or velocity values, without the influence of part shape.

Curves of velocity versus strength which would serve as the basis for these types of evaluations are not identical for all P/M alloys, however, and part shape and type of powders can sometimes affect the measured velocity values. For these reasons, it is recommended that the following considerations be observed when using ultrasonic velocity as a quality control tool:

a. Direction of velocity measurement in a P/M part or test bar should be standarized, preferably in the pressing direction.
b. Different curves of velocity versus strength should be observed for different P/M alloys, and possibly for blends containing additions of low density powders.
c. It may be necessary to use individual velocity versus strength curves for production parts of different shape. The present results show no effect of wall thickness down to at least 1/8", but differences in part length (measurement length) may cause differences in measured velocity value.
d. All quality control curves should be obtained using ultrasonic signals of similar frequency, since effects of signal frequency on measured velocity have not yet been determined for P/M materials.

CONCLUSIONS

The following conclusions can be made from this study:

1. Ultrasonic velocity and resonant frequency measurements are proportional to part strength. The closeness of these relationships are sometimes enhanced by considering a separate curve for specimens of similar density.
2. Ultrasonic velocity and resonant frequency of a sintered part increase with density, sintering temperature and time. Green pressed materials show essentially no effect of density on velocity.
3. Velocity versus density curves change with time and temperature of sintering, with the velocity for a particular density increasing with degree of sintering.
4. Different carbon levels or alloy additions to P/M steel produce different curves of strength versus velocity, due to increasing the strength of the matrix without similar increase in velocity.
5. Ultrasonic velocity can vary with measurement direction in a P/M part.
6. No effect of part (wall) thickness on measured velocity has been observed for atomized iron parts. Neither was any effect of measurement part length (e.g. length versus width direction) on velocity observed for atomized iron bars. Addition of low density iron to atomized iron powder did cause a variation in velocity with part length in the directions of measurement.
7. Velocity measurements may enable the degree of sintering to be determined for production parts of different sizes and shapes. Considerable differences in NDC signals have been found between well sintered and poorly sintered parts.

ACKNOWLEDGEMENT

The authors would like to acknowledge the Metal Powder Industries Federation for allowing publication of Figures 1-18 and 20-22, Tables I-

III, and portions of the text which previously had been published in Progress in Powder Metallurgy, Vols. 37 and 38, references 1 and 2 below. They also wish to acknowledge performance of part of the reported work at Southern Research Institute.

REFERENCES

1. Patterson, B. R., C. E. Bates, and W. V. Knopp, Progress in Powder Metallurgy, 37, 67-79 (1981).
2. Patterson, B. R., K. L. Miljus, and W. V. Knopp, Progress in Powder Metallurgy, 38, 401-10 (1982).
3. R. H. Brockelman, pp.201-24, Perspectives in Powder Metallurgy, Vol. 5, "Advanced Experimental Techniques in Powder Metallurgy," Eds. J. S. Hirschhorn and K. H. Roll, Plenum (1970).
4. Papadakis, E. P. and B. W. Petersen, Materials Evaluation, 37, 76-80 (1979).
5. Patterson, B. R. and C. E. Bates, American Foundrymen's Society Transactions, 89, 369-78 (1981).

ULTRASONIC MEASUREMENTS OF GRAPHITE PROPERTIES

Shaio-Wen Wang
Department of Welding Engineering
The Ohio State University
Columbus, OH 43210

Laszlo Adler
Department of Welding Engineering
The Ohio State University
Columbus, OH 43210

ABSTRACT

Ultrasonic velocity and attenuation measurements were carried out in graphite materials. The graphites are of various grades with varying properties, such as particle size, pore size, pore distribution, oxidation, impregnation, etc. The objectives of this work are to correlate the frequency dependence of the ultrasonic attenuation coefficients and velocities to these various parameters and to correlate ultrasonic measurements to destructive testing. The pore size and distribution can be estimated from ultrasonic measurements. An empirical relationship (suggested by E. R. Kennedy, 13th National Symposium of Fracture Mechanics, 1981) between fracture strength of graphites and ultrasonic parameters has been verified.

INTRODUCTION

Graphite plays an important role in many fields such as nuclear and aerospace because of its ability to absorb neutrons, its resistance to thermal stress and its mechanical strength at high temperature. Fabrication of a graphite suitable for a specific application requires the precise knowledge of the relationship between microstructures and the mechanical properties. Since this relationship has been fairly successfully explained by fracture mechanics concepts [1,2], studies using fracture mechanics are extremely helpful in relating fabrication variables to the eventual product strength. On the other hand, the morphological factors that govern the mechanical strength strongly modulate the ultrasonic elastic wave. Thus, both fracture strength and ultrasonic wave depends similarly on the same for microstructural factors. Therefore, in this study, nondestructive ultrasonic methods are investigated for the evaluation of the fracture parameters of graphite. Fracture strength is then calculated from the Griffith-Irwin equation which is the basis of fracture mechanics in graphite.

THEORETICAL CONSIDERATION

The Griffith model [3] for unstable crack propagation in elastic materials gives

$$\sigma_f = \frac{K_{IC}}{\sqrt{\pi a}} = \left(\frac{E \cdot G_{IC}}{\pi a} \right)^{1/2} \qquad (1)$$

where σ_f = fracture stress, G_{IC} = strain energy release rate, E = Young's modulus, a = radius of critical flaw size, and K_{IC} = fracture toughness = $(E \cdot G_{IC})^{1/2}$. Generally, to meaningfully evaluate each parameter in the equation requires not only a statistically significant number of destructive tests but also a large specimen in each test. If, however, both Young's modulus E, and the inherent defect size, a, could be measured nondestructively, with constant G_{IC} for a graphite grade, the fracture strength can be estimated by the Griffith-Irwin equation. Based on the following theories, the Young's modulus is evaluated from the ultrasonic velocity measurement and the critical defect size is estimated from the ultrasonic attenuation measurements.

BUCH'S MECHANICAL MODEL FOR GRAPHITE [4] - Fracture was modeled as a progressive phenomenon involving microcrack initiation, microcrack arrangement and failure (Fig. 1). A pore was then

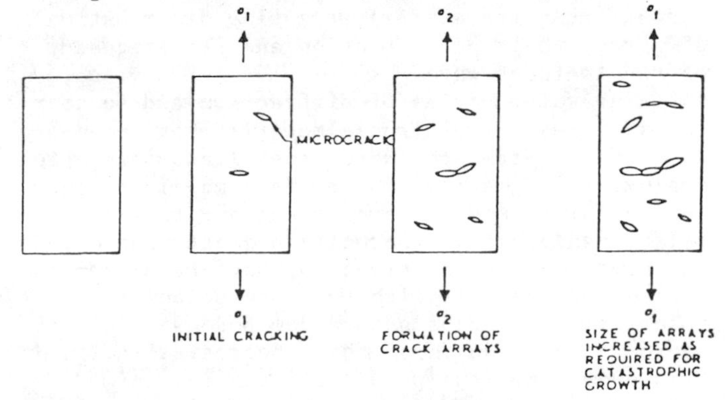

Fig. 1 - Progressive Microcracking Model.

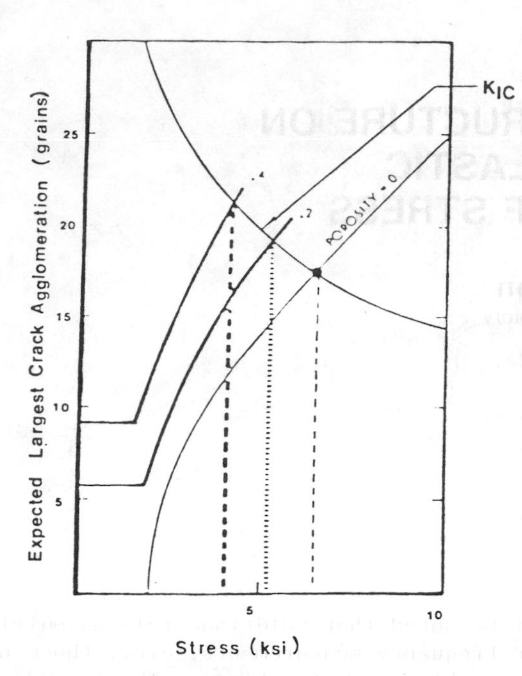

Fig. 2. - Porosity and Fracture Stress for a given K_{IC}. Expected largest crack agglomeration (grains) means the largest crack length that could occur by microcrack agglomeration. Its unit is the size of a grain. Porosity means the percentage of pores existing before stress is applied. Porosity = 0 means zero porosity; porosity = 0.2 means 20% porosity; etc.

considered as a substitute for a cracked grain. Larger pore sizes were modeled as multiple continuous grains. Then, the probability of cracked grains increased. From the statistical calculation, the expected size of the largest microcrack agglomerate (defect size) was larger for higher porosity content at each stress level (Fig. 2).

MULTIPLE SCATTER THEORY - Since graphite has a porous nature, the ultrasonic wave propagation in such an inhomogeneous medium undergoes multiple scattering due to the presence of pores thus reducing the amplitude of the propagating wave by attenuation. The attenuation of waves depends critically on the material properties of the host medium (graphite matrix), the size distribution of pores, their concentration and the frequency of the incident wave.

Several theories of diffraction and scattering of waves in multiphase material have been developed. Here, the more general approach which employs the T-matrix to formulate multiple scattering [5] is used to show the quantitative relationship between porosity and attenuation. The numerical computations [5] have been performed for graphite with the assumption of spherical pore shape and random pore size distribution. The results of this theoretical calculation show that the phase velocity $Re(k_p/K_p)$ decreases as the pore concentration increases, where k_p is the wave number for porous material and K_p is the wave number for the same material with zero porosity. In Fig. 3 the attenuation $Im(k_p/K_p)$ is higher for higher k_pa and higher pore concentration. In Fig. 4, since the multi-

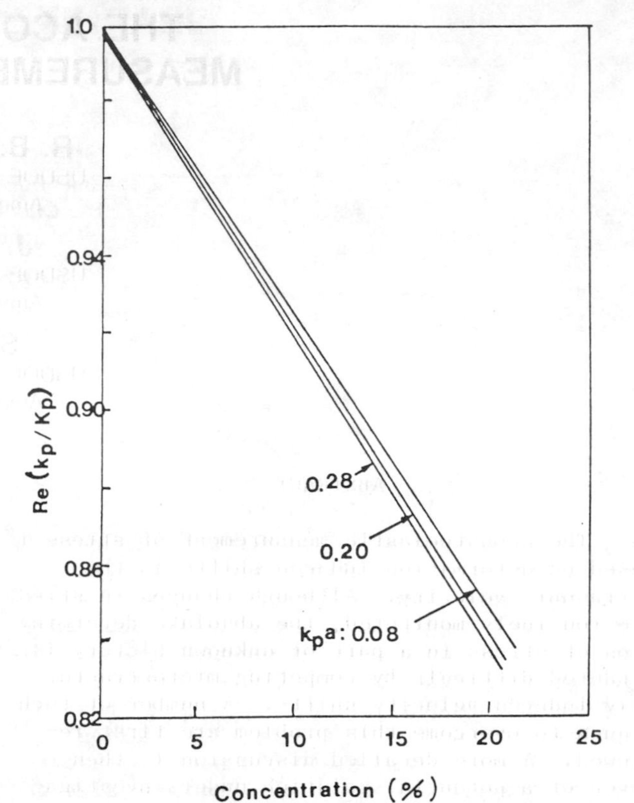

Fig. 3 - Phase Velocity as a Function of Pore Concentration in Graphite for a Gaussian Size Distribution.

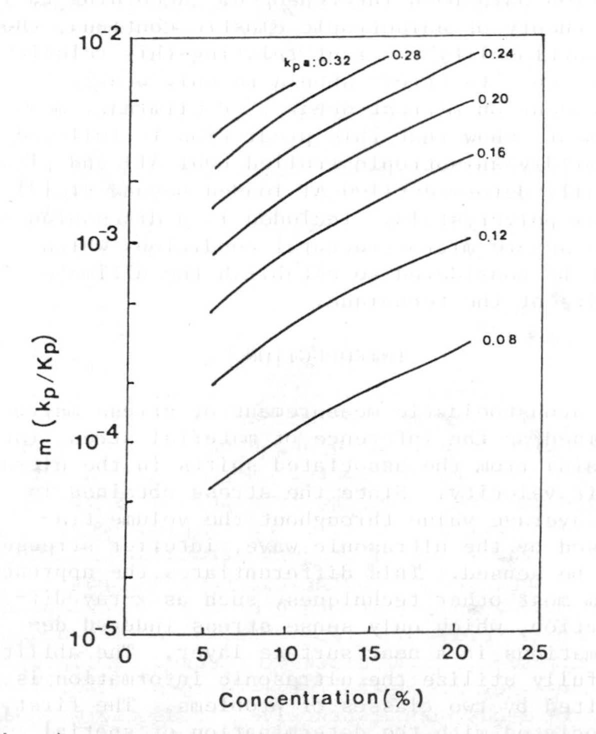

Fig. 4 - Attenuation as a Function of Pore Concentration in Graphite for a Gaussian Size Distribution.

ple scattering theory strongly links pore size with ultrasonic attenuation, it seems feasible to make an estimate of the critical defect size from the attenuation coefficient.

EXPERIMENTAL SET-UP - Both velocity and attenuation coefficient of graphite were measured by immersion, pulse-echo method. A single frequency quartz transducer was coupled to the graphite specimen through an alcohol column. The upper face and side walls of the graphite sample were in contact with air and thus the graphite-air interface was totally reflective. The first two consecutive signals, A and B (Fig. 5), were used to calculate the attenuation coefficient. The peak-to-peak amplitudes of these signals were recorded. Then corrections were made (by a computer program) to account for beam spreading [6] and reflection-transmission energy losses at the specimen boundaries. Fig. 5 shows the origin of the signals used in the attenuation measurements. After reduced by radiation coupling, attenuation in the alcohol and energy loss at boundary, the amplitudes of signal A at the receiving transducer

$$A = K_A\,(S_A)\,e^{-\alpha_1 2l}\,R_1 u_o$$

and of signal B

$$B = K_B\,(S_B)\,e^{-\alpha_1 2l}\,T_1 e^{-\alpha_2 2L}\,(-1) T_2 u_o$$

where $K_i(S_i)$ is the beam spreading correction (Fig. 6), and l and L are the length of alcohol path and graphite sample, respectively. T is the transmission coefficient, R is the reflection coefficient and α_1, α_2 are the attenuation coefficients of the alcohol and the graphite sample. The intrinsic attenuation coefficient which depends only on the graphite properties was then calculated by the following equation:

$$\alpha = \frac{1}{2L} \ln \frac{(A/K_A)T_1T_2}{(B/K_B)R_1} \quad (2)$$

RESULTS

THE ULTRASONIC ATTENUATION COEFFICIENT - Figure 7 indicates that the higher the porosity, the larger the attenuation coefficient. In addition, for graphite samples having a Gaussian pore radius distribution with a mean value at 50 μm, experiment matches the theoretical line of $k_pa = 0.09$. At 1 MHz (the frequency used for the experiment), 0.09 k_pa corresponds to a 48 μm pore radius. Graphites with 55 μm mean pore size match the 0.10 k_pa line which corresponds to 54 μm pore radius (Fig. 8). That is, the larger the pore size, the higher the attenuation coefficient. Furthermore, from the comparison of the experimental data with the theoretical values, either the pore size or the porosity content could be determined if the other parameter is provided.

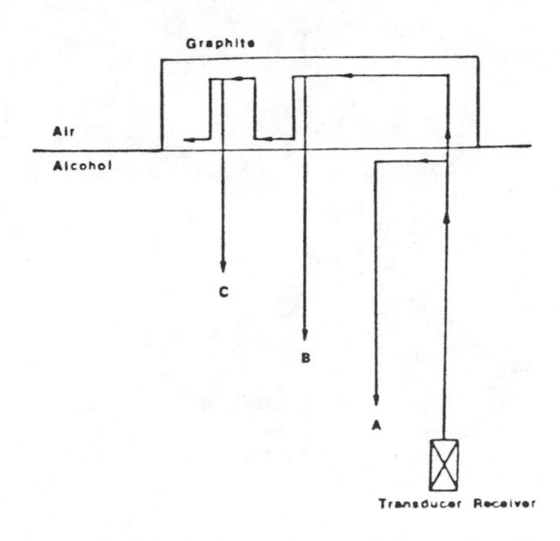

Fig. 5 - The Origin of the Signals Used in the Attenuation Measurements for Grade 65-66 Graphites.

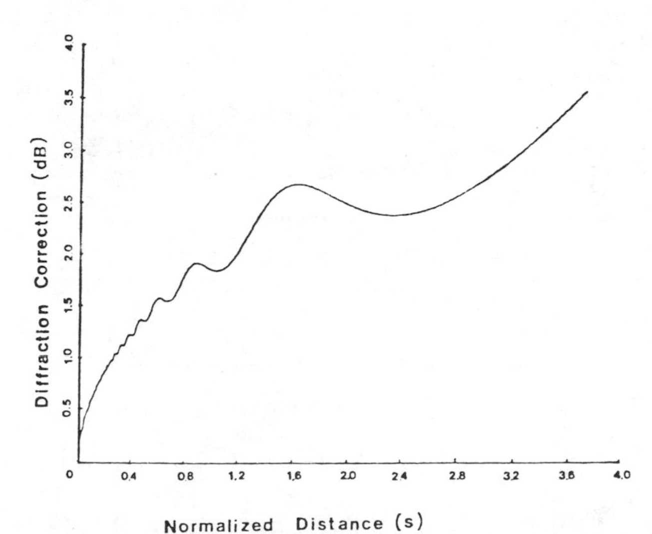

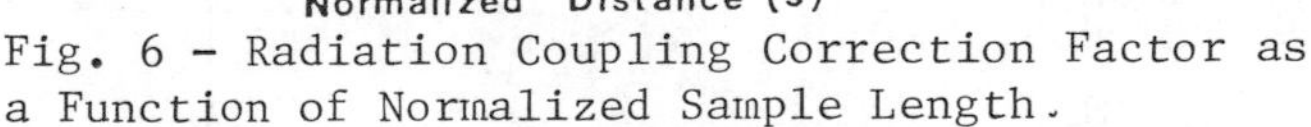

Fig. 6 - Radiation Coupling Correction Factor as a Function of Normalized Sample Length.

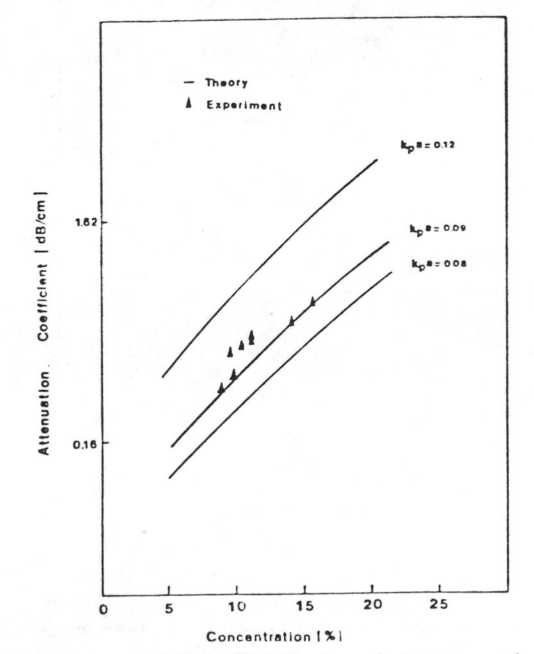

Fig. 7 - Porosity Dependence of Attenuation in Grade 65-66 Graphites

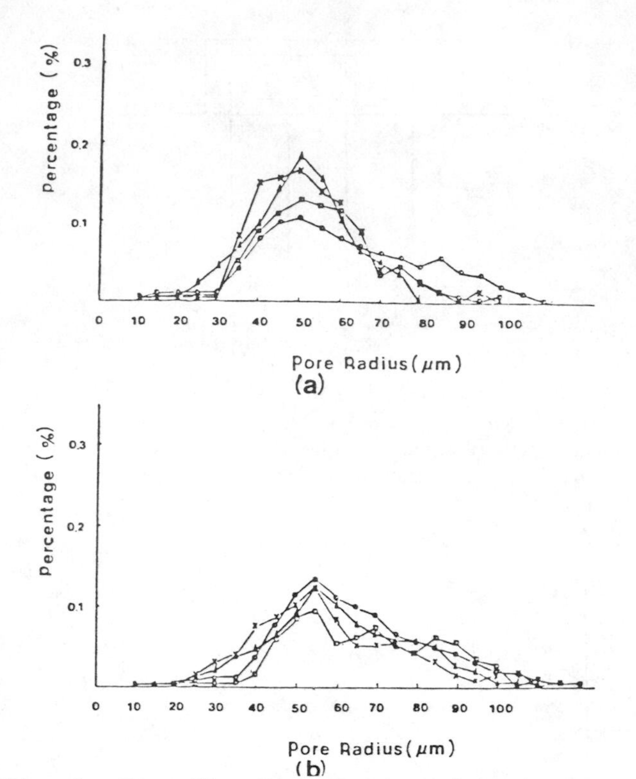

Fig. 8 - Pore Size Distribution of Grade 65-66 Graphites, a) main peak at 50 μm, b) main peak at 55 μm.

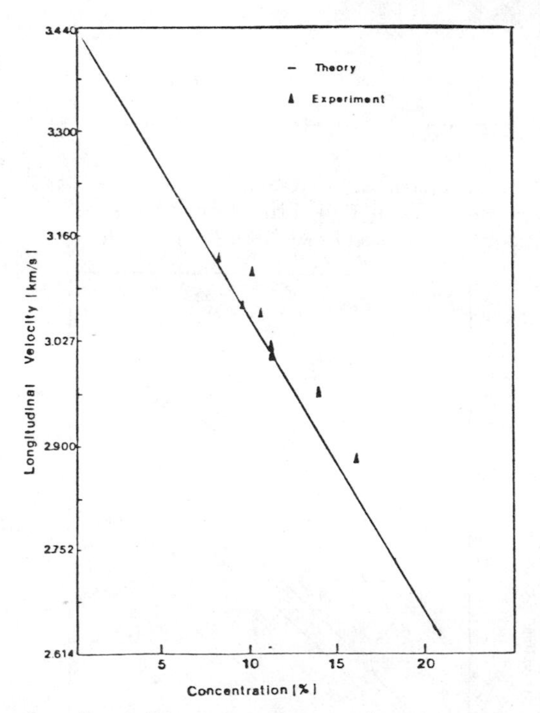

Fig. 9 - Porosity Dependence of Velocity in Grade 65-66 Graphites

THE ULTRASONIC VELOCITY - The decrease in velocity, as the porosity increases, can be explained by the fact that the effective density becomes smaller as the volume, occupied by pores becomes larger. Therefore, besides a determination of Young's modulus, no further information about the microstructure was obtained from the velocity results.

COMPARISON OF THE PARAMETER OF (YOUNG'S MODULUS/ATTENUATION OF COEFFICIENT)$^{1/2}$ TO THE FRACTURE STRENGTH - For graphites within a particular grade (i.e., G_{IC} is constant) the only variable in the Griffith equation is $(E/a)^{1/2}$ (see Eq. (1)). The experimental results indicate that the attenuation coefficient depends on pore size, suggesting that it is reasonable to use the attenuation coefficient to estimate the critical defect length. Also, Young's modulus can be evaluated from the ultrasonic velocity. Therefore, the parameter (Youngs's modulus/ attenuation coefficient)$^{1/2}$ obtained by ultrasonic methods replaces $(E/a)^{1/2}$ in the Griffith equation and from Eqs. (1) and (2) one obtains

$$\sigma_f \simeq C(E/\alpha)^{1/2}$$

where C is a frequency dependent constant (because α is frequency dependent). In Figs. 10 and 11 the fracture strength as a function of the parameter (Young's modulus/attenuation coefficient)$^{1/2}$ is plotted and a linear relationship is found. This implies that once the constant C is found empirically, the fracture strength of a graphite with different porosities could be predicted (from ultrasonic measurements) as long as it is of the same graphite grade. Therefore, nondestructive ultrasonic methods for evaluating material fracture properties not only provide a less expensive alternative to mechanical destructive tests but also suggests a potential procedure for nondestructively estimating product quality. Furthermore, from the standpoint of material science, continued work of this type should contribute to identification and analysis of factors that control toughness and thus aid in fracture technology.

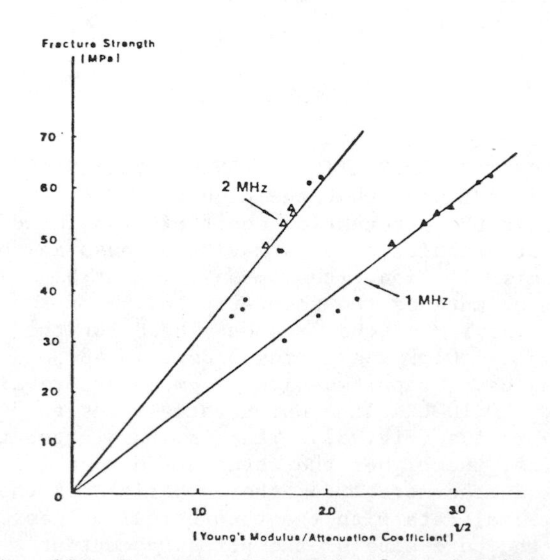

Fig. 10 - Fracture Strength of Grade H451 Graphites as a Function of the Parameter (Young's Modulus/Attenuation Coefficient)$^{1/2}$

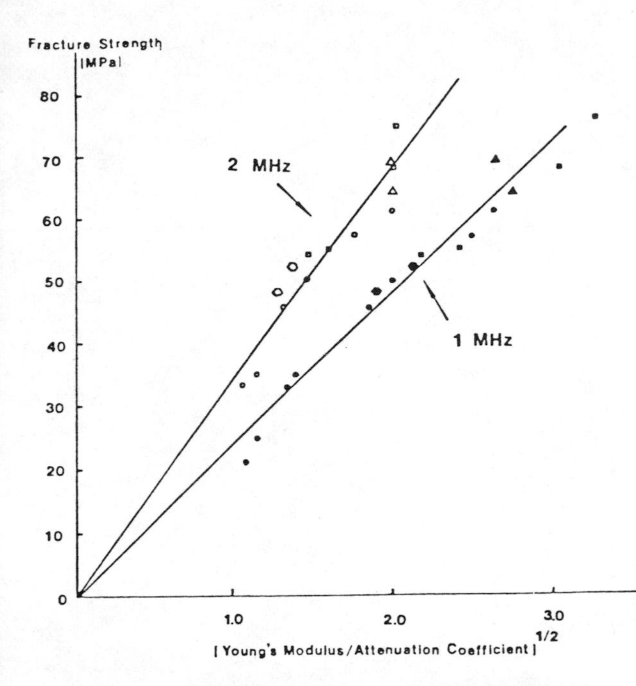

Fig. 11 - Fracture Strength of Grade 2020 Graphites as a Function of the Parameter (Young's Modulus/Attenuation Coefficient)$^{1/2}$

REFERENCES

1. Kennedy, C. R., and W. P. Eatherly, Proceedings, 11th bi-annual conference on carbon, 1973, pp. 304 and 973.
2. Eatherly, W. P., and C. R. Kennedy, ASTM Spec. Tech. Publ. 743 (American Society Testing and Materials, Phila., 1981), p. 303.
3. Griffith, A. A., Transaction, Royal Society of London, 221 (1920).
4. Buch, J. D., ASTM Spec. Tech. Publ. 605 (American Society Testing and Materials, Phila., 1976), p. 124.
5. Varadan, V. K., and V. V. Varadan, Acoust. Electromagnetic and Elastic Wave Scattering--Focus on the T-Matrix Approach, (Pergamon Press, New York, 1979), pp. 3 and 103.
6. Rogers, P. H., and A. L. Van Buren, J. Acoust. Soc. Am., 55, 742 (1974).

ACKNOWLEDGMENTS

Thanks to Dr. C. R. Kennedy and Dr. W. P. Eartherly of the Oak Ridge National Laboratory, Metals and Ceramics Division--who organized this investigation, provided the study materials--graphites and their fracture data, and funded this study through ORNL contract 11X-0-09069V.

III . Residual Stress

RESIDUAL STRAIN MEASUREMENTS IN MICROELECTRONIC MATERIALS

William E. Mayo
Dept. of Mechanics and Materials Science
College of Engineering
Rutgers University
Piscataway, N.J. 08854

Jharna Chaudhuri
Dept. of Mechanics and Materials Science
College of Engineering
Rutgers University, Piscataway, N.J. 08854

Sigmund Weissmann
Dept. of Mechanics and Materials Science
College of Engineering
Rutgers University, Piscataway, N.J. 08854

ABSTRACT

A new x-ray method is presented for determining elastic strains with high precision in single crystal materials. This method is unique not only in its ability to determine the full elastic strain tensor but also its distribution about a strain center with a resolution of approximately 60 μm. It employs the recently developed Computer Aided Rocking Curve Analyzer (CARCA) and is particularly well suited for analysis of thin film structures common to electronic materials. In this paper, the operating principle of the method will be described and potential applications will be presented. Among these are the generation of long-range strain concentrations at dielectric steps, epitaxial interfaces and diffused metal contacts. Also, for epitaxial structures, it is possible to utilize the method to measure the uniformity of the film thickness. As an example of the power and sensitivity of the method, a study of the strain distribution in an InGaAsP epitaxial film on an InP substrate will be presented.

INTRODUCTION

In studies of thin film electronic materials, one area which has not received sufficient attention is the detection and measurement of residual elastic strains. Although many x-ray studies have been performed on these materials (1-11), the resulting strain analyses remained incomplete for two important reasons. First, only a limited number of diffracting planes were examined and thus, the complete strain state could not be determined. For example, Kawamura et al. (6) and Matsui et al. (11) studying InGaAs and InGaAsP on an InP substrate respectively, examined only one symmetric (400) and two asymmetric reflections, (511) and ($\bar{5}$11). From these reflections only two strains (parallel and perpendicular to the film surface) could be determined. Therefore, their data presents only a partial and potentially inaccurate representation of the true strain state. The second shortcomming of the conventional x-ray methods of strain analysis is that these studies assume that strains are uniformly distributed in the plane of the thin film. Although attempts have been made by Rozgonyi and coworkers (1,8), vanMallaert and Schwuttke (12), Wang (5) and Bohg(4) to determine non-uniform strain distributions, these studies are also incomplete since they monitored only a single strain component and thus provided an incomplete description of the strain distribution. It appears therefore, that up to now, there exists no adequate x-ray technique for determining completely the strain state when it is non-uniformly distributed. Recently, Stock, Chen and Birnbaum (13) have utilized an x-ray topographic method based on equiinclination contours to determine the strain concentration surrounding an inclusion. Although this method appears to be an important step toward the development of a complete strain analysis, it too remains incomplete since only four of the six independent strain components could be determined. Nonetheless their work seems to be significant since it demonstrated the relatively large magnitude of the shear strains which in most of the conventional techniques are assumed to be nonexistant.

The purpose of this paper is to present a new x-ray method for analyzing the complete elastic strain tensor and its distribution about a strain center. The method is an x-ray topographic technique utilizing a double crystal diffractometer in a non-dispersive arrangement in which the sample rotation and detector rotation are decoupled. This method is based on the recently developed Computer Aided Rocking Curve Analyzer (CARCA) which was previously used for the analysis and mapping of plastic strains. Minor modifications principally to the computer software, were all that were required to convert the system from an analyzer of plastic strains to an analyzer of elastic strain distributions.

The technique is based in part on the theory of Imura et al. (14) who developed a model for a complete strain analysis which, however, was practically applied only for a uniformly deformed material. These authors showed that by measuring the relative change in d spacings for a number of reflecting planes, it is possible to generate the average strain tensor by a least squares method. Their analysis indicated that at least 7 reflecting planes must be used and further, that there are restrictions about the choice of planes. They suggest that the simplest means for insuring a valid average strain tensor is to use as large a number of independent reflections as possible. In the experimental work, Imura et al. were able to obtain the required large number of reflections in a single exposure by using the x-ray divergent beam method. However, as pointed out previously, this experimental method is not suitable when strains are heterogeneously distributed.

In the method described here, the analysis of Imura et al. is carried out on a point-by-point basis for up to 12 reflections with a resolution of 60 μm. To accomplish this task in a reasonable period of time, the recently developed Computer Aided Rocking Curve Analyzer is utilized. The important feature of the method is the use of a position sensitive detector (PSD) in combination with a parallel monochromatic incident x-ray beam. Since there is a topographic relationship between a point on the diffracted beam and the point on the sample giving rise to that beam, the PSD is capable of providing information simultaneously from up to 1024 regions. The rocking curve analysis is then performed in order to determine the Bragg angle for each corresponding point on the test sample and this in turn is used to generate the strain value at each point.

The operating principle of the method will be described along with a description of the principal sources of error and the correction method developed. Application of the method to an analysis of thin film structures typical in electronic usage will be suggested. Special attention will be given to the strain development due to non-uniform epitaxial deposition of InGaAsP on an InP substrate.

NON-UNIFORM STRAIN ANALYSIS

Imura et al. (14) have considered the calculation of the full strain tensor from the individual strains of 6 or more planes. Their analysis requires a minimum of 6 planes whose normal vectors do not form two sets of mutually orthogonal vectors. This restriction is necessary since the matrix equation contains no unique solution for such non-independent vectors. To avoid this problem, and to minimize errors, it is necessary to use more than six reflecting planes and their corresponding strain values from which a least squares fit to the strain tensor is determined. For such a case, Imura et al have shown that the average strain tensor can be derived by solving the set of equations shown in equation 1. In order to solve this system of equations, a computer program utilizing the Gauss-Jordan elimination algorithm was used (15). Note that if the determinant of the 6x6 matrix is zero, there is no solution for $\langle \varepsilon_{ij} \rangle$. Imura et al. suggested that other than carrying out the computation directly that there exists no a priori method for determining whether any set of indices of reflecting planes will give a determinant value of zero. The simplest way to insure a finite solution for $\langle \varepsilon_{ij} \rangle$ is simply to use a sufficiently large number of planes. The least squares fit to the average strain tensor is given by the system of equations:

$$\begin{bmatrix} [h^2 H^2 s] \\ [k^2 H^2 s] \\ [l^2 H^2 s] \\ [kl H^2 s] \\ [lh H^2 s] \\ [hk H^2 s] \end{bmatrix} = \tag{1.}$$

$$\begin{bmatrix} \langle \varepsilon_{11} \rangle \\ \langle \varepsilon_{22} \rangle \\ \langle \varepsilon_{33} \rangle \\ \langle \varepsilon_{23} \rangle \\ \langle \varepsilon_{31} \rangle \\ \langle \varepsilon_{12} \rangle \end{bmatrix} \mathrm{X} \begin{bmatrix} [h^4] & [h^2k^2] & [h^2l^2] & [h^2kl] & [h^3l] & [h^3k] \\ [h^2k^2] & [k^4] & [k^2l^2] & [k^3l] & [hk^2l] & [hk^3] \\ [h^2l^2] & [k^2l^2] & [l^4] & [kl^3] & [hl^3] & [hkl^2] \\ [h^2kl] & [k^3l] & [kl^3] & [k^2l^2] & [hkl^2] & [hk^2l] \\ [h^3l] & [hk^2l] & [hl^3] & [hkl^2] & [h^2l^2] & [h^2kl] \\ [h^3k] & [hk^3] & [hkl^2] & [hk^2l] & [h^2kl] & [h^2k^2] \end{bmatrix}$$

where

$\langle \varepsilon_{ij} \rangle$ = average strain tensor component,
$[abc] = \Sigma\, a_r b_r c_r \ldots$,
h,k,l = conventional Miller indices,
s = measured strain for hkl reflection,
$H = \sqrt{h^2 + k^2 + l^2}$.

In the original analysis by Imura et al., it was assumed that the strain was uniformly distributed. However, their analysis is equally valid for non-uniform strain distributions provided that the appropriate x-ray diffraction data is obtained on a point-by-point basis. This is accomplished in the present method by utilizing a parallel monochromatic x-ray beam in combination with a position sensitive detector. The PSD is capable of breaking the diffracted beam into segments as small as 60 μm, which because of the topographic nature of the experiment can be traced uniquely to the corresponding diffraction area (200 μm x 60 μm) on the sample. Thus it is possible to measure the relative d spacing changes at each spot on the sample over a relatively large area in a very short period of time. This data is then used as input data for the strain analysis of Imura et al. to determine the non-uniform strain distribution.

EXPERIMENTAL PROCEDURE

The method to be presented here is a modification of the Computer Aided Rocking Curve Analyzer (CARCA). Descriptions of the system hardware and its application have been reported previously (16-20). In all of the previous work with the system, the rocking curve analysis has been used to measure plastic strains and their distributions. However, in the present system, modifications have been made in both the methodology and the data analysis in order to perform the elastic strain distribution studies. Only the modifications to the system and their underlying principles will be discussed here.

The method is based on the double crystal diffractometer in the nondispersive antiparallel arrangement. A monochromatic and nearly parallel x-ray beam is produced by reflection from a (111) oriented Si single crystal. After the $K\alpha_2$ component is removed by a long collimator and slit system, the highly monochromatic beam impinges on the test crystal. The beam is long enough (1 cm long X 200 μm wide) so that it covers the full length of most specimens. The test crystal is then rotated (or rocked) through its reflecting range and its intensity versus angular rotation is recorded. Since the diffracted beam is a topographic image of the irradiated region of the specimen, there is a one to one correspondence between the position in the diffracted beam and the spot giving rise to it. The diffracted beam is then registered by a linear position detector (PSD) mounted parallel to the rotation axis of the sample. When combined with a multichannel analyzer, the diffracted beam can be broken into 60 μm increments. Thus as shown in Figure I, the diffracted beam A' originates from point A on the sample. This holds likewise for pairs BB', CC', and so forth. By stepwise rotations of the specimen, rocking curves are generated simultaneously and independently for each 60 μm segment of the sample. Of course, the peak of the rocking curve is identical to the Bragg angle for that reflection and consequently, peak shifts can be utilized to determine normal strains for the particular reflection.

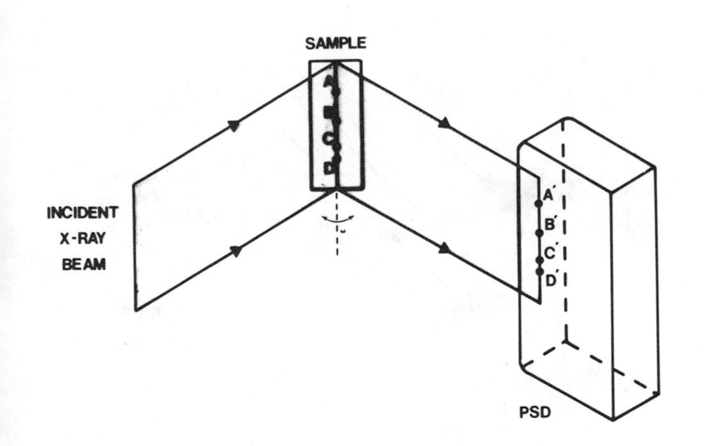

Figure I - Principle of operation of the CARCA elastic strain analyser.

When elastic strains are uniformly distributed in a single crystal, the peak of each rocking curve will come into reflecting position at the same time as shown in Figure II(a). If, however, the strains are non-uniformly distributed about a strain center, then the rocking curves will be shifted in a manner similar to that shown in Figure II(b). Since each rocking curve can be analyzed independently, the local normal lattice strain can be determined readily as a function of position.The analysis is then performed for many

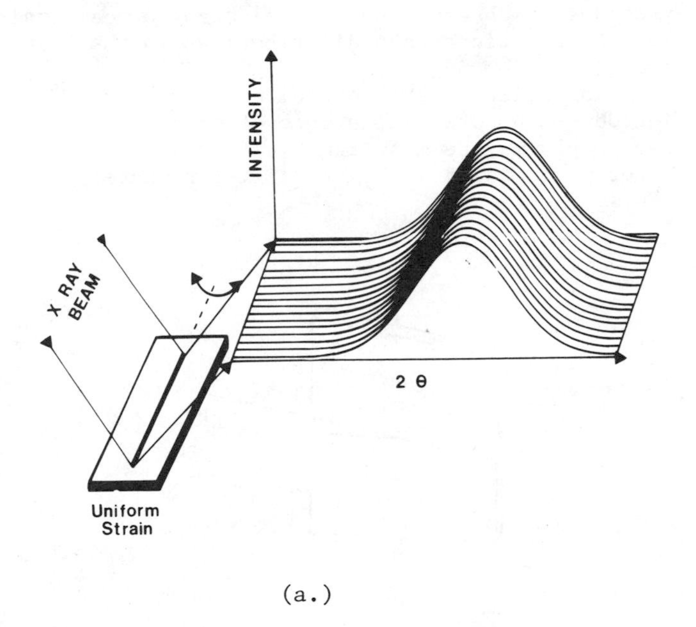

(a.)

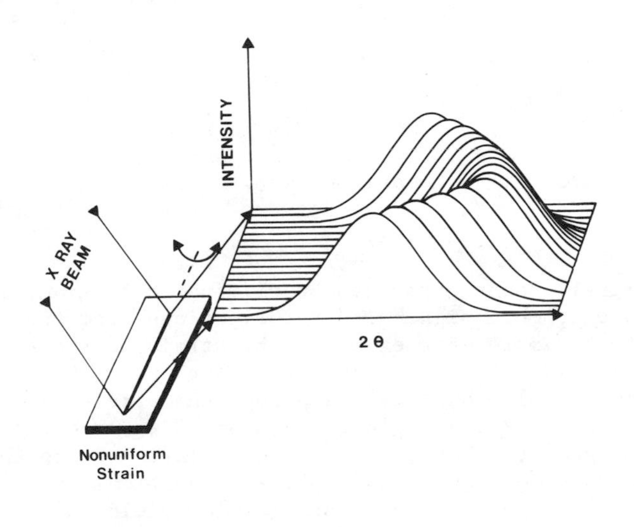

(b.)

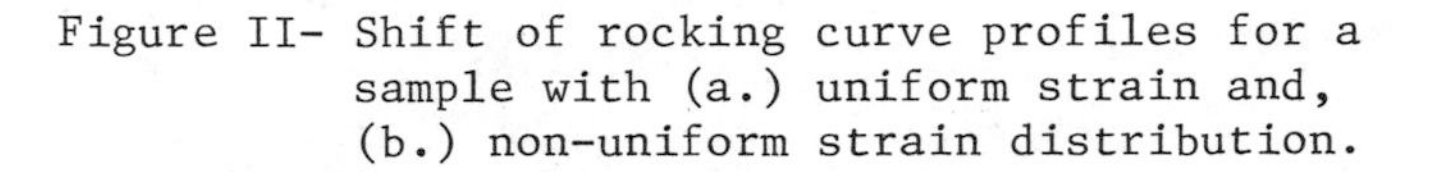

Figure II- Shift of rocking curve profiles for a sample with (a.) uniform strain and, (b.) non-uniform strain distribution.

other (hkl) reflections until the number required by the method of Imura et al. are obtained. The least squares strain tensor is finally calculated for each small probed area on the sample. In this way, the non-uniform strain distribution about a strain center can be obtained.

SYSTEM ACCURACY- In order to improve system accuracy, an unstrained control sample is included in each experimental run. As described in the previous section, the width of the incident x-ray beam is much longer than any of the samples tested and therefore there is sufficient space to include the internal standard. Since the lattice parameter of the standard is known with great accuracy, it provides an absolute and visual means of calibrating the elastic strain distributions in the test sample.

Inclusion of the control sample however, introduces a potential source of error, since the two sample may be misaligned with respect to each other as depicted in Figure III. Fortunately, this

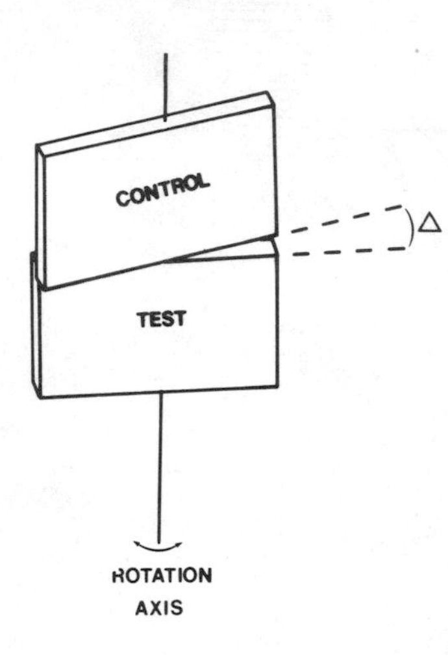

Figure III- Schematic drawing of mounting misalignment between control and test samples.

misalignment error can be eliminated by examining one or more higher order reflections from the test sample. For example, the strain on the (200) and (400) planes of the test sample must be identical. Thus, by measuring the corresponding observed Bragg angles, Θ^{O}_{200} and Θ^{O}_{400}, the misalignment error ,δ, can be calculated. Since the strains on (200) and (400) are identical, a simple application of Bragg's law yields:

$$\frac{\lambda}{2d^{t}_{200}} = \sin(\Theta^{O}_{200} - \delta)$$

$$\frac{\lambda}{2d^{t}_{400}} = \sin(\Theta^{O}_{400} - \delta)$$

$$d^{t}_{400} = d^{t}_{200}/2 \qquad (2.)$$

where,

λ = x-ray wavelength,

d^{t}_{hkl} = true d spacing for (hkl).

By rearranging, the misalignment error, δ, is given by

$$\tan\delta = \frac{\sin\Theta^{O}_{200} - 1/2\sin\Theta^{O}_{400}}{\cos\Theta^{O}_{200} - 1/2\cos\Theta^{O}_{400}} \qquad (3.)$$

Similar equation can be developed for even higher order reflections, such as (600), (800), etc.. In practice, the (600) and (400) reflections are more desirable to use than the (200)/(400) combination owing to the much greater accuracy inherent in back reflection measurements. During a single experiment, up to 60 simultaneous measurements of δ are taken, corresponding to the 60 points on the test sample. Thus, δ can be determined with a high degree of accuracy.

THIN FILM THICKNESS MEASUREMENTS- Most of the structures investigated here consist of one or more thin film layers deposited onto a substrate. When the thin film is an epitaxial layer, it is possible to determine its thickness by examining the ratio of intensities from the epitaxy and the substrate. This method is non-destructive and permits a high resolution mapping of the film thickness over the entire area of the specimen.

Figure IV depicts kinematic diffraction from a layered structure where both single crystals are in diffracting position at nearly the same Bragg angle. This situation is most likely encountered in the case of an epitaxial structure, for example (100) (In,Ga),(As,P) on an InP substrate.

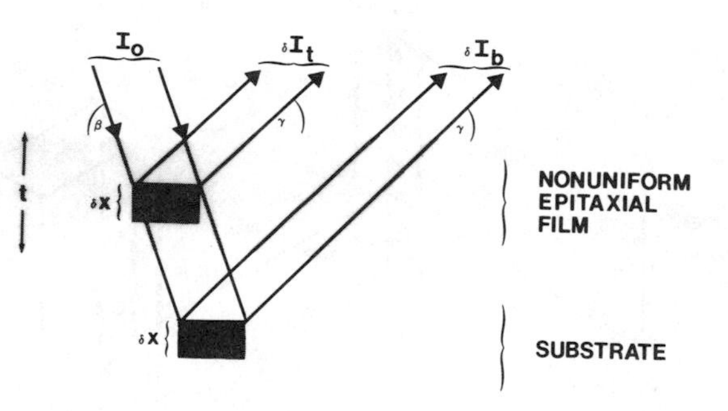

Figure IV - Simultaneous diffraction from an epitaxial thin film and the underlying substrate.

The diffracted intensity from the top layer, I_t, is given by (21):

$$I_t = \frac{I_0 b}{\sin\gamma} \int_{x=0}^{x=t} e^{-\mu x\left(\frac{1}{\sin\gamma} + \frac{1}{\sin\beta}\right)} dx \qquad (4.)$$

where,

I_0 = incident x-ray intensity,
b = scattering power,
μ = mass absorption coefficient of epitaxy,
γ = x-ray beam incidence angle,
β = diffraction angle,
t = film thickness.

Similarly, the diffracted intensity from the bottom layer, I_b, is given by:

$$I_b = \frac{I_0 b}{\sin\gamma} e^{-\mu t\alpha} \int_{x=t}^{x=\infty} e^{-\mu' \alpha x} dx \qquad (5.)$$

where,

μ' = mass absorption of substrate,

$\alpha = \frac{1}{\sin\gamma} + \frac{1}{\sin\beta}$,

$e^{-\mu t\alpha}$ = absorption of x-ray beam by epitaxy.

By taking the ratio of the intensities, the film thickness can be determined by the equation

$$\frac{I_t}{I_b} = \frac{\mu'}{\mu} \left[\frac{1 - e^{-\mu\alpha t}}{e^{-(\mu+\mu')\alpha t}} \right] . \qquad (6.)$$

The above equation assumes kinematical diffraction conditions and does not take into account atomic scattering factor differences between the two layers. Since the chemical composition of the epitaxy is different from the substrate, the atomic scattering factor will be different for the two layers due to substitutional disorder. For this specific caes, however, due to the compensating nature of the substitutions (Ga→In; As→P) the effect is relatively small for high Bragg angle reflections. If however, lower angle reflections are utilized, the differences in scattering factor must be taken into account.

Equation 6 can be solved numerically to provide a local measurement of the film thickness. By repeating the measurements at adjacent sites on the sample, a complete thickness profile can be readily generated with a spatial resolution of 60 μm. In the current studies, these profiles were correlated with the generation of a non-uniform strain state, and these thickness measurements turned out to be very important in elucidating the origin of the residual elastic strain distribution.

APPLICATION OF THE METHOD

The method is currently being applied to the study of III-V semiconductor materials which are of great interest for application as optical devices operating in the range of 1.0-1.6 μm. This range matches the optimum frequency of glass fibers in optical transmission systems. These studies are part of an extensive project to determine the origin of residual elastic strains in thin film structures deposited onto an InP substrate.

There are currently four separate problems being examined with the CARCA strain analyzer. These represent a broad spectrum of processes utilized in device construction- namely; epitaxial deposition, passivation, metallization and the use of buffer layers. No attempt will be made here to try to clarify the role that each processing step plays in controlling the strain state since this would be a formidable undertaking. Rather, it is the purpose of this paper to demonstrate the feasibility of applying the present method to problems of interest to device designers, and to provide them with quantitative information which none of the other common x-ray methods can deliver.

ELASTIC STRAINS IN EPITAXIAL InGaAsP - The quarternary alloy InGaAsP has been extensively investigated for potential applications in double heterostructure lasers. It is an especially desirable system since the lattice parameter can be matched to the substrate by chemical control, thus producing a device which is potentially strain-free. However, there still exists a difference in thermal expansion coefficients which ultimately leads to the generation of residual elastic strains upon cooling from the deposition temperature. Thus, Bisaro et al.(22) estimate that a quarternary film matched at room temperature to the InP substrate, would be deposited at 650°C with a mismatch as large as 0.06%.

Another potential source of elastic strain is the variation in composition through the film thickness. Auger analyses performed by Matsui et al.(11) clearly showed that the group V elements, P and As, vary in a complimentary manner through the thickness while the In and Ga concentrations remained constant. These variations were attributed to the LPE growth conditions rather than to any solid state diffusion. In any case, large variations in lattice parameter will result from the non-uniform composition profiles.

A third source of strain is the variation in film thickness arising from convection currents in the melt during the epitaxial growth. Yamazaki and coworkers (23) have shown that growth rates at the edge of the substrate can be more than an order of magnitude higher than in the adjoining regions. Even under well controlled melt conditions, some variation in film thickness is still to be expected. This non-uniformity will lead to the generation of large shear strains(24) which are due to the local accomodation of the non-uniform normal strains in adjacent regions of uneven thickness. Moreover, it is expected that these strains are nonuniformly distributed.

In light of all of the effects described above, it is unlikely that the simple description of the strain state in thin epitaxial films commonly accepted is a valid one.This description envisions that the film is entirely uniform both in thickness and in composition, is totally free of shear strains, and is characterized by a simple tetragonal distortion normal to the film surface (6,11). In the ensuing discussion, contrary evidence is presented for the existance of a much more complicated strain distribution. During the course of these studies, the full strain tensor is determined as a function of position on the sample, and it is clear that the conventional strain description is indeed a faulty one.

A sample of InGaAsP approximately 1.9 μm thick was deposited onto an InP substrate by a conventional LPE method (25). The x-ray characterization studies consisted first, of Lang topographic examinations to determine if misfit dislocations were present, and were followed by the strain analysis using the CARCA method. The Lang topographic studies were performed with Mo Kα radiation to insure transmission through the substrate but the strain studies were conducted with Cu $K\alpha_1$ radiation in order to maximize absorption in the thin epitaxial film. The Cu $K\alpha_1$ radiation is also ideal for the thin film thickness measurements since it resulted in epitaxy-to-substrate intensity ratios between 0.6 and 4.5.

Results of the thickness measurements show that the uniformity is poor. It varies by as much as 45% over the region examined as shown in Figure V. For convenience, the film thickness has been normalized to the maximum value observed, t_{max} = 2.63 μm. For these measurements, only the

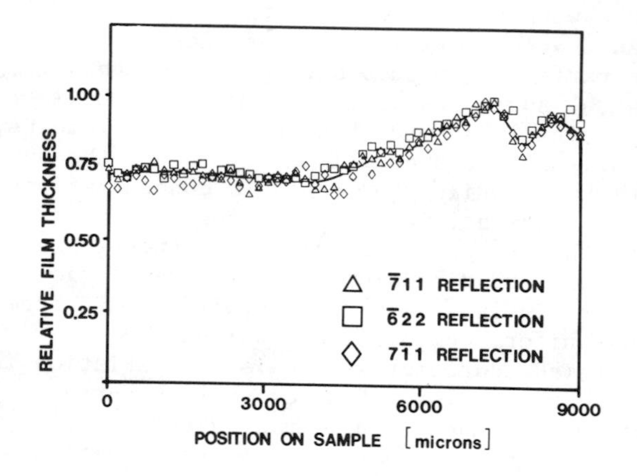

Figure V - Normalized thickness measurements along the epitaxial film surface utilizing x-ray intensity measurements.

reflections with the highest Bragg angles were used, viz. $\bar{7}11$, $\bar{6}22$ and $7\bar{1}1$, thus insuring near-normal incidence of the x-ray beam with a resulting strong diffracted beam from the substrate. Other reflections with smaller Bragg angles were not utilized since the diffracted intensity from the substrate was too small for a reliable estimate of the film thickness. Thus, in Figure V three sets of data are shown for the film thickness as a function of position on the sample, and these sets of data are in very good agreement. It should be noted that film measurements by the conventional optical fringe technique (26) indicate a thickness of 1.90 μm , or 0.72 on the scale of Figure V. This independent measurement of the film thickness is in excellent agreement with the present results for an equivalent region of the sample. However, the x-ray technique is more versatile since it is capable of scanning large areas non-destructively.

Elastic strains in the thin film are quite large and their distribution is related to the non-uniformity in film thickness. The full strain tensor has been obtained at 180 μm increments along the length of the film. Smaller increments (as low as 60 μm) are possible, but they provide no new information beyond the 180 μm step size. The normal strains from up to 10 reflections were used for these studies, and included the 200, 400, 600, $3\bar{1}1$, $\bar{6}22$, $\bar{5}11$, $\bar{7}11$, $5\bar{1}1$, $\bar{7}\bar{1}1$ and $6\bar{2}2$ Bragg reflections. An example of the measured strain state present in the film and referred to the {100} crystallographic axes is presented in Table 1. Note the presence of large shear strain

Table 1 - Representative strain state in InGaAsP thin film on InP

Strain (%)

ε_{11}	ε_{22}	ε_{33}	ε_{23}	ε_{13}	ε_{12}
-0.077	0.175	0.074	-0.217	-0.137	0.033

components indicating a complex state of stress. From data similar to that in Table 1, the principal strains and their directions have been derived. As shown in Figure VI, the thin film is

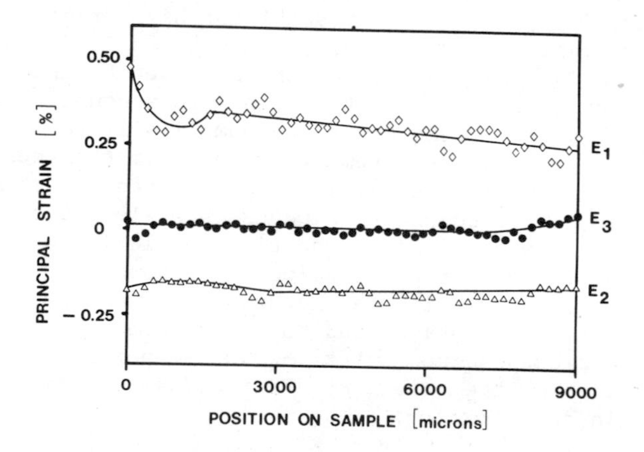

Figure VI - Principal strains in the epitaxy as a function of position on the sample. The non-zero principal strains lie in the $(\bar{1}11)$ plane.

essentially in a state of plane strain as evidenced by the fact that e_3 is approximately zero except near one of the film edges. The principal directions of the two non-zero strains generally lie in the ($\bar{1}11$) plane, which is a slip plane in III-V materials. There is some rotation of this plane of non-zero principal strains, but it is usually within 10^0 of ($\bar{1}11$).

The data presented in Figure VI indicate that the magnitude of the principal strains depends upon the film thickness. This relationship is presented in Figure VII in which only the maximum principal strain is plotted versus film thickness. It indicates a general reduction in the magnitude of the strain with increasing film thickness. The origin of this effect is not known at this time and is under further investigation. Two possible explanations for this effect are

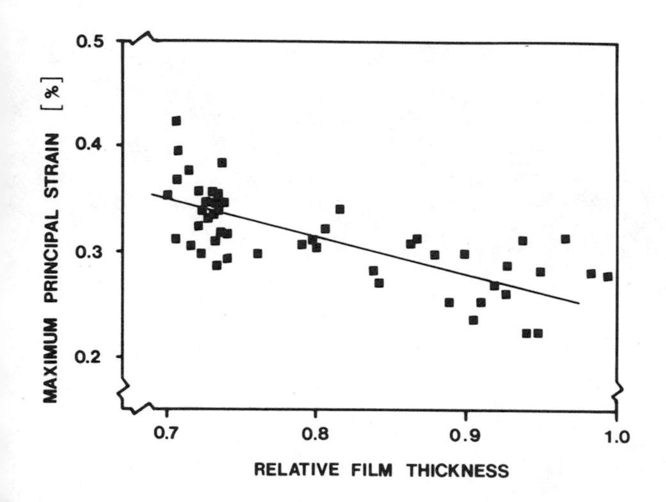

Figure VII - Relationship between the maximum principal strain, e_1, and the film thickness as measured by the present x-ray technique.

proposed here. In the first model, misfit dislocations are preferentially generated at the edge of the sample beneath the thickest region of the film in accord with the observations of Yamazaki et al (23). These dislocations in turn act to relieve the elastic strains in the immediate vicinity. A second possibility lies in the inherent ability of thicker films to accomodate the interfacial misfit strains by a relaxation through the film thickness, without generation of misfit dislocations. X-ray Lang topography is being undertaken to resolve this issue by directly imaging the dislocations and correlating their densities with film thickness. These studies however, are only in an initial phase and will not be reported here.

SUMMARY

A new x-ray method has been described which is capable of measuring the non-uniform elastic strain distribution in single crystal materials. This method is more powerful than existing x-ray techniques since it is capable of determining the complete strain state whereas conventional methods determine only one or two strain components. Since the strain can be determined at spatial increments as small as 60 μm , the method is very useful for examining the strain distribution around a strain center. Also, by comparing diffracted intensities from the substrate and the epitaxy, the uniformity of the film thickness can be determined.

The method has been applied to elucidating the strain distribution in an InGaAsP thin film grown by liquid phase epitaxy on a [100] InP substrate. It was found that the general strain state was much more complex than previous investigators have reported. The following results were found:

1. the principal strains and their directions revealed a state of near-plane strain lying in the ($\bar{1}11$) plane;
2. the magnitude of the maximum principal strain is related to the epitaxial film thickness;
3. the non-uniformity in the film thickness was a major source for the uneven strain distribution; and
4. the strain decreased with increasing film thickness.

ACKNOWLEDGEMENTS

The authors are indebted to Dr. Subash Mahajan of Bell Laboratories for supplying all of the samples in the present studies. The authors also gratefully acknowledge the support of the National Science Foundation, Division of Materials Research, Ceramics Program under grant DMR8104985 and the Rutgers Research Council .

REFERENCES

1.) Rozgonyi, G.A. and T.J.Ciesielka, Rev.Sci. Inst.,44(1973)1053.
2.) Datsenko,L.I.,A.N.Gureev,N.F.Korotkevich, N.N.Soldatenko and Yu.A.Tkhorik, Thin Solid Films,7(1971)117.
3.) Kamins,T.I. and E.S.Meieran, J.Appl.Phys., 44(1973)5064.
4.) Bohg,A.,Phys.Stat.Sol.(a),46(1978)445.
5.) Wang,C.C. and S.H.McFarlane III, Thin Solid Films,31(1976)3.
6.) Kawamura,Y. and H.Okamoto, J.Appl.Phys., 50(1979)4457.
7.) Angilello,J.,F.d'Heurle,S.Peterson and A. Segmuller, J.Vac.Sci.Tech.,17(1980)471.

8.) Rozgonyi,G.A. and D.C.Miller, Thin Solid Films, 31(1976)185.
9.) Oe,K.,Y.Shinoda and K.Sugiyama, Appl.Phys. Lett.,33(1978)962.
10.) Ewing,R.E. and D.K.Smith, J.Appl.Phys., 39(1968)5943.
11.) Matsui,J.,K.Onabe,T.Kamejima and I.Hayashi, J.Electrochem.Soc.,126(1979)664.
12.) vanMellaert,L. and G.H.Schwuttke, J.Appl. Phys.,43(1972)687.
13.) Stock,S.R.,H.Chen and H.K.Birnbaum,Proc. US-France Cooperative Science Seminar, Snowmass,Co.,1983(inPress).
14.) Imura,T.,S.Weissmann and J.J.SLade Jr., Acta Cryst.,15(1962)786.
15.) Carnahan,B.,H.A.Luther and J.O.Wilkens, "Applied Numerical Methods",John Wiley, (1969)560.
16.) Liu,H.Y.,W.E.Mayo and S.Weissmann, Mat.Sci. Eng.,(in press).
17.) Yazici,R.,W.E.Mayo,T.Takemoto and S.Weissmann, J.Appl.Cryst.,16(1983)89.
18.) Mayo,W.E.,H.Y.Liu and J.Chaudhuri, Proc. High Speed Growth and Characterization of Crystals for Solar CElls, Port St.Lucie,Fl. 1983(in press).
19.) Mayo,W.E. and S.Weissmann, Proc.US-France Cooperative Science Seminar,Snowmass Co., 1983(in press).
20.) Weissmann,S.,Z.H.Kalman,J.Chaudhuri,R.Yazici and W.E.Mayo, in "Residual Stress and Stress Relaxation", ed. by E.Kula and V.Weiss, Plenum Press,1982(501-517).
21.) Cullity,"Elements of X-Ray Diffraction", Addison Wesley,1978,134.
22.) Bisaor,R.,P.'erenda and T.P.Pearsall,Appl. Phys.Lett.,34(1979)100.
23.) Yamazaki,S.,Y.Kishi,K.Nakajima,A.Yamaguchi and K.Akita, J.Appl.Phys.(in press).
24.) Serenbrinsky,J.H., Solid State Elec.,13 (1970)1435.
25.) Mahajan,S.,Bell Telephone LAboratories, Murray Hill N.J.,Private Communication.
26.) Chin,B.,Bell LAboratories,Private Commun.

EFFECTS OF MICROSTRUCTURE ON THE ACOUSTOELASTIC MEASUREMENTS OF STRESS

R. B. Thompson
USDOE - Ames Laboratory
Ames, Iowa 50011

J. F. Smith
USDOE - Ames Laboratory
Ames, Iowa 50011

S. S. Lee
USDOE - Ames Laboratory
Ames, Iowa 50011

ABSTRACT

The acoustoelastic measurement of stress is based on deformation induced shifts in the ultrasonic velocity. Although changes in stress are routinely monitored, the absolute determination of stress in a part of unknown history is rendered difficult by competing microstructurally induced velocity shifts. A number of techniques to overcome this problem are first reviewed. A more detailed discussion is then given of a potential solution under investigation by the authors. This involves measuring the difference in the velocities of two shear waves whose directions of propagation and polarization have been interchanged. According to the theory of anisotropic elastic continua, the proportionality constant relating this velocity difference to stress should be only weakly dependent on microstructure. Preliminary measurements show that this prediction is followed in mildly anisotropic (rolled 6061 Al) and plastically deformed (1100 Al loaded beyond yield) metal polycrystals. Included is a discussion of more severe microstructural conditions which must be considered to establish the ultimate limits of the technique.

INTRODUCTION

The acoustoelastic measurement of stress may be defined as the inference of material stress (or strain) from the associated shifts in the ultrasonic velocity. Since the stress obtained is the average value throughout the volume traversed by the ultrasonic wave, interior stresses can be sensed. This differentiates the approach from most other techniques, such as x-ray diffraction, which only sense stress induced deformations in a near surface layer. The ability to fully utilize the ultrasonic information is limited by two classes of problems. The first, associated with the determination of spatial variations and tensor components of stress fields, will not be discussed here. In passing, it will be noted that solutions rely on multiple beam or frequency measurements, e.g., those used in tomographic reconstructions. The second class of problems, uncertainties associated with microstructural variations, is the subject of this paper.

The acoustoelastic measurement of stress is based upon an assumed linear relationship between velocity and stress which can be characterized by a slope and intercept. One problem is that both of these are dependent on microstructure. Figure 1 illustrates the slope variability for a series of aluminum and steel alloys, assumed isotropic, whose third-order elastic constants have been reported in the literature.[1,2] Three commonly used configurations for stress measurement are considered. Parts (a) and (b) deal, respectively, with measurements of the velocity of longitudinal waves propagating parallel and perpendicular to stress. Part (c) treats the shear wave birefringence technique, in which the difference in velocities of two shear wave propagating perpendicular to the stress is measured. One of these is polarized perpendicular to the stress while the other is polarized parallel to the stress. In each case, third order elastic constants reported from the literature[1,2] have been used to compute the indicated figure of merit. No attempt at an error analysis of the original data has been made. The microstructural variations in the alloys leads to significant scatter, which is most pronounced in part (b), whose scale is compressed by a factor of ten. For comparison, the much smaller scatter in the shear moduli of the alloys is shown in part (d).

Figure 2 gives a finer look at the problem by comparing the responses of two 6061 aluminum alloys of identical nominal composition and differing only in thermo-mechanical treatment.[3,4] The measurement geometry is that of longitudinal waves propagating perpendicular to an applied tensile stress, as was treated in Fig. 1b. In the sample cut from bar stock, a linear relationship between velocity and stress

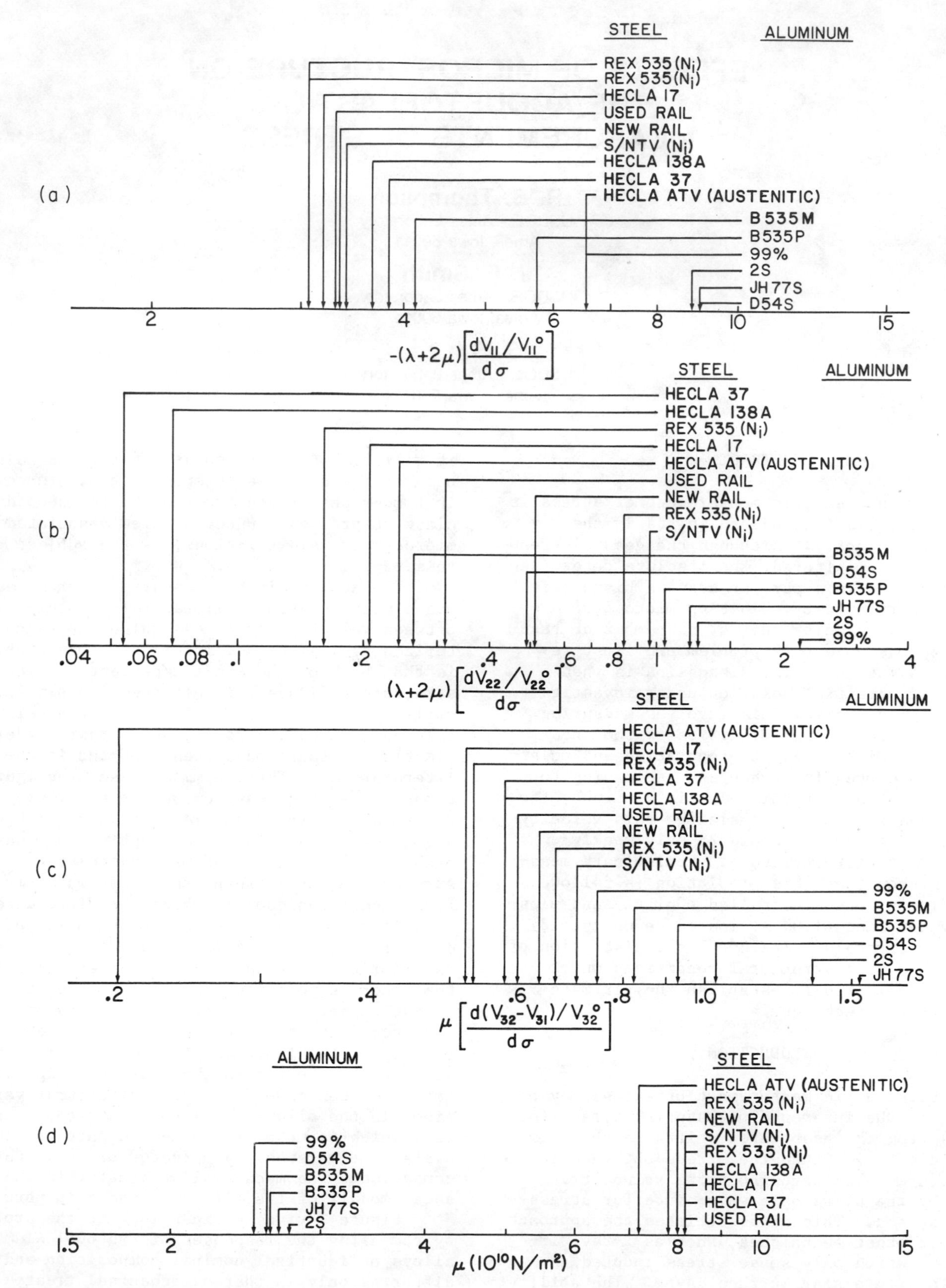

Fig. 1. Microstructure influences on common acoustoelastic stress measurement techniques in aluminum and steel.

a. Longitudinal waves propagating parallel to stress.
b. Longitudinal waves propagating perpendicular to stress (note different scale).
c. Shear wave birefringence.
d. Shear modulus to illustrate smaller variability of the linear elastic properties.

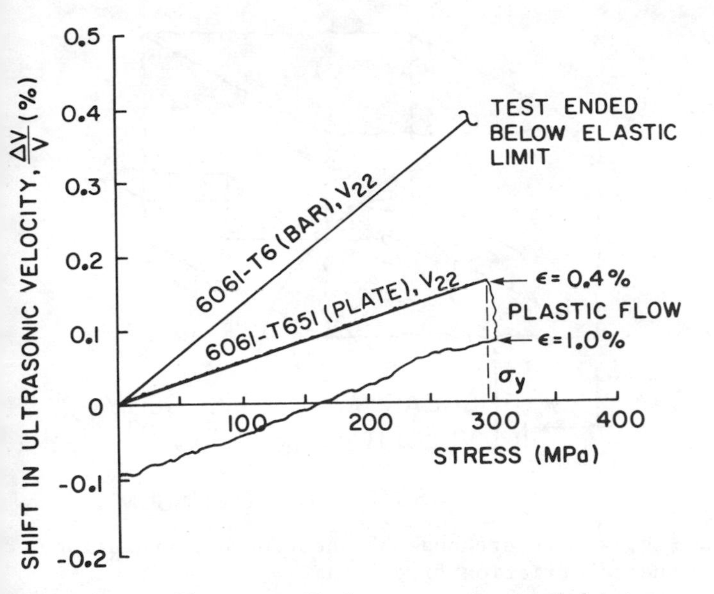

Fig. 2. Influences of thermo-mechanical history and plastic deformation on longitudinal wave (propagating perpendicular to stress) acousto-elastic response of 6061 Al.

was observed over the entire range of the test. In the plate sample, initial loading also produced a linear shift in velocity, but with only about 50% of the slope of the bar data. Furthermore, after tensile yielding, the velocity declined rather than increased. Finally, during unloading, it continued decreasing but with a different slope. It is clear from these data that inference of stress from a single measurement of velocity, with no prior knowledge of microstructure or deformation history, would be impossible.

A number of solutions to various aspects of this problem have been proposed. In certain alloys, it has been suggested that the linear relationship between velocity and stress persists beyond yield. Such a case has been demonstrated in 2024 aluminum by Fisher and Herrman[5] for longitudinal waves propagating perpendicular to the stress. After bending a beam shaped sample beyond yield, stresses were predicted from the ultrasonic velocity using acoustoelastic constants derived in the elastic regime. The predictions agreed well with those obtained from elastic-plastic stress calculations. Husson, Bennet, and Kino[6] made stress predictions on the basis of Rayleigh wave velocities in the vicinity of a circumferential weld in a 304 stainless steel pipe. Again the results were in good agreement with expectations. Bray has also suggested that the variation in the acoustoelastic constants for longitudinal waves propagating parallel to stress is sufficiently small in railroad steels[7] to allow field measurements of the thermally induced stresses which are occasionally responsible for rail buckling. Inspection of Fig. 1 shows that, in fact, the variability of the acoustoelastic effect for this case is smaller than for other alloys or configurations. More extensive measurements on a wider range of samples would be required to fully evaluate whether the required accuracy would be achieved by this approach.

Additional nondestructive information is commonly used to help distinguish velocity shifts induced by stress from microstructural effects. Salama[8] has made use of the temperature dependence of the ultrasonic velocity. In a series of aluminum and steel alloys, he has found the temperature dependence to be much less influenced by stress than the velocity itself. Furthermore, the fractional changes of the temperature dependence of the velocity are typically greater than those of the velocity itself. Other approaches under investigation are based on obtaining additional information from the measurement of the ultrasonic attenuation or the frequency dependence of the velocity. Successful demonstrations of measurements of stress in real structural components such as saw blades have been reported.[9]

A special case of the use of additional information involves simultaneous measurements of several wave velocities and combining these in such a way as to be independent of microstructure. One approach has been discussed in a review article by Allen, Cooper, Sayers, and Silk.[10] The basic objective is to measure combinations of elastic constants which are independent of texture. Consider a situation, as illustrated in Fig. 3, in which the velocities of a longitudinal wave and two transverse waves, propagating in the same directions, can be measured. It has been shown that the sums of the squares of these three velocities is the constant $(C_{11} + 2C_{44})/\rho$ for any direction in an assembly of cubic crystallites.[11] Here C_{11} and C_{44} are elastic stiffnesses and ρ is the density. It also has been found that stress does influence this parameter. Thus, a change in this parameter provides a texture independent indication of the existence of stress.

This approach requires a precise measurement of velocity through the thickness of a plate. The necessary precision in time measurement is readily obtained. However, the corresponding determination of thickness may be impossible due to part geometry. Hence, a second texture independent combination of velocities is required as the basis for eliminating the unknown thickness. Candidates have also

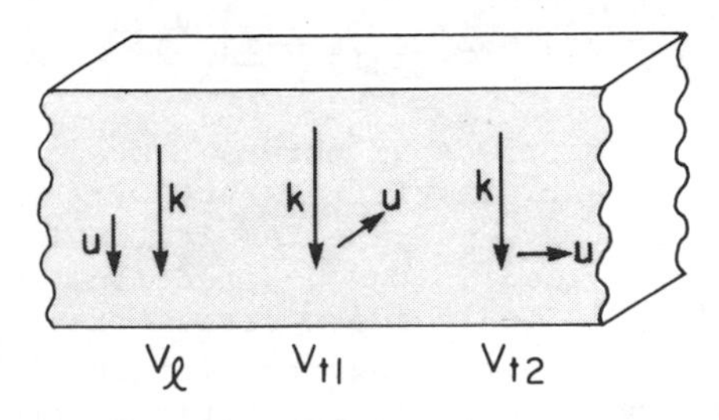

Fig. 3. Three orthogonal waves, propagating through the thickness of a plate, which provide independent information about stress and texture.

been identified by Allen et al.,[10] although they do require a preknowledge of the texture axes (but not the magnitude of preferred orientation). A number of questions remain regarding this technique. For example, the possibility exists that, even though the stress free value of those combinations of measurements are texture independent, the slopes of their stress variations may depend on texture. Also, the effects of plastic deformation have yet to be established.

Alternate approaches to the texture independent combinations of velocities have been independently investigated by the authors[12-14] and by King and Fortunko.[15-18] In each case, the specific configurations were conceived by the investigators on the basis of the nonlinear equations of continuum mechanics. However, historically, the underlying physical principles had been previously identified, though not experimentally demonstrated. These underlying ideas, and their recent experimental demonstrations, are reviewed in the next section.

A more complete discussion of residual stress measurement techniques may be found in a recent review article by Pao, Sachse, and Fukuoka.[19]

INFERENCE OF STRESS FROM THE ANGULAR DEPENDENCE OF THE SHEAR VELOCITY

In 1940, Biott[20] noted that "The propagation of elastic waves in a material under initial stress is fundamentally different from the stress free case...[it] follows laws which cannot be explained by elastic anisotropy or changes in elastic constants". The point was discussed in greater detail in 1965 by Thurston,[21] who stated that "A transverse wave propagates faster in the direction of tension than in the perpendicular direction..., just as in a stretched sting. The difference in ρv^2 for the two waves equals the tensile stress." The possibility of using this effect for the separation of the effects of stress and texture was proposed in 1981 by MacDonald,[22] who suggested that "A comparison of wave speeds... indicates how to separate the effects of stress and texture." The two shear waves to which this prediction applies, having directions of polarization and propagation interchanged, are illustrated in Fig. 4. The theory can be reduced to the form

$$\sigma_{ii} - \sigma_{jj} = 2\, C_{ijij} \left(\frac{V_{ij} - V_{ji}}{V_{ij}^{\circ}} \right) \qquad (1)$$

where σ_{ij} is a principal stress component, C_{ijij} is a shear elastic constant, V_{ij} the velocity of a wave propagating in the i-direction and polarized in the j-direction, and the superscript "o" indicates a stress free value. Equation (1) is a rigorous prediction of the nonlinear theory of elastic continua. It is only necessary that the i and j directions support pure modes (wave polarization perpendicular to propagation direction).

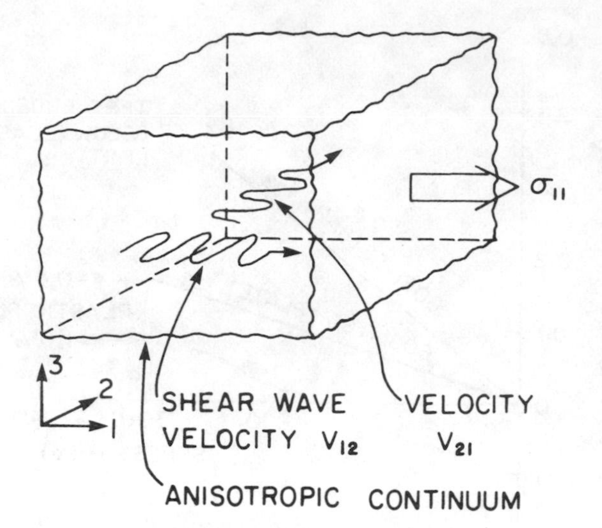

Fig. 4. Interchange of shear wave propagation and polarization directions.

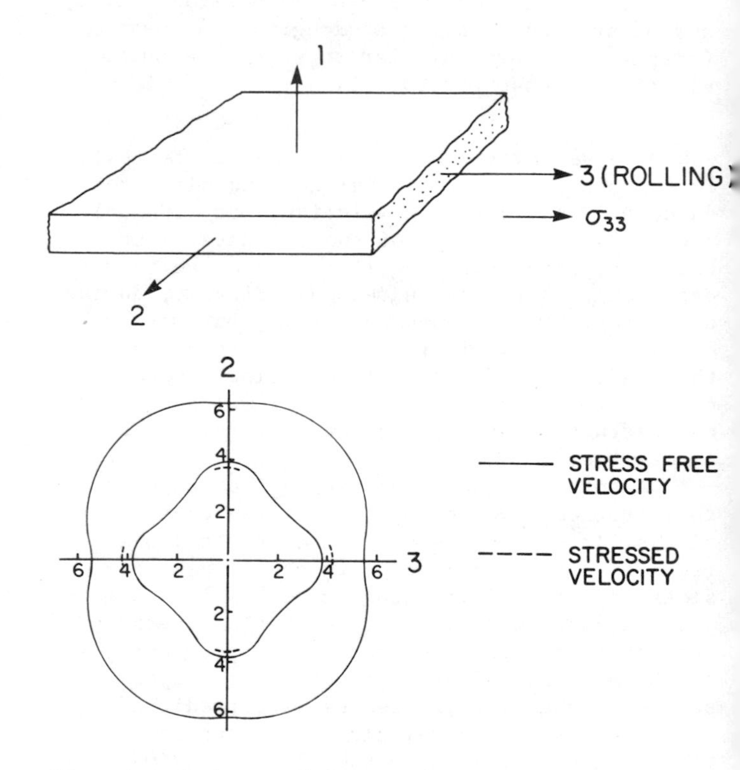

Fig. 5. Conceptual basis of technique.
Top: Definition of coordinate system with respect to a rolled plate of orthorhombic symmetry.
Bottom: Hypothetical angular dependences of longitudinal (outer curve) and shear (inner curve) velocities in plane parallel to plate surface. Degeneracy splitting by stress indicated by dashed curves.

An important practical example in which this is satisfied is a rolled metal plate placed in uniaxial tension along the rolling direction, as illustrated in Fig. 5. Rolled plates are generally elastically anisotropic because of the existence of some preferential orientation of the grains. In many cases, the anisotropy closely approximates an orthorhombic symmetry (three mutually perpendicular mirror planes) characterized by nine independent elastic constants. In such a material, the velocity of ultrasonic waves will depend on direction. The sketch at the bottom of Fig. 5 illustrates the behavior of longitudinal and horizontally polarized shear (SH) waves propagating in the plane of the plate (neglecting for the moment the effects of the plate surfaces). The anisotropy causes the velocities of longitudinal waves propagating in the 2 and 3 directions to be different. However, symmetry requires that the SH modes propagating in these two directions have equal velocities, given by $\sqrt{C_{66}/\rho}$ where C_{66} is the appropriate anisotropic elastic constant in abbreviated notation and ρ is the density. This equality is a consequence of the fact that both waves involve the same tensor component of shear.

The presence of stress alters this situation. Suppose a uniaxial stress σ_{33} is applied in the 3-direction. From a symmetry point-of-view, the material is still orthorhombic since the three orthogonal mirror planes remain. However, the continuum theory predicts [12,13,20-22] that the two SH velocities are no longer equal. As indicated by dashed lines, the velocity in the direction of stress becomes greater than that in the perpendicular direction. The stress is proportional to this difference and is given by Eg. (1).

Experiments have been performed to test the degree to which this prediction of the theory of continuous elastic media applies to polycrystalline metals.[13-14] Figure 6 conceptually illustrates the configuration. Two samples were studied, a 6061-T6 Al plate and an 1100-H14 Al plate, both as received from commercial vendors. The ultrasonic velocity for shear waves polar-

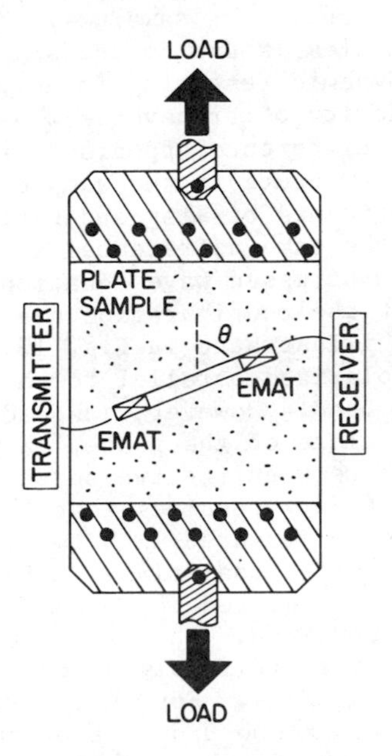

Fig. 6. Experimental configuration. EMAT spacing is fixed by a spacer.

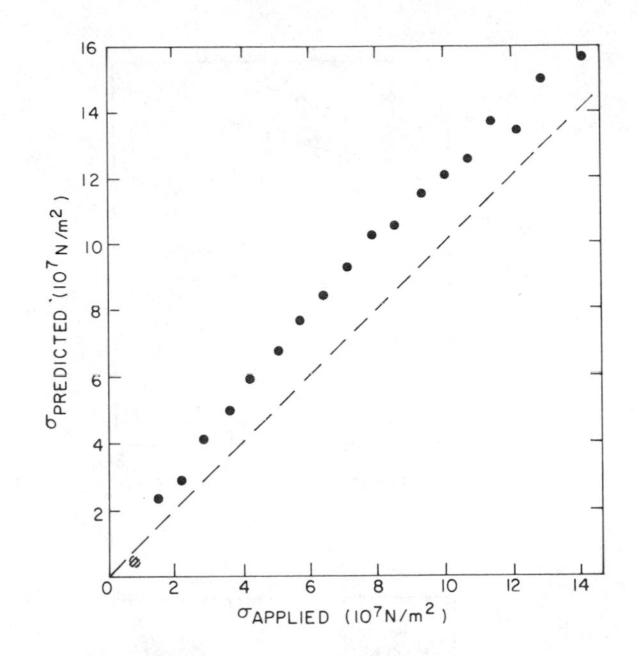

Fig. 7. Predicted versus applied stress in 6061-T6 Al plate.

ized in the plane of the plate (SH waves) was measured by exciting the transmitter with a tone burst and monitoring, with a time interval counter in the averaging mode, the arrival time of a particular cycle at the receiver. Shifts in velocity as a function of both applied load and angle of propagation with respect to that load were recorded.

The 6061 Al plate was selected to test the ability of the technique to suppress the effects of weak anisotropy. Measurements of an unstressed plate show a maximum SH wave velocity change of 0.15% between waves propagating along the rolling direction and at 45° with respect to that direction.[13] Although small, this value is comparable to the stress induced velocity shifts As shown in Fig. 7, application of Eq. (1) did, in fact, allow stress to be deduced in the presence of this anisotropy. It should be emphasized that there are no adjustable parameters in this comparison. C_{66} was determined from the measured density and wave speed.

The experiments in the 1100 Al plate were intended to assess the influence of plastic deformation on Eq. (1). As shown in Fig. 8a, the plate was loaded in tension beyond yield, with gripping failure occurring at a maximum elongation of 0.6%. The time delays for the shear waves propagating both parallel and perpendicular to the loading direction are shown in Fig. 8b. Note that (a) the relationship between time delay and stress becomes nonlinear as soon as yielding occurs and (b) the delays observed during unloading are different than those observed during loading. Recall that propagation distance is fixed by the EMAT spacing (see Fig. 6) so that the delay shifts are propor-

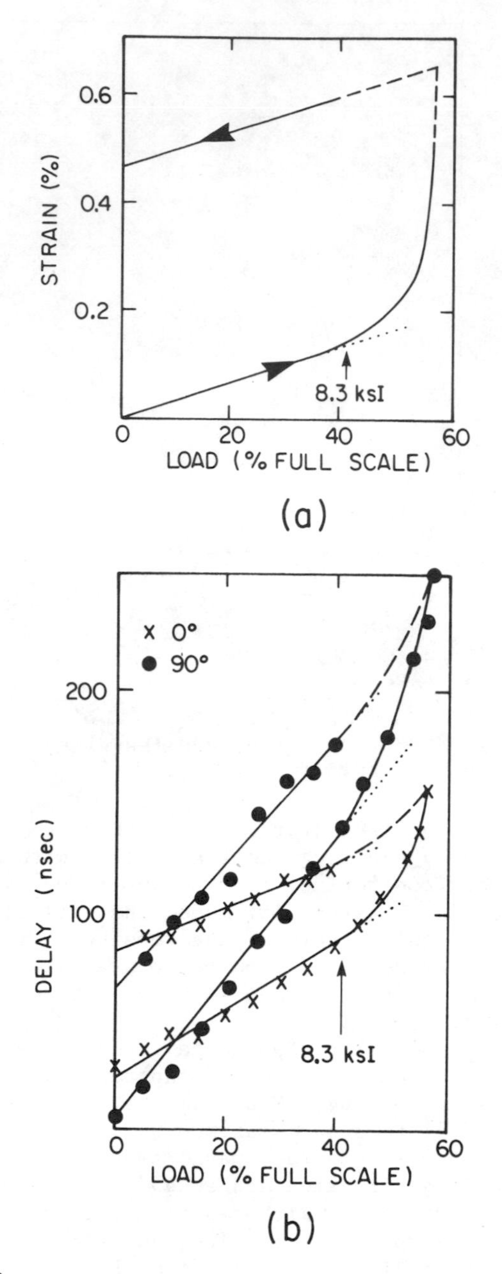

Fig. 8. Experimental data for 1100-H14 Al plate.

a) Strain versus load.
b) Ultrasonic delay versus load for 0° and 90° waves. In each case the lower portion of the curve was traced during loading and the upper portion was traced during unloading.

tional to material velocity shifts independent of length changes. These points again illustrate the microstructurally related variabilities which make prediction of stress from a single velocity measurement unreliable.

In contrast, the prediction of stress does not show either of these features, as illustrated in Fig. 9. The predictions during loading and unloading are essentially the same, and no break in the curve occurs at the yield point. The small systematic deviation between theory

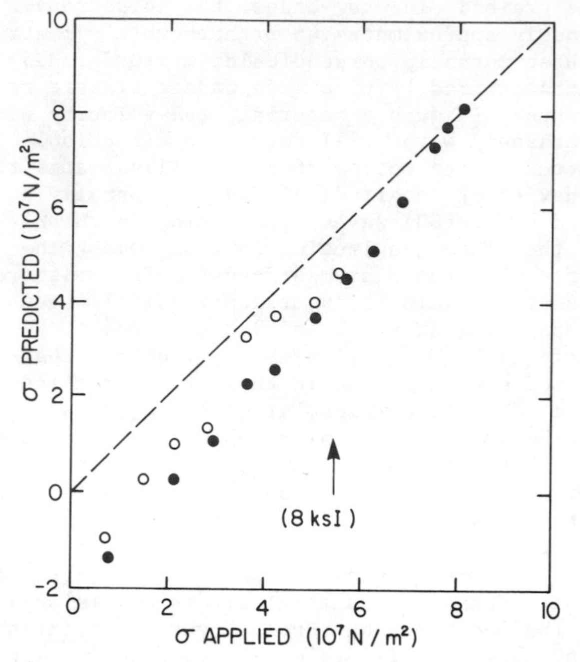

Fig. 9. Predicted versus applied stress in 1100-H14 Al plate. Solid points occured during loading, open points occurred during unloading.

and experiment at low loads is believed to be associated with the bending of the plate and will be explored in greater detail in the future.

The results on both samples are consistent with the apparent microstructural insensitivity of Eq. (1). The physical interpretation of this result is believed to be analogous to wave propagation in a taut string.[13,21,22] The absolute velocity of either wave is influenced by microstructure in the same way. The difference in velocities is due to the solid equivalent of the dynamic restoring force which determines the velocity of transverse waves on a taut string. This difference appears to be insensitive to microstructure in the samples studied.

The experiments of King and Fortunko[15-17] are based on the same differences in the stressed and unstressed wave equation. However, the details of the configuration are somewhat different, as illustrated in Fig. 10. Again, shear waves polarized parallel to the surface of the plate are used. However, instead of propagating in the plane of the plate, they bounce back and forth between the two surfaces at an angle. Their theory requires that four different velocities be measured with two values of Θ in each of two orthogonal planes. In addition, a calibration is required on an unstressed region of the specimen. The limit of $\Theta = 90°$ corresponds to the previously described approach of the present authors. For that particular angle, certain unknowns drop out of the equations. The number of measurements required reduces from 4 to 2, and the need for an unstressed calibration vanishes.

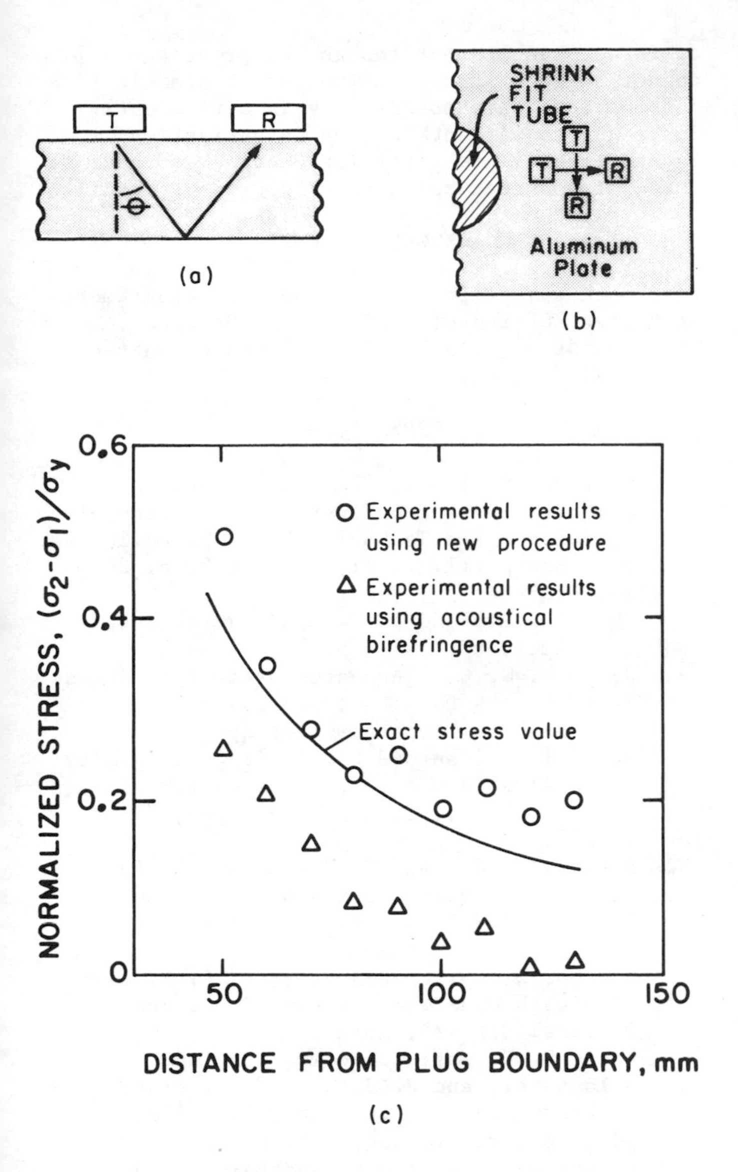

Fig. 10. Approach of King and Fortunko.
a. Waves propagating at angles with respect to surface.
b. Geometry used in measurement of shrink fit stresses.
c. Comparison of theory and experiment for shrink fit stresses.

As an experimental test of the approach of King and Fortunko, the easily calculated stresses around a shrink fit plug were measured. Figure 10 shows the excellent agreement between theory and experiment and the fact that the through-thickness results are superior to those obtained with the more traditional shear wave birefringence technique. In more recent results, King and Fortunko have found 5% absolute agreement between theory and experiment for the $\Theta = 90$ case.[18]

The approaches of Allen et al.,[10] King and Fortunko,[15-18] and the present authors[12-14] all have the common objective of seeking combinations of measurements which are independent of texture. In some cases the approach of Allen et al. may be the simplest to implement experimentally since it only requires waves propagating through the thickness of a plate. However, although the influences of variabilities in the second order elastic constants of crystalline aggregates are suppressed, the possibility of changes in the slope of measurement versus stress, due to variations in third order elastic constants, remains. The approaches of King and Fortunko and of the authors require non-normal measurements and hence may be most easily applied to parts of greater lateral dimension. However, a more complete suppression of microstructural effects might be expected. In all of the techniques, considerably more data is required to allow a full definition of potential and limitations. Some of the material questions which must be answered in the evaluation of the approach of the present authors are defined in the next section.

MICROSTRUCTURAL INFLUENCES

Equation (1) is an unambiguous prediction of the nonlinear theory of elastic continua. The only material constant which appears is the <u>second-order</u> elastic constant, C_{ijij}. This is known to depend on microstructure, but does so to a much lesser extent than the third-order constants which determine the slope of the velocity-stress relationships for most acoustoelastic techniques. This can be seen by comparing the spread in Fig. 1d for the shear modulus with that in Figs. 1a - 1c for the acoustoelastic effect in different orientations. From a knowledge of density and a measurement of shear wave velocity, C_{ijij} can be determined experimentally on any piece. The stress dependence of C_{ijij} can usually be neglected in this determination. Hence Eq. (1) would appear to greatly reduce microstructural influences on stress determinations.

The appropriateness of the assumption of a homogeneous, continuous, elastic media must be carefully examined. Real structural metals are polycrystalline, having inhomogeneities on the scale of the grain size. They may also exhibit anelastic behavior. The possible influence of such conditions will be discussed below. In that discussion, it will be assumed that the material is homogeneous on a macroscopic scale, i.e. that the grain size distribution and deformation does not vary over the scale of the measurement.

Figure 11 introduces some of the important factors associated with grain size. The elastic anisotropy of single crystals, coupled with change in orientations at grain boundaries, creates an inhomogeneous medium in which the ultrasonic velocity is frequency dependent. Recent calculations of the resulting dispersion have been performed by Hirsekorn[23,24] and by Stanke and Kino.[25] As shown at the top of Fig. 11 for the case of aluminum, the result is a decrease in velocity as the wavelength approaches the grain radius, a.

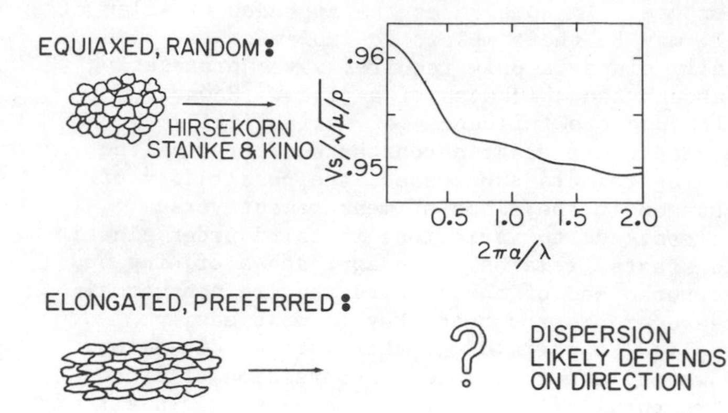

Fig. 11. Effects of Grain Structure on Dispersion. (Note $1/\lambda$ is proportional to frequency).

These dispersion calculations have been performed for a polycrystal consisting of a random orientation of equi-axed grains. Although dispersion exists, it will be the same in all directions. Hence one might intuitively imagine that, when the difference in two velocities is computed as required for application of Eq. (1), the dispersion effects would be eliminated and the technique would still prove successful. On the other hand, if there were preferred orientation and/or elongated grains, the dispersion would be expected to depend on direction; the difference in the two velocities, even in the absence of stress, would represent a limit on the accuracy of the technique. Such errors could be minimized by the use of low frequencies (long wavelengths) in the measurement.

The effects of plastic deformation must also be considered. The original derivation of Eq. (1) is based on a purely elastic analysis. The analysis must be reconsidered in the presence of plastic deformation. Several continuum models for the effects of plastic deformation on acoustoelasticity have recently been reported.[26-28] Among the conclusions are the predictions that the acoustoelastic effect depends on both plastic strains as well as initial stresses. However, in the specific configuration shown in Fig. 4 and governed by Eq. (1), the plastic strain terms vanish. Hence the elastic-plastic continuum theories do not appear to suggest an error associated with plastic deformation. More work is required to treat the effects of dislocations from a microscopic point-of-view. The vibrating string model for ultrasound-dislocation[29] interactions could form the basis of such studies.

It should be noted that the experiments reported in the previous section were in a favorable regime with respect to the potential errors. Aluminum crystals have a relatively weak elastic anisotropy so that dispersion effects are small. At the measurement frequencies of 500 kHz, the wavelength of 6mm is much larger than typical grain sizes, further suppressing dispersion. Higher frequency measurements in steels might produce considerably different results. Although the present models do not suggest strong influences of plastic deformation, this possibility is sufficiently important to warrant further investigation. Future experiments are being designed to address these microstructural issues in greater detail.

ACKNOWLEDGEMENT

This work was performed for the U.S. Department of Energy, Office of Basic Energy Sciences, Division of Materials Sciences, under contract No. W-7405-Eng-82.

REFERENCES

1. Hearmon, R. F. S., "Landolt-Börnstein, Numerical Data and Functional Relationships in Science and Technology", K.-H. Hellwege, ed., Group III, V. 11, Chapter 2, p. 265, Springer-Verlag, Berlin, (1979).
2. Egle, D. M., and D. E. Bray, J. Acoust. Soc. Amer. 60, 741 (1976).
3. Johnson, G. C., "Acoustoelasticity: Stress Measurements Using Ultrasonics", PhD Dissertation, Department of Applied Mechanics, Stanford University, University Microfilms, Ann Arbor, Michigan (1979).
4. Johnson, G. C., J. Appl. Mech. 48, 791 (1981).
5. Fisher, M. J., and G. Herrmann, "Review of Progress in Quantitative NDE 3", D. O. Thompson and D. E. Chimenti, eds. Plenum Press, New York (in press).
6. Husson, D., S. D. Bennett, and G. S. Kino, "Rayleigh Wave Measurement of Surface Residual Stress", ibid.
7. Bray, D. E., private communication.
8. Salama, K., and J. J. Wang, "New Procedures in Nondestructive Testing", P. Höller, ed., Springer-Verlag, Berlin, 539 (1983).
9. Schneidner, E., and K. Goebbels, ibid, 551.
10. Allen, D. R., W. H. B. Cooper, C. M. Sayer, and M. G. Silk, "Research Techniques in Nondestructive Testing", V. 6, R. S. Sharpe, ed., Adademic Press, London, 151 (1983).
11. Reynolds, W. N., J. Phys. D. 3, 1798 (1970).
12. Thompson, R. B., J. F. Smith, and S. S. Lee, "Review of Progress in Quantitative NDE 2", D. O. Thompson, and D. E. Chimenti, eds., Plenum Press, New York, 1339 (1983).
13. Thompson, R. B., S. S. Lee, and J. F. Smith, "Review of Progress in Quantitative NDE 3", D. O. Thompson and D. E. Chimenti, eds., Plenum Press, New York (in press).
14. Thompson, R. B., S. S. Lee, and J. F. Smith, Appl. Phys. Letters, to be published Feb. 1, 1984.
15. King, R. B., and C. M. Fortunko, "1982 Ultrasonics Symposium Proceedings", IEEE, New York, 885 (1982)
16. King, R. B., and C. M. Fortunko, "Review of Progress in Quantitative Nondestructive Evaluation 2", D. O. Thompson and D. E.

Chimenti, eds., Plenum Press, New York, 1327 (1983).
17. King, R. B., and C. M. Fortunko, J. Appl. Phys. 54, 3027 (1983).
18. King, R. B., and C. M. Fortunko, "Review of Progress in Quantitative NDE 3", D. O. Thompson and D. E. Chimenti, eds., Plenum Press, New York (in press).
19. Pao, Y.-H., W. Sachse, and H. Fukuoka, "Physical Acoustics", V. 17, W. P. Mason and R. N. Thurston, eds., Academic Press, New York (in press).
20. M. Biott, J. Appl. Phys. 11, 522 (1940).
21. Thurston, R. N., J. Acoust. Soc. Amer. 37, 348 (1965).
22. MacDonald, D. E., IEEE Trans. on Sonics and Ultrasonics, SU-28, 75 (1981).
23. Hirsekorn, S., J. Acoust. Soc. Amer. 72, 1021 (1982).
24. Hirsekorn, S., J. Acoust. Soc. Amer. 73, 1160 (1983).
25. Stanke, F. E., and G. S. Kino, "Review of Progress in Quantitative NDE 3", D. O. Thompson and D. E. Chimenti, eds., Plenum Press, New York (in press).
26. Johnson, G. C., J. Acoust. Soc. Am. 70, 591 (1981).
27. Johnson, G. C., "Review of Progress in Quantitative NDE 1", D. O. Thompson and D. E. Chimenti, eds., Plenum Press, New York, 77 (1982).
28. Pao, Y.-H., and Udo Gamer, "A Theory of Acoustoelasticity for Elasto-Plastic Solids," Report No. 5121, Materials Science Center, Cornell University, Ithaca, New York (1983).
29. Granato, A., and K. Lücke, J. Appl. Phys. 27, 583 (1956).

DETERMINATION OF STRESS GENERATED BY SHRINK FIT

N. Chandrasekaran, Y. H. Wu and K. Salama
University of Houston
Houston, Texas 77004

ABSTRACT

The temperature dependence of ultrasonic velocity was used to determine the stress generated in a disc made of A533B steel. The stress is introduced by shrink fitting a rod of the same steel in a hole of slightly smaller size. The measurements were made using longitudinal and shear waves propagating along the thickness of the disc in the temperature range between room temperature and -40°C. In addition, shear velocities were measured with the polarization adjusted parallel to and perpendicular to the radius on which the stress measurements were performed. The results show that at all locations, the ultrasonic velocity decreases linearly as the temperature is increased and that the temperature dependence of ultrasonic velocity can be determined to ±2%. The results also indicate that the temperature dependence of the longitudinal velocity remains unchanged as the radial distance on the disc is changed, which agrees with predictions of elasticity theory. The temperature dependences of the shear velocities, however, show considerable variations with radial distance, reflecting the changes in the radial and the tangential components of stress as a function of the radial distance.

INTRODUCTION

Ultrasonic methods seem to hold the best promise for the nondestructive evaluation of bulk residual stresses in both crystalline and non-crystalline materials[1]. Several methods based on the measurements of ultrasonic velocities[2,3] or their temperature dependences[4] are currently being developed and show promising possibilities for their applications in stress measurements. In order for these methods to provide quantitative determination of stress, specimens with well-known stress distributions should be available for reference. Hsu et al.[5] have studied the distribution of ultrasonic longitudinal velocity in a shrink-fit assembly and found a good agreement between the velocity measurements and the calculated stress distribution.

The temperature dependences of ultrasonic velocities are due to the anharmonic nature of the solid, and therefore can be used for stress measurements. Experiments performed on aluminum[6], copper and steel[7] show that, in the vicinity of room temperature, the ultrasonic longitudinal velocity decreases linearly with temperature, and the slope of the linear relationship varies considerably when the specimen is subjected to stress. The results also indicate that the temperature dependence of ultrasonic velocity is insensitive to variations in alloy chemical composition which represents a major difficulty for the use of direct velocity measurements in the determination of residual stresses[8].

In order to further develop the temperature dependence method for field applications, the radial distribution of the slope of ultrasonic velocity vs. temperature has been studied in a shrink-fit assembly made of ASTM A533B steel. The study is made using longitudinal and shear waves polarized parallel as well as perpendicular to radial direction. The results show that the temperature dependence of ultrasonic velocity remains unchanged throughout the radius of the stressed specimen, which agrees with the predictions of elasticity calculations. The behavior of the temperature dependence of shear waves, however, is found to have less agreement with elasticity theory.

EXPERIMENTAL

SPECIMEN PREPARATION - Carbon steel ASTM A533B was used in the present research. This steel is generally used in the manufacture of pressure vessels in the nuclear industry. A specimen of ring-plug type was machined from this steel. The dimensional details of the ring

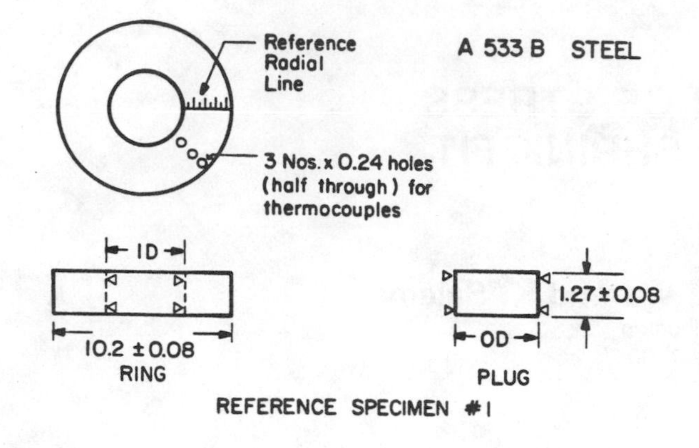

Fig. 1 Dimensional details of reference specimen #1. ID = 1.27 cm; OD = 1.275 cm.

specimen are shown in Fig. 1. The figure shows the difference in the dimension of outer diameter of the plug and the inner diameter of the ring before plugging. This difference is the extent of diametrical interference. The mating surfaces of this specimen was carefully machined to within 0.0005 in. tolerance which was met at only three fourths of the circumference. At the worst point the deviation from circularity was 0.002 in.

The interference fit was made by simultaneously cooling the plug and heating the ring, and thereby creating a temperature differential enough to compensate for the difference in dimension between the ring and the plug. The plug was immersed in liquid nitrogen for about 5 minutes while the ring was heated to about 600°F for about half an hour. This temperature difference was found to be adequate for plugging.

TEMPERATURE CONTROL SYSTEM - The temperature control system is designed to enclose the specimen and the measuring accessories to ensure a stablized temperature during the time required for the velocity measurement. The desired subambient temperature is obtained by controlling the flow rate of liquid nitrogen to the insulated enclosure. After the specimen is cooled to the desired low temperature, the supply of nitrogen is stopped. The warming rate is adjusted so that the rise in temperature is steady at about 0.25°K/minute. This rate is achieved by using a low heat input light source within the enclosure. All the measurements were made in the range of 220°K to 260°K, which produced a steady warming rate for the system. The actual temperature of the specimen was measured by means of a Copper-Constantan thermocouple attached directly to the specimen. The thermocouple along with a potentiometer provides the measurement of the temperature to an accuracy of ±0.1K.

VELOCITY MEASUREMENT SYSTEM - Velocity measurements in the present investigation were performed by the pulse-echo-overlap method, which has been fully described by Papadakis. A block diagram of the apparatus is shown in Fig. 2. In this system, a pulse of approximately 1 microsecond in duration and of variable pulse repetition rate is generated by the r.f. pulsed oscillator and is impressed on a 0.25 inch in diameter quartz transducer of a fundamental frequency of 10 MHz. The transducer is acoustically bonded to the specimen by means of a thin layer of viscous Nonaq stopcock grease.

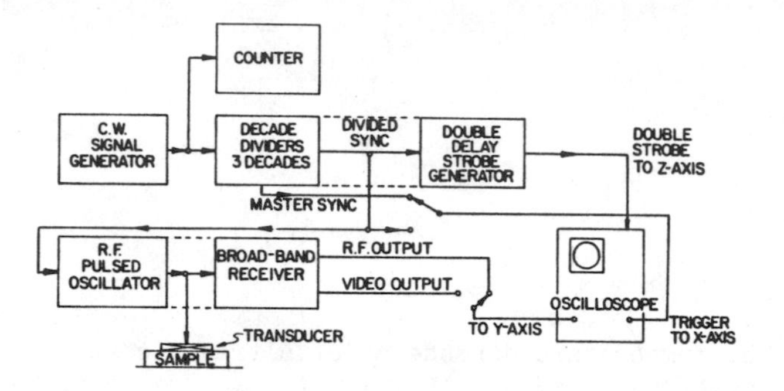

Fig. 2 Pulse-echo-overlap system.

After a conventional series of echoes is displayed on the screen of the oscilloscope, two echoes are visually superimposed by adjusting the frequency of the CW oscillator so that the period of this frequency is exactly equal to the round trip time of the ultrasonic pulse in the specimen. The frequency f at the exact cycle for cycle match, is then recorded directly from the electronic counter and used to compute the ultrasonic velocity using the relationship $V = 2\ell f$, where 2ℓ is the total path length.

Figure 3 shows the transducer holder used in the measurements of velocity. The spring supported plunger served as the inner conductor of the coaxial cable which carries the rf signals. These signals are transmitted from the plunger by a piece of teflon coated wire to a BNC receptacle mounted directly on the transducer holder. A clamping force of 30 to 50 Newtons is found to be sufficient to produce a uniform thin layer bond between transducer and specimen.

RESULTS

The velocities of longitudinal and shear waves propagating through the thickness of the specimen are measured at various locations along that radial reference line. The measurements of the shear velocity are made when the waves are polarized parallel as well as perpendicular to the reference line. The results of these measurements are representative to the stress distribution in the entire shrink-fit

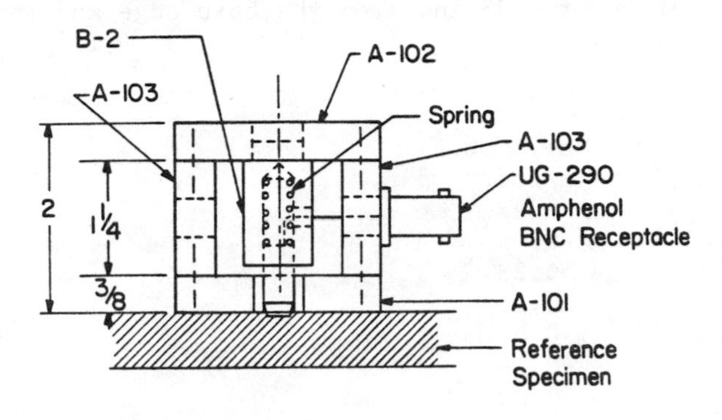

Fig. 3 Transducer holder used for the measurement of ultrasonic velocity on reference specimens.

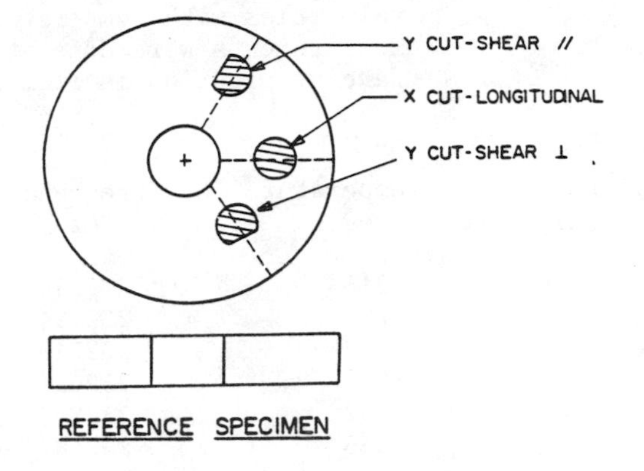

Fig. 4 Specimen used in the ultrasonic velocity measurements.

specimen since the triaxial stress generated by the interference fit is radially symmetric. The velocity measurements are also performed at various stages of manufacturing of the specimen. A shear wave polarized parallel to circumferential direction is hereinafter called shear-parallel and a shear wave polarized along the radial direction is called shear-perpendicular, as shown in Fig. 4. The measurements are made at about six locations, along the reference radial line.

The results of typical examples of the temperature variations of longitudinal, shear-parallel and shear-perpendicular velocities on specimen with the hole plugged are shown in Table (1). These measurements are made at a distance of 0.25 in. from the edged ofthe hole. Table (1) also shows the velocity deviation from the best fit line and the values of the dV/dT. Fig. 5 shows plots of the results of these three examples. From this figure it can be seen that the velocity of ultrasonic waves varies linearly with temperature and the slopes of the straight line of best fit are different.

The velocities of longitudinal and shear-parallel and shear-perpendicular are measured as a function of temperature before and after stresses are introduced in specimen. These measurements are made along the same radial reference line. The results of the temperature dependences of ultrasonic velocities prior to making the interference fit between the plug and the ring, form the baseline data for the study of stresses generated by shrink fit. Separate baseline data is obtained for the longitudinal, the shear-parallel, and the shear-perpendicular velocities for the specimen.

Figure 6 shows the variation of the temperature dependence of ultrasonic longitudinal velocity as a function of radial distance from the hole. The plot shows the variation of $(dV/dT)_{\ell}$ before and after stresses are introduced. The results of the temperature dependences of ultrasonic shear velocities for the same two cases are shown as a function of radial distance from the hole in Figs. 7 and 8. Figure 7 shows the variation of $(dV/dT)_{\parallel}$ while Fig. 8 shows the variation of $(dV/dT)_{\perp}$. In all cases the ultrasonic waves are propagated in the axial direction.

DISCUSSION

ACCURACY OF MEASUREMENTS - In order to estimate the repeatability and the accuracy of the measuring technique the measurements of ultrasonic velocity as a function of temperature were repeated for the cases of longitudinal, shear-parallel and shear-perpendicular. The experiments were repeated on several different dates using different wave propagation modes and polarizations. The results of these experiments are tabulated in Table (2). The table also shows the standard deviation, ρ, based on the results of the four sets of measurements performed. The standard error, S_m, is computed using the relationship of eqn (1), to obtain the error in determining the temperature dependence of ultrasonic velocity, dV/dT[9]

$$S_m = \rho/\sqrt{n} \tag{1}$$

where n is the number of measurements and ρ is the standard deviation of an infinite number of measurements.

In this investigation, dV/dT is measured only once for each particular type of wave propagation. Therefore, equation (1) becomes,

$$S_{m\ell} = \rho \tag{2}$$

where $S_{m\ell}$ is the standard error for a single measurement. Using the results in table (2) and equation (2), $S_{m\ell}$ is found to be ±1.7%.

Table (1) - Variations of ultrasonic longitudinal, shear (parallel) and shear (perpendicular) velocities with temperature in reference specimen #1, after the interference fit. Measurements were done at the same location of 0.25 in. from the hole edge and the wave propagation is from surface to surface.

Pol.	Temperature Deg K	Frequency Hz	Velocity m/sec	Vel. Dev. m/sec
Long	215	232889	5927.2	-0.05
	220	232794	5924.8	0.28
	222	232749	5923.7	0.11
	227	232654	5921.2	-0.09
	232	232540	5918.3	-0.19
	235	232487	5917.0	0.22
	238	232439	5915.8	-0.09
	251	232163	5908.7	-0.15
	257	232065	5906.2	-0.32
	260	231995	5904.4	0.05
Shear (Par.)	211	127776	3252.0	0.38
	218	127680	3249.6	0.11
	221	127628	3248.2	0.03
	223	127599	3247.5	-0.29
	226	127561	3246.5	-0.31
	234	127432	3243.3	-0.24
	243	127324	3240.5	-0.20
	251	127151	3236.5	0.04
	260	127003	3232.3	0.05
	262	126957	3231.2	0.12
Shear (Perp.)	210	127853	3254.0	0.09
	214	127790	3252.4	0.04
	221	127668	3249.3	-0.05
	225	127611	3247.8	-0.28
	230	127523	3245.6	-0.26
	233	127473	3245.3	-0.35
	236	127415	3242.8	-0.38
	241	127317	3240.3	-0.12
	246	127245	3238.5	-0.04
	249	127187	3237.0	-0.08

This accuracy in the temperature dependence measurements agrees very well with the accuracy of ±1.6% determined by Barber[10] in his study on aluminum.

It can also be noted from Table (2) that at zero applied stress, the values of dV/dT for shear-parallel and shear-perpendicular are equal within the accuracy of measurements. This means that at zero applied stress, dV/dT of shear waves in the steel investigated is texture insensitive.

CALCULATION OF STRESS DISTRIBUTIONS IN A SPECIMEN - Figure 9 shows the calculated distributions of the circumferential, the radial and the axial stress variations along a radial line in the specimen used in this investigation. The stress distributions are calculated using the relationships,

$$\sigma_t = \frac{E}{2}\,\frac{b}{c^2}(1 + c^2/r^2)\;\delta \tag{3}$$

$$\sigma_r = \frac{E}{2}\,\frac{b}{c^2}(1 - c^2/r^2)\;\delta \tag{4}$$

$$\sigma_a = \nu(\sigma_t + \sigma_r) = \frac{E}{2}\,\frac{b}{c^2}\nu\delta \tag{5}$$

where σ_t is the tangential stress, σ_r is the radial stress and σ_a is the axial stress. E is the Young's modulus, δ is the extent of interference, ν is the Poisson's ratio, b is the outer

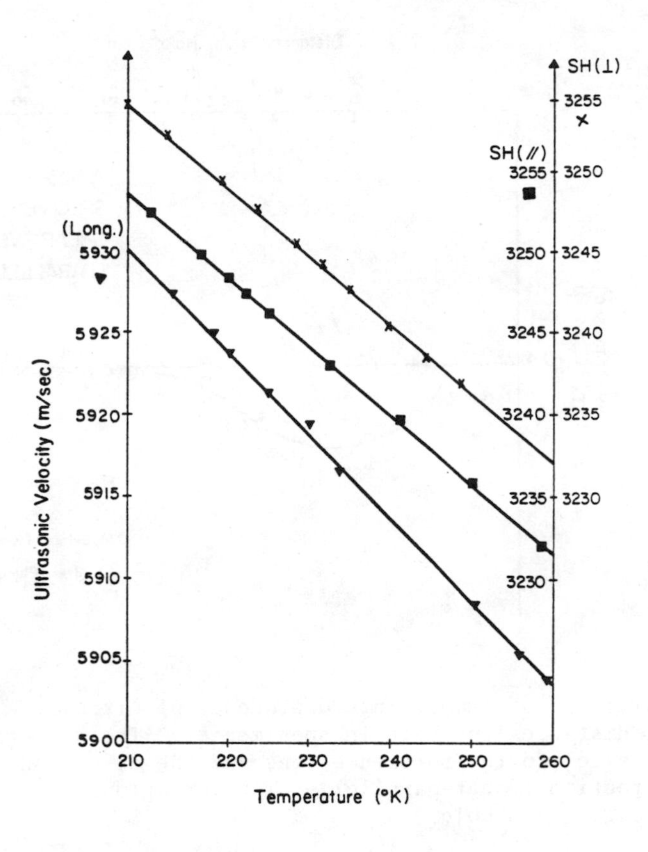

Fig. 5 Typical variations of ultrasonic longitudinal, shear (parallel) and shear (perpendicular) velocities with temperature at 0.64 cm. from the hole in reference specimen

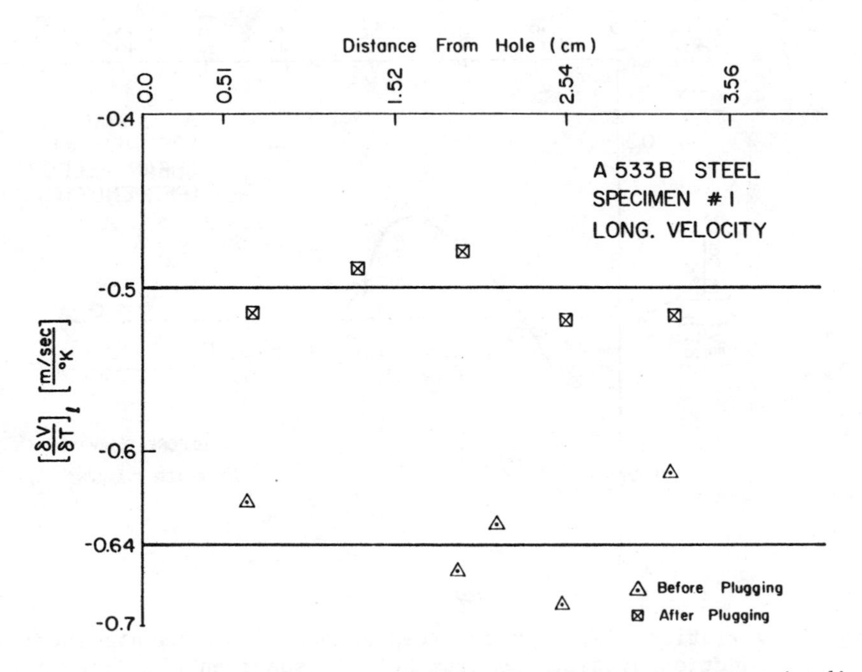

Fig. 6 Variations of temperature dependence of ultrasonic longitudinal velocity as a function of distance from the hole in specimen #1. The propagation is in axial direction. The variations are shown before and after plugging the hole.

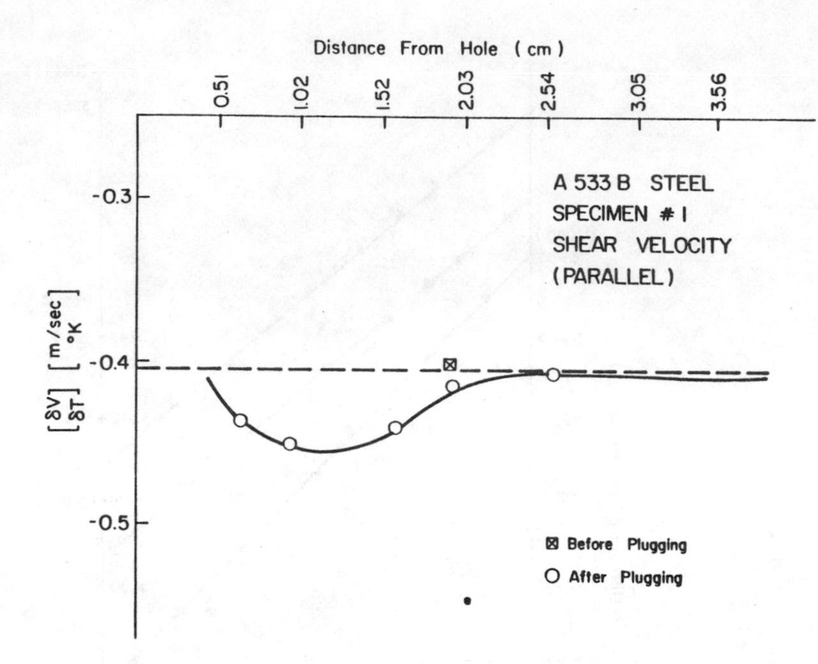

Fig. 7 Variation of temperature dependence of ultrasonic shear velocity as a function of distance from hole in specimen #1. The direction of polarization is parallel to the reference line and the direction of propagation in the axial direction (shear-parallel). The variations are shown before and after plugging the hole.

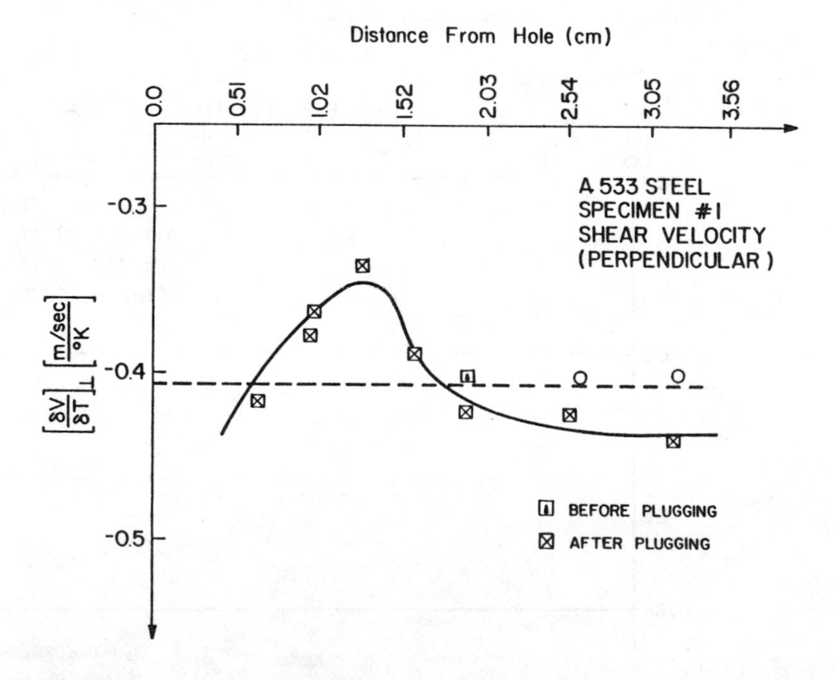

Fig. 8 Variation of temperature dependence of ultrasonic shear velocity as a function of distance from hole in specimen #1. The direction of polarization is perpendicular to the reference line and the direction of propagation is in the axial direction (shear-perpendicular). The variations are shown before and after plugging the hole.

Table(2) Variations in (dV/dT) obtained on reference specimen using longitudinal, shear (parallel), and shear (perpendicular) waves.

Date	Condition of spec.	Type of Wave	dV/dT	% Dev.
4/20/82	As recd.	Longitudinal	-0.640	1.84
4/03/82	As recd.	Longitudinal	-0.652	0.0
7/02/82	As recd.	Longidutinal	-0.613	0.0
9/23/82	As recd.	Longitudinal	-0.629	2.54
6/05/82	Machined	Shear (Par.)	-0.401	0.0
6/18/82	Machined	Shear (Per.)	-0.407	1.47
7/03/82	Machined	Shear (Par.)	-0.474	0.0
7/14/82	Machined	Shear (Per.)	-0.483	0.86
12/30/82	Hole Made	Shear (Par.)	-0.443	0.0
12/22/82	Hole Made	Shear (Per.)	-0.445	0.44

Std. Dev. = 0.0172
= 1.7%

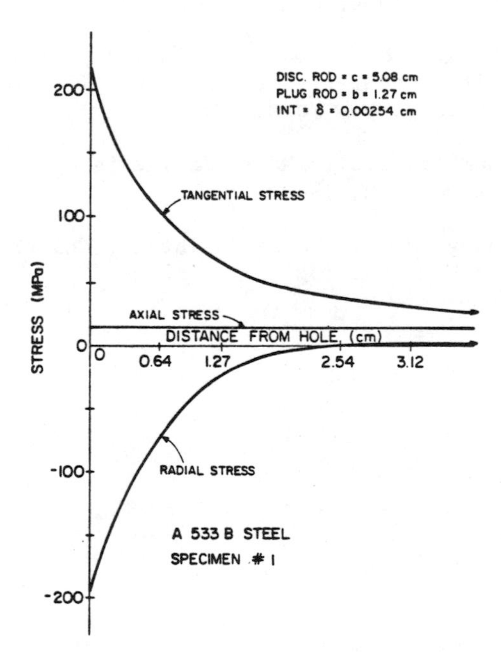

Fig. 9 Theoretical variations of tangential, radial and axial stresses as a function of distance from the hole in specimen #1.

diameter of the plug and c is the outer diameter of the disc. The distributions shown in Fig. 9 are calculated using the dimensions shown in Fig. 1 and a radial interference of 0.001 in. From equations (3), (4) and (5) it can be seen that the shape of the stress distribution is not affected by the actual values of interference, though the magnitude of the stress components depends on the extent of that quantity.

TEMPERATURE DEPENDENCE OF LONGITUDINAL ULTRASONIC VELOCITY - Figure 6 shows the variations of the temperature dependence of ultrasonic longitudinal velocity, as a function of radial distance from the hole before and after specimen was plugged. In these measurements longitudinal waves are propagated along the axial direction of the specimen. From this figure it can be noted that both before and after plugging, dV/dT remains constant along the radial direction of the specimen. Also from the figure, it is seen that $(dV/dT)_\ell$ remains constant at values between -0.64 m/s.k and -0.63 m/s.k. This value is very close to the value of -0.645 obtained by Salama and Wang[11] on a stress-free specimen of A533B steel. This equality indicates that the amount of the axial component of residual stress in specimen in the as-received condition should be very small.

Furthermore, one can see from Fig. 6 that after stresses are introduced, the axial component of stress also remains constant along the radial distance from the hole. This qualitatively agrees with eqn. (5) where σ_a does not depend on radial distance from the hole. It is difficult, however, to obtain quantitative values for the corresponding stress component since $(dV/dT)_{\ell}$ measurements are affected by the axial component of the stress, which is parallel to the direction of wave propagation, and calibration curves available are only for the case of stress applied in a direction perpendicular wave propagation.

TEMPERATURE DEPENDENCE OF ULTRASONIC VELOCITY SHEAR-PARALLEL AND SHEAR-PERPENDICULAR - It can be seen from equations (4) and (5) and Fig. 9 that the radial and circumferential stresses vary inversely with the square of the radial distance from the hole. Figure 7 shows the relationship between the temperature dependence of ultrasonic shear velocity $(dV/dT)_{||}$ and the radial distance, when the polarization is parallel to the radial line. From this figure, one can see that the data points for the distances ranging from 1.27 cm to 3.74 cm may be represented by an inverse square relationship. However, the data points less than 1.27 cm show a decrease in dV/dT as the distance decreases. This decrease might be due to the plastic deformation in that region resulting from the introduction of the plug into the hole. A similar relation between $(dV/dT)_{\perp}$ and the radial distance is obtained when the polarization is aligned perpendicular to the radial line. These results are shown in Fig. 8. In this figure the behavior of the temperature dependence of ultrasonic shear velocity is also similar to that observed when the polarization is parallel to the reference line.

Again, it is difficult to convert these results of $(dV/dT)_{SH}$ to quantities of tangential and radial components of stress. The measurements of dV/dT in the radial and the tangential directions will be affected by the component of stress parallel to propagation, in addition to that of the component perpendicular to propagation. Since the calibration curves available are only for the case of stress applied in a direction perpendicular to the wave propagation, it will be difficult to quantitatively evaluate the corresponding values of stress from the values of $(dV/dT)_{SH}$.

ACKNOWLEDGEMENT

This work is supported by the Office of Naval Research under Contract N00014-82-K-0496 and the Electric Power Research Institute under Contract T107-2.

REFERENCES

1. Proceedings of a Workshop on Nondestructive Evaluation of Residual Stress, NTIAC-72-2 (1976).

2. Kino, G.S., D.M. Barnett, N. Grayeli, G. Hermann, J.B. Hunter, D.B. Ilic, G.C. Johnson, R.B. King, M.P. Scott, J.C. Shyne and C.R. Steele, Journal of Nondestructive Evaluation 1, 67 (1980).

3. Gordon, B.E. Jr., ISA Trans. 19(2), 33, 1980.

4. Salama, K. and C.K. Ling, J. Appl. Phys. 51(3), 1505, 1980.

5. Hsu, N.N., T.M. Proctor, Jr. and G.V. Blessing, J. of Testing and Evaluation 10 (5), 230, 1982.

6. Chern, E.J., J.S. Heyman and J.H. Cantrell, Proceedings IEEE Ultrasonic Symposium, p. 960 (1981).

7. Salama, K. and J.J. Wang, Translated in Germany of New Procedures in Nondestructive Testing (Proceedings), p. 539, 1983.

8. James, M.R. and O. Buck, Critical Reviews in Solid State and Materials Sciences, p. 61 (1980).

9. Spigel, M.R., Schaum's Outline of Probability and Statistics, McGraw-Hill Book Company, New York, 1975.

10. Barbar, G.C., Master's Thesis, University of Houston, 1982.

11. Wang, J.J., Master's Thesis, University of Houston, 1981.

EVALUATION OF INTERFACIAL STRESSES FROM THE AMPLITUDE OF REFLECTED ULTRASONIC SIGNALS

D. K. Rehbein
Ames Lab, USDOE
Ames, IA 50011

J. F. Smith
Ames Lab, USDOE and Department of Material Science & Engineering
Iowa State University
Ames, IA 50011

D. O. Thompson
Ames Lab, USDOE and Department of Engineering Science & Mechanics
Iowa State University
Ames, IA 50011

ABSTRACT

In seeking to develop a technique for quantitatively evaluating interfacial stresses, it has been found that the amplitude of a reflected ultrasonic signal furnishes a useful measure. At present, the accuracy of measure is undefined because no way has been found to evaluate the true interfacial stress. Rather, the stress against which the amplitudes have been calibrated is a calculated stress based upon a two-dimensional point by point loading with an elastic-plastic model. There is semiquantitative accord between theoretical predictions and normalized data for reflection coefficient versus stress level. These results are at fixed frequency with comparable surface roughness. Variation of either frequency or surface roughness further affects the reflection coefficient. An automated scanning system has been assembled to provide longitudinal and circumferential scanning and stress evaluation capability for ultrasonic study of cylindrical interfaces.

INTRODUCTION

In materials applications a wide variety of interfacial discontinuities occur: e.g., cracks, laminates, coatings, etc. Nondestructive techniques for characterizing these interfaces need to be developed, and the present investigation was undertaken to determine whether ultrasonic techniques could provide a method for evaluating stress levels at clamped interfaces. Theoretical analyses by Haines [1] and Thompson et al. [2] treat such interfaces as having a microscale roughness so that the intimacy of contact between surface asperities is stress sensitive. In essence, these models predict that a stress increase results in an increase in area of efficient sonic contact so that, with increasing stress, a greater fraction of an incident sonic signal is transmitted and a lesser fraction is reflected. The Haines and Thompson approaches are in complete agreement but differ in notation so that the symbolism associates materials parameters in different combinations, albeit with the same overall effect. In the present state of development both approaches treat material on either side of an interface as being the same substance.

The Haines' formalism results in the following expressions for the transmission and reflection coefficients:

$$T = 2/[2 + \frac{i\alpha\bar{r}f}{S/p_m}] \qquad (1)$$

$$R = -[\frac{i\alpha\bar{r}f}{S/p_m}]/[2 + \frac{i\alpha\bar{r}f}{S/p_m}] \qquad (2)$$

In these expressions S/p_m represents the normalized applied stress with p_m being the pressure associated with lateral flow of the material at the interface and relating linearly the area of sonic contact to the applied stress, f is the frequency, $\bar{r}$ is the average radius of the regions of contact and depends upon surface roughness as well as upon the normalized applied stress, α is a composite of material parameters including Young's modulus and acoustic impedance, and i is the imaginary root of -1. It is readily apparent that the requisite conditions for energy conservation

$$|T|^2 + |R|^2 = 1 \qquad (3)$$

and for amplitude conservation

$$T - R = 1 \qquad (4)$$

are met at the interface. These expressions predict that the fraction of an ultrasonic signal which is reflected from an interface should (a) decrease with increasing stress, (b) increase with increasing frequency, and (c) increase with increasing surface roughness. The Thompson formalism results in the same predictions.

For the present investigation, clamping stresses were generated by the martensitic 'shape memory' transformation [3] in Nitinol (NiTi) couplers. These couplers were fabricated in the form of cylindrical sleeves by the Raychem Corp. and contained a small percentage of iron so that the transformation temperature was near -55°F. A longitudinal cross section through a coupler which has been cut in half is shown in Fig. 1; four raised lands may be seen and these are the loci of the maximum clamping stresses. The pattern of usage for the couplers follows the general pattern: (a) fabrication at ambient temperature to desired dimensions, (b) refrigeration to temperatures below the martensitic transformation temperature followed by deformation to produce transverse expansion, (c) storage at low temperature until time for installation, and (d) installation around tubing to be joined where the transformation during the return to room temperature causes reversion to the fabricated shape and produces the desired clamping stress.

Fig. 1. Longitudinal cut through a coupler showing the positions of the four lands.

PROCEDURE

In practice the inside of the clamped tubing is not accessible so it was decided to look at the reflection characteristics of clamped interfaces rather than transmission characteristics. Ultrasonic measurements were made in water tanks with specimens in which short sections of 316 stainless steel tubing were clamped in Nitinol couplers; water was excluded from the inner surfaces of the tubing by plugging the free ends of the clamped tubes. A focussed x-cut quartz transducer was placed so that an ultrasonic pulse of longitudinal waves could be impinged at normal incidence to the cylindrical surface of a coupler. The transducer could be translated along the length at constant distance from the coupler. Real time echo patterns from three characteristic positions along such translational traces are shown in Fig. 2. In this figure, echo patterns are from position A which is midway between two lands, from position B which is immediately adjacent to a land, and from position C which is a land position. These correspond to the A, B, and C positions in the sketch of a loaded coupler in Fig. 3.

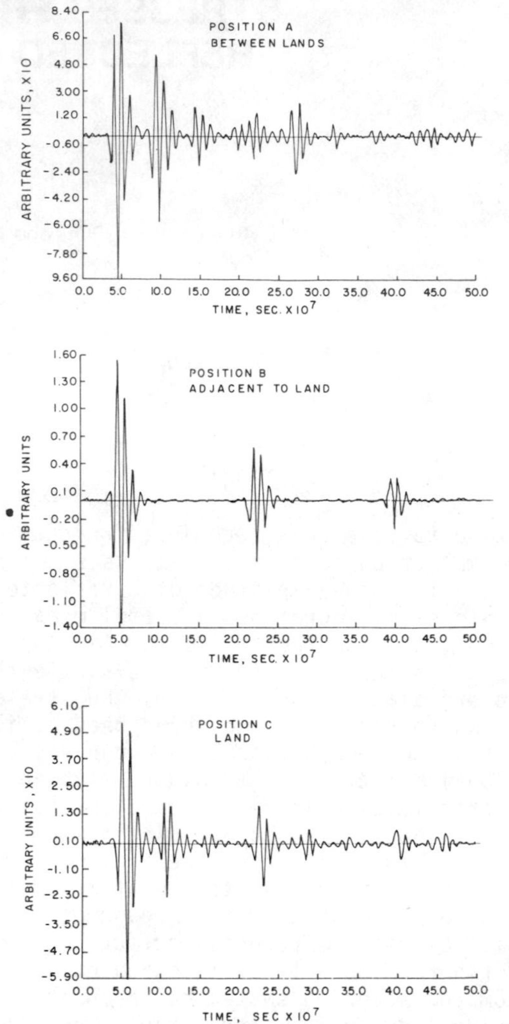

Fig. 2. Ultrasonic echo patterns from area between lands (Position A), adjacent to land (Position B), and on a land (Position C). Position B illustrates sonic disbond between the coupler and tubing.

Examination of the echo patterns shows that position B has an ordinate scale which differs by a factor of ten from the scales of the other two positions. It may also be noted that the echo pattern from position B has far fewer echoes than either position A or position C. These echoes correspond to first, second, and third reflections from the inner surface of the coupler with zero transmission into the 316 stainless steel tubing. Such a sonic disbond requires total reflection and accounts for the high echo amplitudes from this position. X-radiographs [4] clearly show that indentation by the lands produces pronounced warpage of the outer surface of the 316 stainless steel tubing on either side of the land. The existence of these disbonds is useful in that they provide an

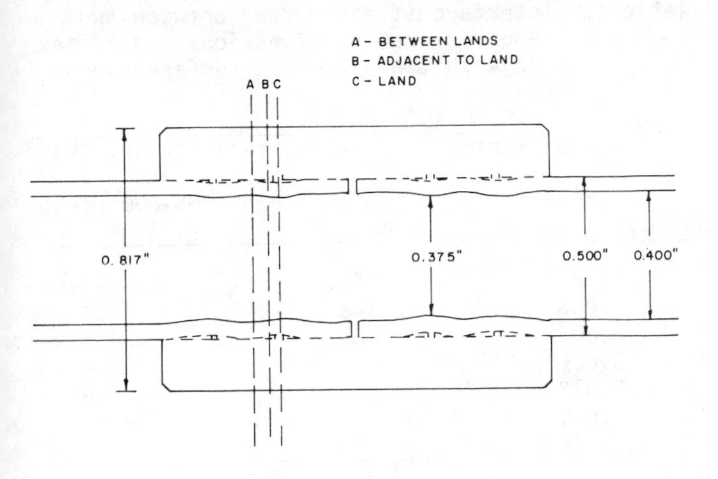

Fig. 3. Sketch of a loaded coupler using 'as received' tubing. Note the locations of Positions A, B, and C.

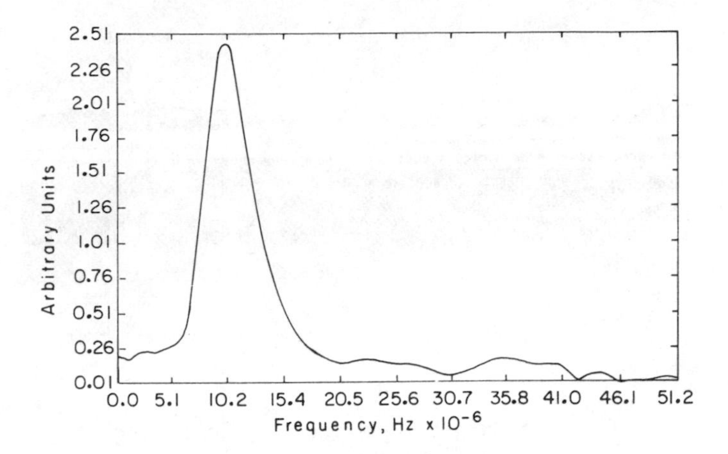

Fig. 4. Frequency spectrum from the Fourier transform of the first reflection at the disbond Position B.

internal standard for calibrating echo amplitudes. At positions A and C, the first echoes correspond to reflections from the coupler-tubing interface, the smaller second echoes correspond to reflections from the inside wall of the coupled tubing, and successive echoes correspond to a variety of longer sonic paths. Close examination shows that the first echo from position A has the least transit time, position B an intermediate transit time, and position C the greatest transit time. This is in accord with expected differences in path length due to the differing thicknesses of the coupler at these points. Interest has focussed on the first echo and, for utilization, this echo was in all cases transformed by Fourier analysis [5]. A typical plot of the frequency spectrum resulting from such a transformation is shown in Fig. 4. This spectrum shows a maximum at approximately 10 MHz with a bandwidth of about 5.6 MHz. The frequency at maximum and the bandwidth have been found to be relatively insensitive to either stress level or surface condition but, in contrast, the amplitude at the maximum has been found to correlate strongly with interfacial stress level and surface roughness so as to provide a useful gauge.

That a clamping stress is exerted by the couplers is evident from the following dimensional information. A sampling of six couplers were brought through their martensitic transformations to room temperature with no tubing inserted. Micrometer measurements showed that, regardless of angular or longitudinal position, the outside diameters of all couplers were consistent at 0.8154±0.0001". Inside diameters between successive pairs of lands were between 0.4830" and 0.4840" with an average of 0.4833". At the lands themselves, inside diameters were found to vary from 0.4610" to 0.4640" with an average value of 0.4615". For comparison with these inside diameters, the outside diameter of the 316 stainless steel tubing in the 'as received' condition was 0.502" with a wall thickness of 0.050". Thus the inside diameters of the transformed couplers and the outside diameter of the inserted tubing are incommensurate so that a hoop stress must exist when tubing is inserted into a coupler.

Stress levels were varied in a series of coupler-tubing combinations by reducing the outside diameter of the 'as received' tubing by centerless grinding to produce an initial series with a variety of undersizes ranging down to 0.019" undersize and with short range RMS surface roughness of 15 μinch as measured by a DECTAC profilometer. Matched pairs of 'as received' or undersized tubing were then clamped together with Nitinol couplers. The inside diameters of the coupled tubing were then measured along distance increments from the end of the coupler to the center; these measurements were made with the strain gauge probe which is shown in Fig. 5. The measured contours of the inside diameters are shown for the complete series of tubes in Fig. 6; the plots end at the center of the coupler because of the mirror symmetry at that locus.

Stress levels at the coupler-tubing interfaces were evaluated with a computer program based upon a two-dimensional point by point loading with an elastic-plastic model [6]. The measured inside diameters of the loaded couplers provided the boundary condition for determining the stress limit. The program was loaded with the appropriate elastic moduli and yield strengths, and these were such that the deformation in the Nitinol couplers were minimal with their outside diameters remaining essentially fixed. The program then simulated the inward contraction of the Nitinol couplers by incrementing the stress levels and computing the deformation in the enclosed tubing. The incrementing was repeated until the deformation in the tubing resulted in

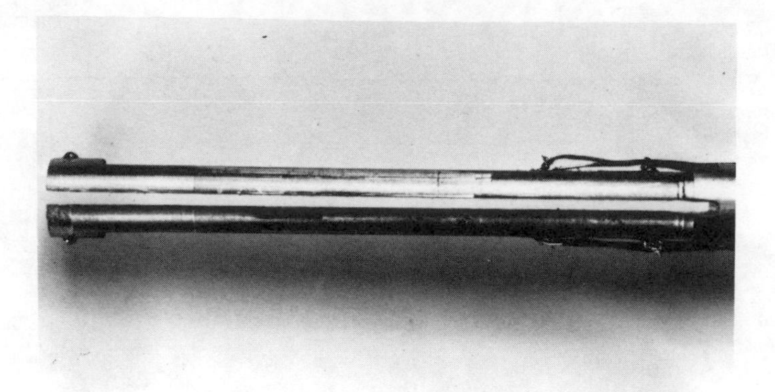

Fig. 5. Strain gauge device for measuring inside diameters of coupled tubing.

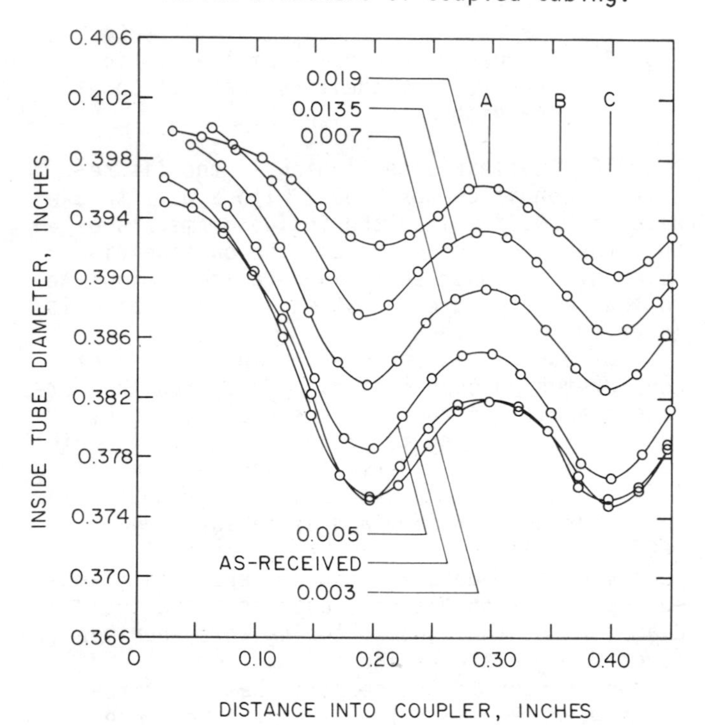

Fig. 6. Inside diameters of tubing in loaded condition with varying initial outside diameters of 15 μinch roughness.

a contour for the inside diameter which matched the experimentally determined contour. The computed interfacial stress levels for this condition were then assumed to be a good approximation for the actual interfacial stress levels.

This whole process was repeated for a second set of coupler-tubing combinations with one significant difference. This difference was that, after the centerless grinding, the tubing was further abraded with 80 grit sandpaper to increase the surface roughness. This second set of tubing had a short range RMS roughness of 600 μinch. The computed interfacial stresses for the position midway between two lands (position A) are tabulated for both coupler-tubing sets in Table I.

Table I. Interface stress midway between two lands for tubing of various undersizes and with two different surface roughnesses.

Tubing Undersizing	Interface Stress, (ksi)	
inches	RMS=15 μinch	RMS=600 μinch
As-Rec	45.5	----
0.003	48.4	43.1
0.005	38.1	34.5
0.007	27.7	22.4
0.013	13.4	5.9
0.019	8.3	----

RESULTS

Echo patterns were taken from each coupler in the two sets of coupler-tubing combinations. In each case the echoes were Fourier transformed and normalized with respect to the echo at the sonic disbond (position B), $R = A_I/A_{I,disbond}$. Such normalization compensates for any variation in experimental parameters which arise from time variations in circuit response, sample mounting, etc. It was found that the normalized reflection coefficients at the land positions (position C) were, in agreement with the comparatively small variation in the computed stress levels, relatively insensitive to the undersized tubing. In contrast the normalized reflections from positions midway between the lands (position A) showed a pronounced and consistent reduction in am-

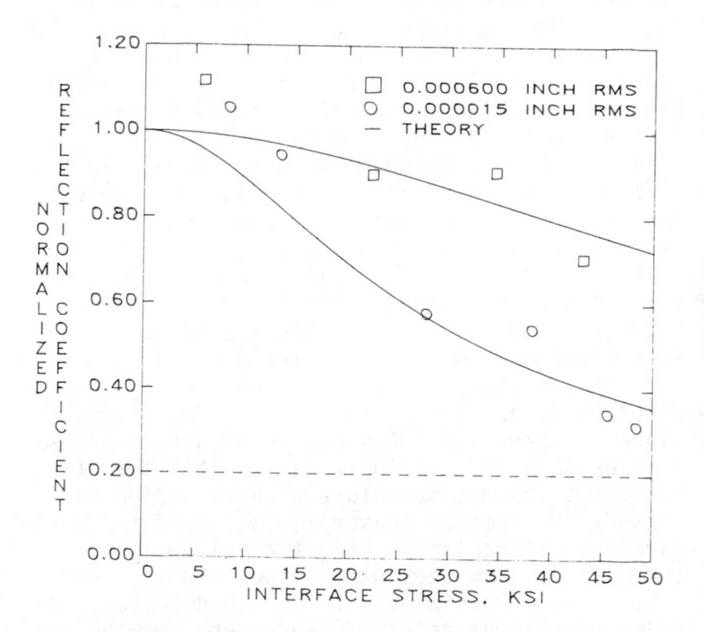

Fig. 7. Normalized reflection coefficients vs. calculated interfacial stresses for the two sets of tubing with individual points representing experimental data and with solid lines indicating theoretically predicted trends.

plitude with increasing undersize of the coupled tubing. Normalized reflection coefficients from position A are plotted against computed stress levels for the two sets of couplers in Fig. 7. In these figures the solid lines represent reflection coefficient vs. stress as implied from Eq. 2 with the composite product of the parameters $\alpha\bar{r}fp_m$ being evaluated for each set from the experimental echo amplitude from the 0.005" undersized tubing. The dashed line toward the bottom of the figures is the inherent reflection coefficient that would remain due to the difference in acoustic impedance [7] between Nitinol and stainless steel even if there were complete atom to atom contact throughout the interface. The dif-

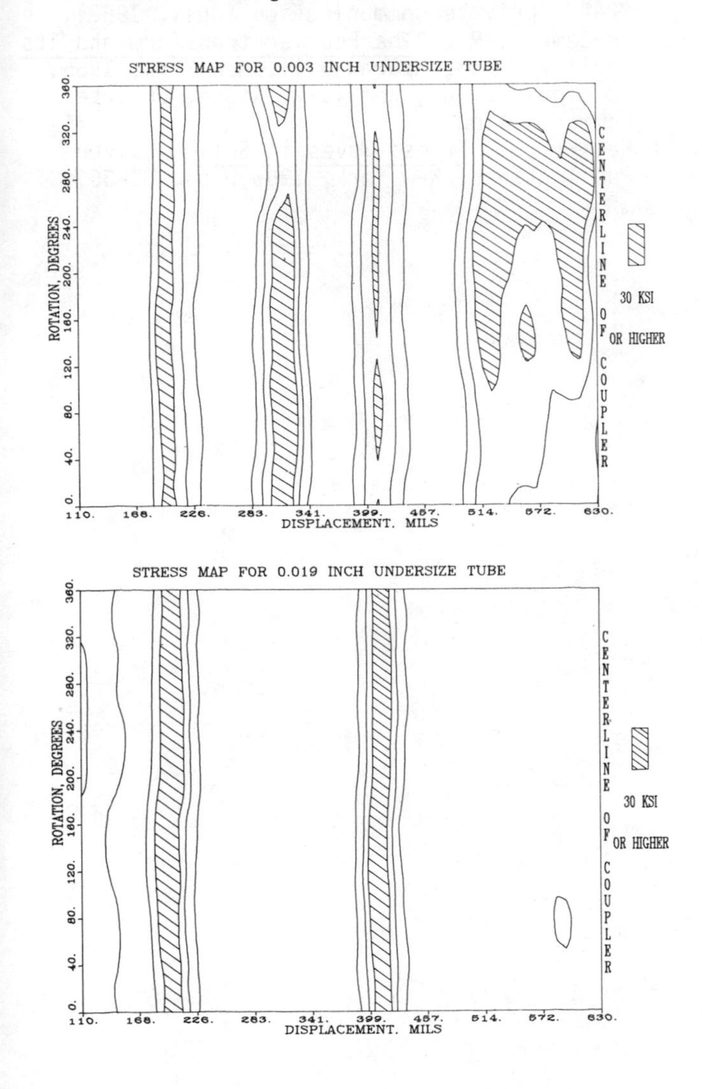

Fig. 8. Comparison of stress contours for one-half the complete tubing-coupler interface for 0.003" and 0.019" undersize tubing, both with 15 μinch surface roughness and in the loaded condition. Stress contours are plotted in multiples of 10 ksi. Note that the land positions in both plots are quite evident but the stress level between lands for the 0.019" undersize tubing has all but vanished.

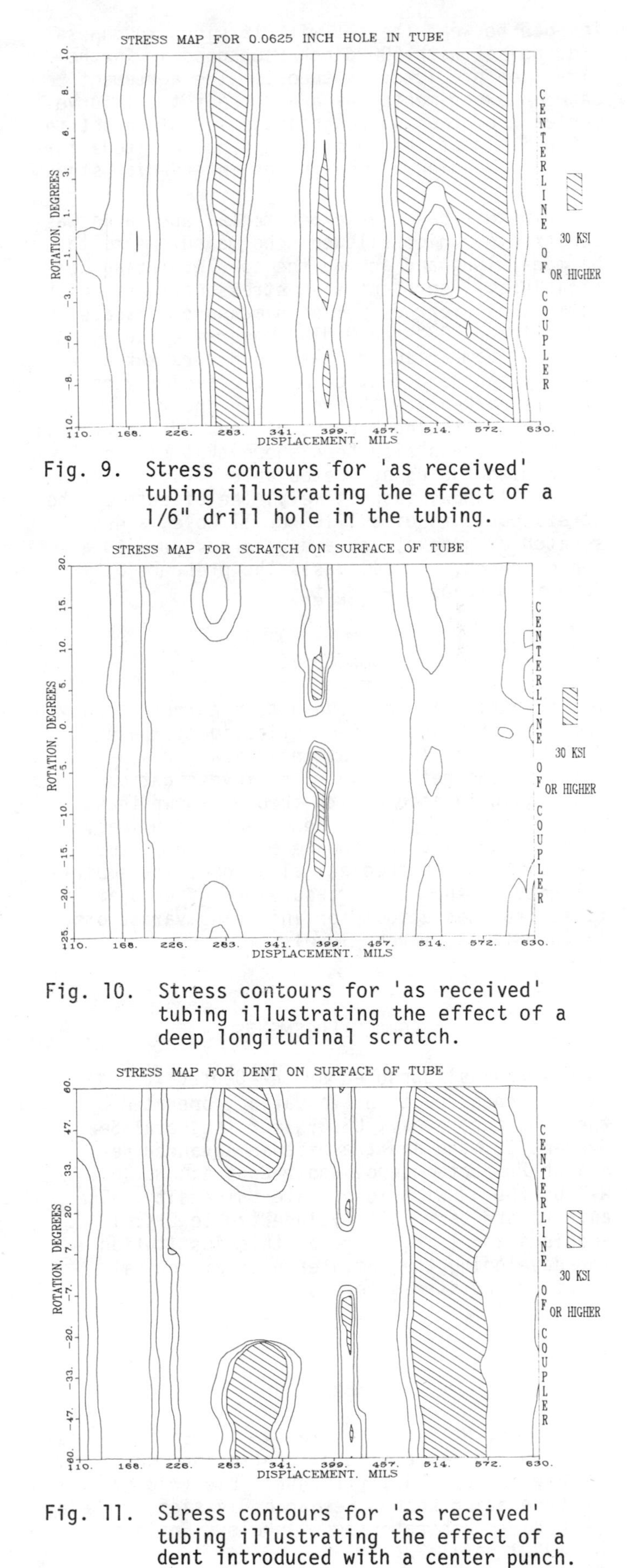

Fig. 9. Stress contours for 'as received' tubing illustrating the effect of a 1/6" drill hole in the tubing.

Fig. 10. Stress contours for 'as received' tubing illustrating the effect of a deep longitudinal scratch.

Fig. 11. Stress contours for 'as received' tubing illustrating the effect of a dent introduced with a center punch.

ference between the set with 15 μinch roughness and the set with 600 μinch roughness is in the direction predicted by theory. The agreement between experimental data and theoretical curve for each of the two roughnesses is sufficient to establish that the normalized echo amplitude furnishes a meaningful measure of interfacial stress level.

On this basis an experimental apparatus was constructed which allowed echo measurements to be made along and around the coupler-tubing combinations. Representative stress contours which have been inferred from such measurements are shown for 0.003" and 0.019" undersize tubing in Fig. 8. Comparison of these contours makes it readily apparent that clamping efficiency decreases markedly with increasing undersize of the coupled tubing because of reduced interfacial stress. It has also been shown that hidden flaws in coupled tubing are detectable. Figures 9, 10 and 11 show, respectively, the effects on the stress contours of a 1/16" drill hole, a deep scratch or groove, and a dent produced with a center punch. In all cases the presence of the defect is readily evident.

CONCLUSION

The investigation has shown that current theory predicts at least semi-quantitatively the effect of stress levels at interfaces on amplitude of echo reflection. The investigation has, again in accord with theory, shown that roughness has a pronounced effect on intimacy of sonic contact and thence on echo amplitude. Even without the theoretical accord, the empirical results show that measurement of echo amplitudes provides a tool for inferring variations in interfacial stress levels.

ACKNOWLEDGEMENTS

This work was sponsored by the Center for Advanced Nondestructive Evaluation, operated by the Ames Laboratory, USDOE, for the Naval Sea Systems Command and the Defense Advanced Research Projects Agency under Contract No. W-7405-ENG-82 with Iowa State University. The authors are particularly indebted to Drs. T. J. Rudolphi and T. R. Rogge of this institution for developing the computer program for calculating the stress levels.

REFERENCES

1. Haines, N. F., "The theory of sound transmission and reflection at contacting surfaces", Report RD/B/N4711, (Central Electricity Generating Board, Research Division, Berkeley Nuclear Laboratories, Berkeley, England, 1980) pp. 1-22.
2. Thompson, R. B., B. J. Skillings, L. W. Zachary, L. W. Schmerr, and O. Buck, "Effects of crack closure on ultrasonic transmission", Review of Progress in Quantitative Nondestructive Evaluation 2A, D. O. Thompson and D. E. Chimenti, Eds. (Plenum Press, New York, 1983), pp. 325-343.
3. Jackson, C. M., H. J. Wagner, and R. J. Wasilewski, "55-Nitinol--The alloy with a memory: Its physical metallurgy, properties, and applications", Report NASA-SP 5110 (National Aeronautics and Space Administration, Washington, DC, 1972)), pp. 1-35.
4. Goff, J. F., Head, Materials Division, Naval Surface Weapons Center, Dahlgren, Virginia 22448, private communication (July, 1982).
5. Bracewell, R., "The Fourier transform and its applications", McGraw-Hill, New York, 1965.
6. Rudolphi, T. J., and T. R. Rogge, private communication.
7. Kolsky, H., Stress Waves in Solids (Dover Publications, New York, 1963), pp. 31-36.

SOLVING RESIDUAL STRESS MEASUREMENT PROBLEMS BY A NEW NONDESTRUCTIVE MAGNETIC METHOD

Kirsti Tiitto
American Stress Technologies, Inc.
Pittsburgh, PA. 15242

ABSTRACT

Nondestructive measurement of residual stresses has conventionally been restricted to the utilization of x-ray diffraction method that yields absolute stress values under certain physical conditions. This method is, however, accompanied by a number of drawbacks such as shallow depth of measurement and high cost. It is also very difficult to use it to the continuous monitoring of residual stresses in industrial processes.

In this paper, magnetoelastic interaction was used to evaluate the residual stresses through analyzing the statistics of magnetization transitions (Barkhausen noise). These transitions were excited by a controlled magnetic field and detected electromagnetically. The average amplitude of the signal generated by the transitions was used to characterize the transition statistics. It has been shown that this amplitude is very sensitive to the residual stresses of the material and can be applied to the nondestructive elastic stress measurements in ferromagnetic materials.

This paper will present the results of residual stress measurements using the above principle in several practical applications. The results show the potential of this method both in laboratory for static measurements and in process control for dynamic, continuous monitoring of the stress level. The advantages as well as the disadvantages of this method over the conventional ones are discussed.

INTRODUCTION

Residual stress measurement by utilizing the magnetoelastic method is based on the principle of magnetoelastic interaction between magnetostrictive and elastic lattice strains. Shortly described this appears as follows. If a piece of ferromagnetic steel is magnetized, it will elongate in the direction of the applied magnetic field; and conversely, if the same piece is stretched by an applied load, it will be magnetized in the direction of the load. The same occurs with compression, except that the resulting magnetization now occurs at 90 degrees to the direction of compressive load.

To conveniently employ the magnetoelastic interaction for stress investigation, another physical phenomenon is needed. This is the Barkhausen effect[1], the series of abrupt changes or jumps in the magnetization of a steel when the magnetizing field is gradually altered. This procedure is further illustrated in Fig. 1. The triangular magnetizing field applied to the specimen forces its magnetic induction to change along the hysteresis loop in small jumps, whereby the magnetic noise to be analyzed is generated.

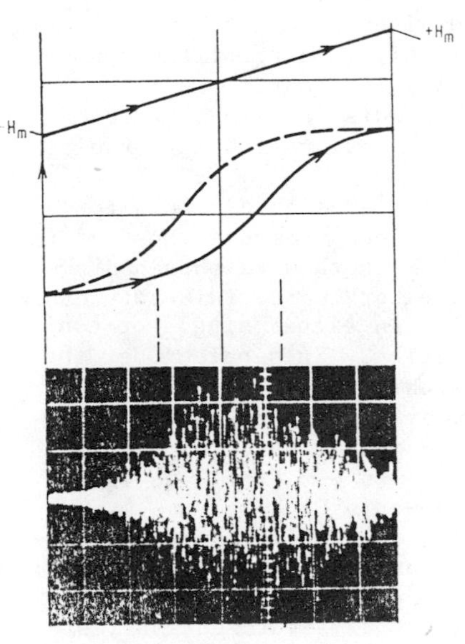

Fig. 1 - Applied magnetizing field, specimen magnetic induction and magnetic noise generated in the sensor for one magnetizing halfcycle.[2]

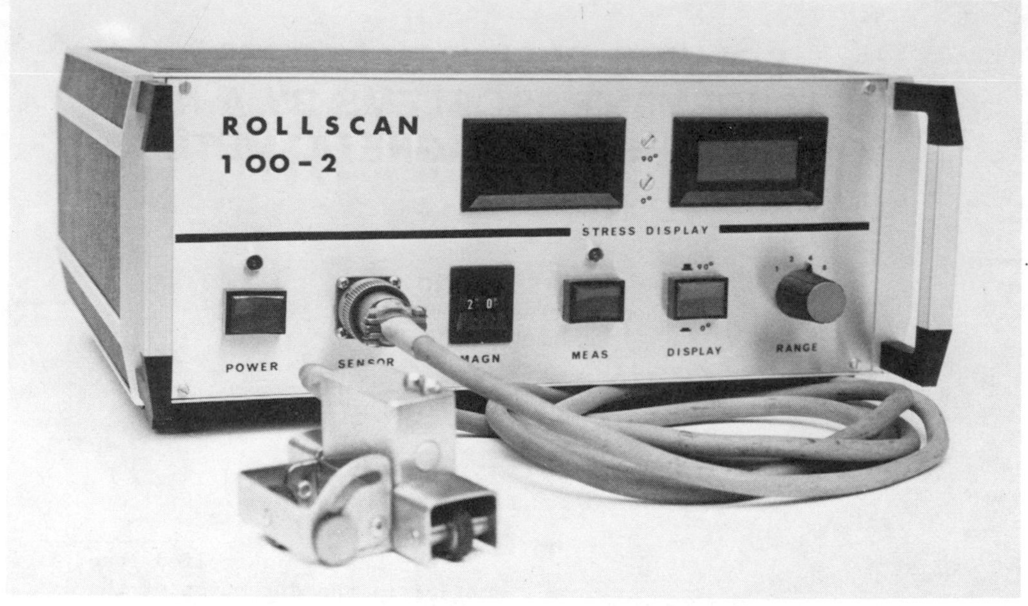

Fig. 2 - The instrument used for dynamic measurements.

Combining these two phenomena leads to a qualitative stress indication so that an increasing tensile stress is accompanied by a growing "Barkhausen noise" level, and an increasing compressive stress by a diminishing noise level[3,4]. More accurate quantitative data can be obtained through the use of calibration procedures with known loads.

EXPERIMENTAL PROCEDURE

INSTRUMENTATION - Commercially available instrumentation was used in this work. It includes a sensor to excite and detect the magnetic noise, and a central unit to control the sensor and process the signal. The central unit automatically evaluates and displays in a numerical form the measurement result, called magnetoelastic parameter here.

The instrumentation available allows for both static and dynamic measurements. Most of the results reported in this work were obtained by using the static laboratory unit. This unit can measure point by point in either single or continuous mode of operation. In single mode, the result is obtained within four seconds. For some measurements, the dynamic model equipped with a sensor that allows the measurements on a moving surface up to speeds 1 m/s (~3 ft/s) was used. This instrument is shown in Fig. 2.

The measurements were conducted nondestructively on the sample surface without any special surface preparation. Depending on the sensor configuration, the area of measurement varies from 1 mm^2 (0.0016 $in.^2$) to 250 mm^2 (0.4 $in.^2$). The depth of measurement depends on the frequency range of noise signal and can hence be varied. For the range of 500-10000 Hz used in the standard instrument, it is 0.2 - 0.3 mm (0.008 - 0.012 in.). The so-called high frequency option was employed in order to study surface layers less than 0.025 mm (0.001 in.) in thickness.

CALIBRATION TESTS - The magnetoelastic parameter as measured in this work represents the stress level of the material in relative units. Since an absolute stress value is desirable in some applications, calibration tests were conducted to study the relation between this parameter and applied stress in different materials. Three steels with yield strength ranging from 248 N/mm^2 (36 ksi, designation SAE 1020) to 710 N/mm^2 (103 ksi, designation AISI 9310) were used in these tests. The calibration samples were typically 12 mm (0.5 in.) wide, 102 mm (4in.) long and 6 mm (0.25 in.) thick. They were first stress relieved at 650 C (1200 F) for 2 to 3 h and cooled slowly to room temperature before calibration. These test pieces were then strained in tension and compression within the elastic limit by using a tensile machine together with strain gauges. Simultaneously, the magnetoelastic parameter was recorded in the direction of the load. Depending on the operation mode of the measuring unit, the tests could be performed either by stepwise or continuously changing loads. At single type of operation, approximately 12 measurements were made in tension and 4 to 6 in compression.

WELD SEAM SAMPLES - T-section samples with two weld beads running parallel to each other were constructed. The material was SAE 1020 with a yield strength of 248 N/mm^2 (36 ksi). The plate thickness was 6.4 mm (0.25 in.) and width 57.2 mm (2.25 in.). The samples were 254 mm (10 in.) long.

The stresses were measured in the as-welded and stress relief annealed condition. The magnetoelastic parameter was evaluated by the static unit with 3 mm (0.12 in.) distance between the measurement points on the bottom surface of the samples. The direction of measurement was both perpendicular and parallel to the weld seams. To obtain the actual stress values of these weld seams, calibration was conducted.

ROLLING MILL ROLLS - Several forged steel back-up rolls (high carbon steel with Cr and Mo) were evaluated by using both the static and dynamic units. These rolls were 114 cm (45 in.) to 140 cm (55 in.) in diameter and approximately 102 cm (40 in.) long. Some of the rolls were new in a finish ground condition, some were used several times in the mill and redressed by various amounts between the mill campaigns.

With the static unit, the rolls were measured across the roll length at spacings of 50 mm (2 in.) in longitudinal and transverse directions to the roll axis. A whole traverse along the roll length was evaluated with the dynamic unit. Both longitudinal and transverse directions were measured simultaneously with a 2-channel unit and the results were recorded.

GRINDING BURN SAMPLES - Injector lobes (Cr-Mo alloy steel) of diesel engine camshafts were measured with the static unit coupled with a dynamic sensor and a recorder in order to detect grinding burns. These burns were generated by abusive grinding during the final machining. Since the lobe surfaces before grinding were induction hardened, they exhibit high compressive stresses which may be changed to tensile stresses by abusive grinding. To confirm the results, the burns were also revealed by nital etching.

RESULTS AND DISCUSSION

CALIBRATION CURVES - The calibration curves obtained for three different steels with yield strength from 248 N/mm^2 (36 ksi, ferrite-perlite structure) to 710 N/mm^2 (103 ksi, martensite structure) are shown in Fig. 3. It can be seen that the curves are close to linear from approx. one third of yield stress in compression to approx. two thirds of yield stress in tension. Close to the yield stress, each of the calibration curves saturates. This phenomenon is especially pronounced for the steel with highest yield strength.

The calibration curves of Fig. 3 show that the magnetoelastic parameter is very sensitive to a change in stress. The experimental curves can now be used to convert the magnetoelastic parameter into actual stress values in materials under inspection.

RESIDUAL STRESSES OF WELD SEAMS - The calibrated residual stresses of the T-section samples in the as-welded and stress relieved condition are shown in Figs. 4a and b. It was found that the as-welded sample has high values of the magnetoelastic parameter and hence high tensile stresses close to the yield stress parallel to one of the weld seams, Fig. 4a. The other seam has lower parallel stresses; it was most probably welded first and "stress relieved" during the welding of the second seam.

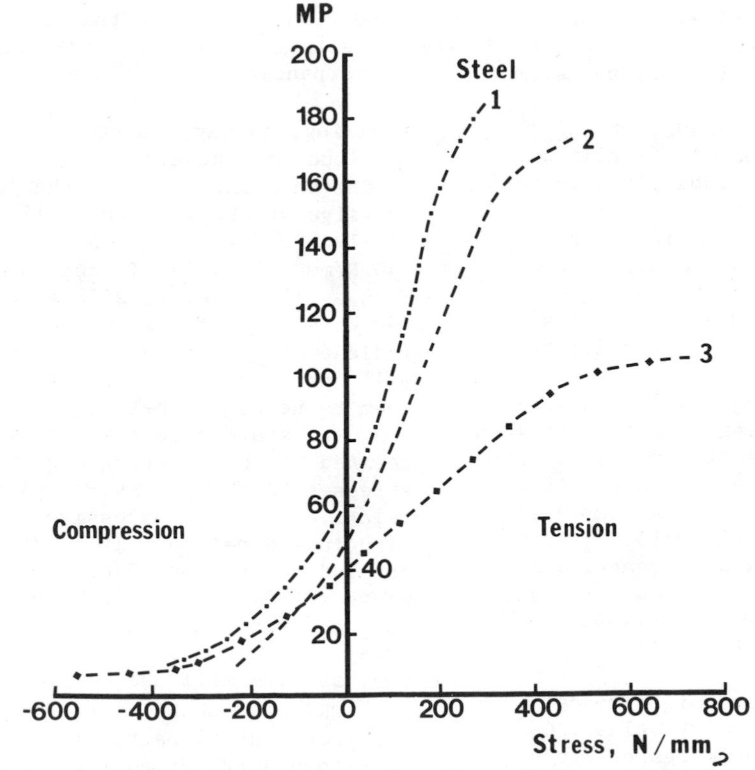

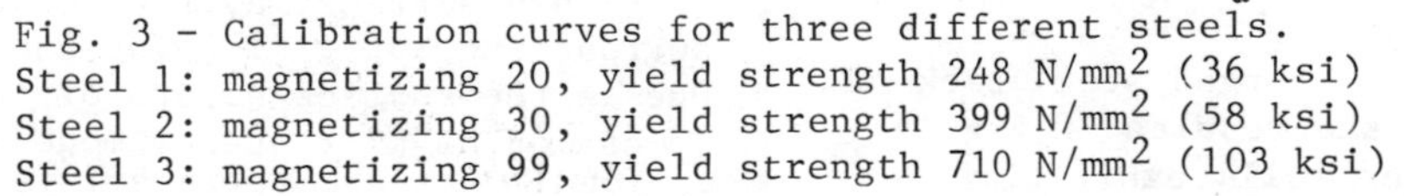
Fig. 3 - Calibration curves for three different steels.
Steel 1: magnetizing 20, yield strength 248 N/mm^2 (36 ksi)
Steel 2: magnetizing 30, yield strength 399 N/mm^2 (58 ksi)
Steel 3: magnetizing 99, yield strength 710 N/mm^2 (103 ksi)

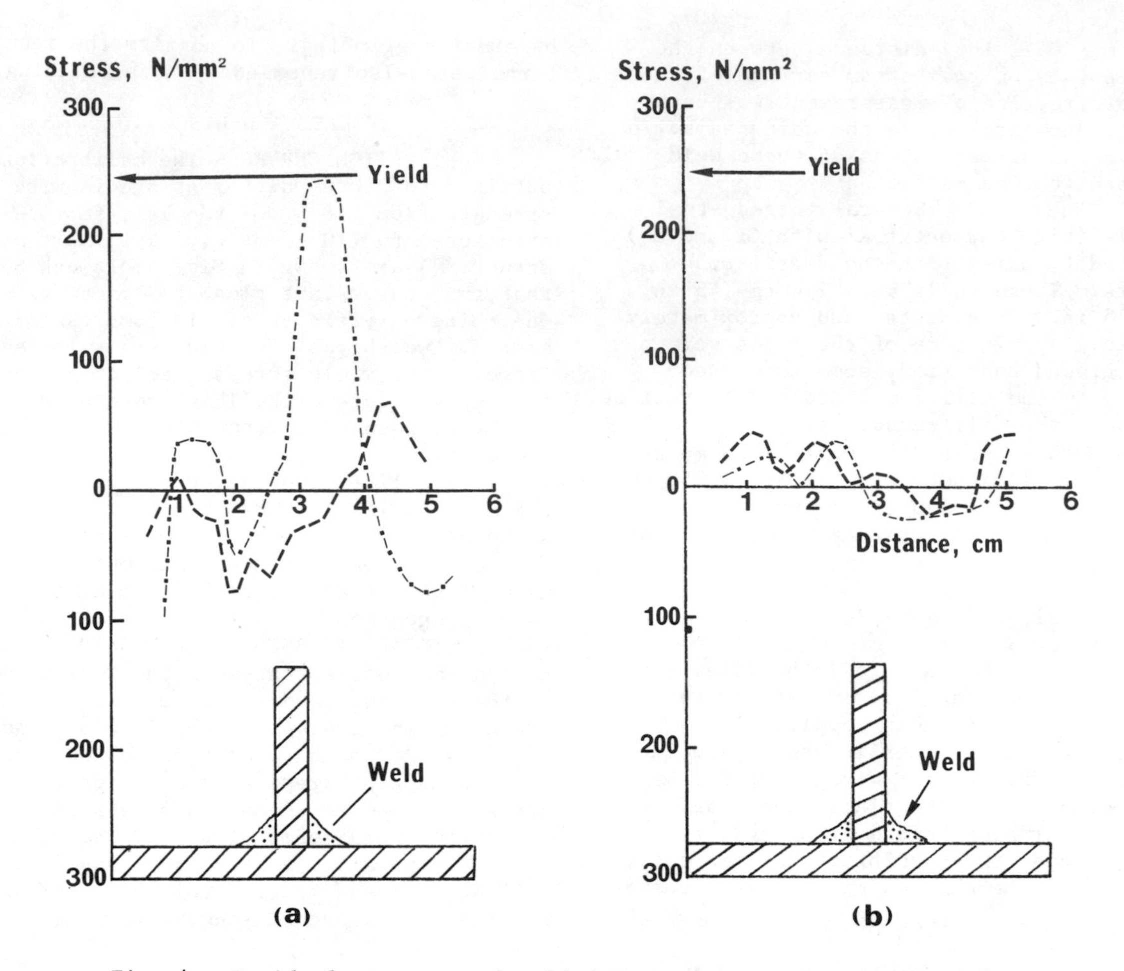

Fig. 4 - Residual stresses of welded T-section samples in the (a) as-welded and (b) stress relief annealed condition. Material : SAE 1020.
-·-·- parallel to the seams --- perpendicular to the seams

During stress relief annealing, the high tensile stresses mentioned above were eliminated, and the stress levels appear to be close to zero in both directions, Fig. 4b.

SPALLING OF BACK-UP ROLLS - Fig. 5 shows results of measurements on a good forged steel back-up roll (1) successfully used in the mill, on a new roll (2) already showing big stress variations and on a used roll (3) exhibiting severe spalling.

It was found that the magnetoelastic parameter MP for roll 1 after turning operation is less than about 15 in both the transverse and longitudinal directions. According to the calibration data obtained, this corresponds to a compressive stress higher than 550 N/mm^2 (80 ksi). After mill use slightly higher MP values were measured on roll 1, i.e. the compressive stresses especially in the transverse direction were decreased. The lowest compressive stress found on roll 1 was 275 N/mm^2 (40 ksi).

Roll 2 in Fig. 5 is a new roll already showing large variations of the MP values. On some locations, the transverse stresses are as low as 6.9 N/mm^2 (10 ksi).

During testing several forged steel back-up rolls, it was found that the results in Fig. 5 for roll 3 are typical for rolls exhibiting spalling, namely a large difference between the longitudinal and transverse MP values. According to the calibration data, the longitudinal compressive stresses on the roll surface are relatively high, ranging from 410 N/mm^2 (60 ksi) to 240 N/mm^2 (35 ksi). In contrast, MP values as high as 130, i.e. tensile stresses of 410 N/mm^2 (60 ksi), were found in the transverse direction indicating a drastic change in stress during the mill operation. Also, the transverse stresses seem to be highly heterogeneous.

The above results suggest that spalling is related to the occurrence of high tensile stresses in the transverse direction and perhaps relatively high compressive stresses in the longitudinal direction. It is difficult, however, to conclude on the basis of the existing data whether the increase in MP values associated with the occurrence of spalling is brought about by erratic stress distributions in the roll before use, abusive mill use, or both.

GRINDING BURNS - Grinding burns in hardened fabricated steel parts typically consist of a very thin surface layer under tension, the surrounding material being under compression. Due to the surface nature of the defect, the so-called high frequency mode was used. This option limits the surface layer evaluated to a

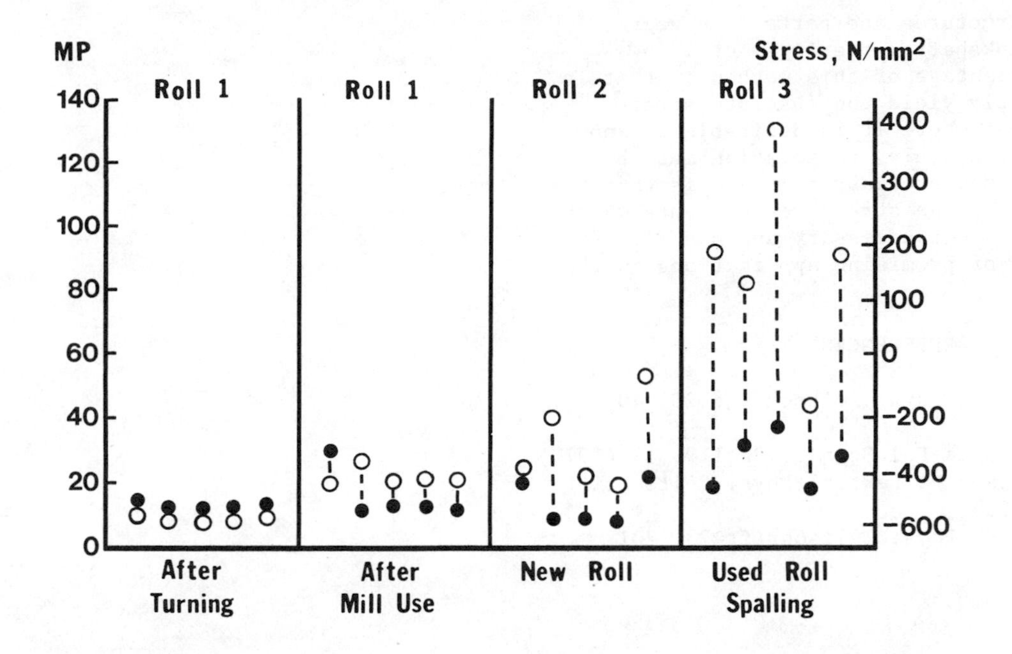

Fig. 5 - Magnetoelastic parameter (MP) values and the corresponding calibrated residual stresses on the surface of three forged steel rolling mill rolls in different conditions. o transverse stresses, ● longitudinal stress.

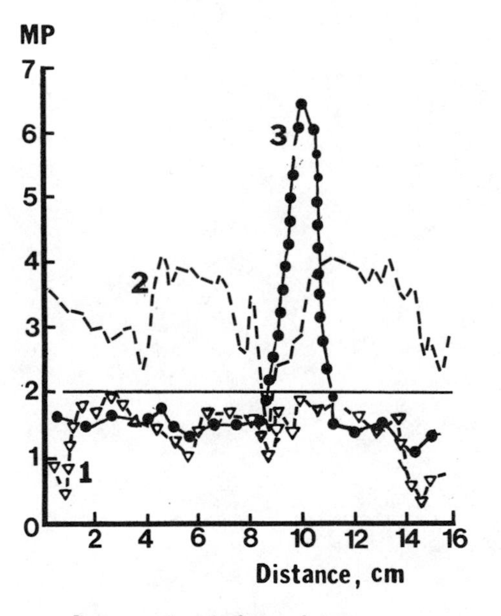

Areas of grinding burns:

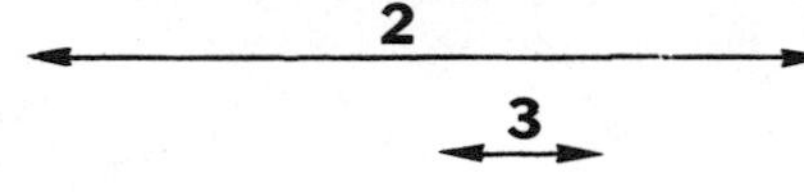

Fig. 6 - Evaluation of grinding burns on diesel engine camshafts.
The line drawn at MP=2 separates sound material (MP<2) from grinding burn areas (MP>2).
Sample 1: no grinding burns
Sample 2: grinding burns nearly all over the injector lobe
Sample 3: severe local grinding burn

thickness less than 0.025 mm (0.001 in.).

The results given in Fig. 6 were obtained in a continuous, dynamic mode of measurement of the static unit coupled with a dynamic sensor and chart recorder. Among the data established, the grinding burns of Fig. 6 were selected to represent relatively severe burns on the injector lobes of a diesel engine camshaft. The signal intensities are approximately twice (curve 2) and four times (curve 3) of that obtained from the sound material (curve 1). These burns could readily be observed also by chemical etching method.

The above results suggest that the magnetoelastic high frequency measurement technique provides a NDT method for automatic inspection of grinding burns.

CONCLUSIONS

The results obtained in this work have shown that the magnetoelastic method can be used to evaluate (i) the residual stresses associated with weld seams in mild steels, (ii) the stress condition of forged steel back-up rolls and hence predict the occurrence of spalling, and (iii) grinding burns on the lobes of diesel engine camshafts. It was shown that this method can give a qualitative and in many cases a quantitative indication of the stress condition in ferromagnetic materials.

The main advantage of this method is that the measurements can be conducted nondestructively on both static and moving samples. Also, no special surface preparation is needed. Lightweight portable battery-driven units can easily be taken to the field. Special sensor constructions already available also enable the measurements on complicated structures and parts like gears, camshafts, crankshafts, bearings, etc.

The disadvantage of this method is that it does not directly yield the absolute stress values. In cases where it is desirable to know the stresses accurately, calibration must be conducted. The magnetoelastic method is thus best suited to comparative stress measurements where high accuracy is not necessary and has already found a number of promising applications in the industry.

REFERENCES

1. Barkhausen, H., Phys. Zeitschrift 20, 401 (1919)
2. Tiitto, S., Acta Pol. Scand., Ph 119, 19 (1977)
3. Gerlach, H. and P. Lertes, Phys. Zeitschrift 22, 568 (1921)
4. Zschiesche, K. Phys. Zeitschrift 23, 201 (1922)

IV . Secondary Processing and Deformation

MICROSTRUCTURAL EVALUATION OF A FERRITIC STAINLESS STEEL BY SMALL ANGLE NEUTRON SCATTERING

S. Kim
Dept. of Materials Science and Engineering
Northwestern University
Evanston, IL 60201

J. R. Weertman
Dept. of Materials Science and Engineering
Northwestern University
Evanston, IL 60201

S. Spooner
Solid State Div.
Oak Ridge National Laboratory
Oak Ridge, TN 37830

C. J. Glinka
Reactor Div.
National Bureau of Standards
Washington, D.C. 20234

V. Sikka
Metals and Ceramics Div.
Oak Ridge National Laboratory
Oak Ridge, TN 37830

W. B. Jones
Sandia National Laboratories
Albuquerque, NM 87185

ABSTRACT

A study is under way to evaluate the feasibility of using small angle neutron scattering (SANS) to detect microstructural changes which result from high temperature service and are likely to be associated with a degradation in strength. The material studied is Fe9Cr1Mo modified by the addition of small amounts of V and Nb. The major changes detected by the neutrons are alterations in size, number and composition of the carbides. SANS measurements of samples aged 5000 h at various temperatures showed that aging at 538°C produces a maximum in the scattering cross section. Aging at this temperature also maximizes hardness. Size distributions of the carbides derived from inversion of the SANS data show that 538° results in the largest number of small carbides. Higher aging temperatures lead to appreciable carbide coarsening. SANS measurements of samples deformed at 649°C showed that microstructural changes can be detected after only a few hours of fatigue. High voltage electron microscopy studies have shown that the major microstructural changes produced by high temperature exposure are carbide coarsening, subgrain development, and a drastic decrease in dislocation density.

INTRODUCTION

An alloy which is exposed to elevated temperatures for prolonged periods is liable to undergo microstructural alterations which degrade its mechanical behavior. Clearly it would be highly desirable to have a nondestructive method available to monitor these changes. Small angle neutron scattering (SANS) offers a number of advantages as a technique for carrying out such monitoring. Features like precipitates, carbides or voids in the size range from a few nanometers to about one micrometer can be detected by SANS. Because of the penetrating power of neutrons, specimens 5 to 10 mm in thickness can be investigated. The potential exists for extracting a large amount of information from a curve of scattering cross section vs scattering vector, e.g., the number and size distribution of the scattering entities. However engineering alloys frequently contain more than one type of scattering center, and the separation of the contribution of each to the scattering curve requires the use of a complementary technique such as TEM. Articles on SANS and its application to the study of metallurgical systems are given in references [1-3].

The present paper describes the use of SANS to investigate microstructural changes in a ferritic stainless steel produced by exposure to elevated temperatures. Both simple aging without loading and the added effects of deformation were considered.

EXPERIMENTAL DETAILS

The steel investigated in the present study is Fe9Cr1Mo modified by the addition of small amounts of the strong carbide formers V and Nb. This alloy has been proposed as a replacement for the austenitic steels 304 and 316 in certain power generation applications. An extensive effort to characterize this material is being carried out as part of the Advanced Alloy Program of Oak Ridge National Laboratory [4,5]. The chemical composition of the steel is given in Table 1.

TABLE 1. Chemical Composition

C	0.08 - 0.12	Ni	0.20 max.
Cr	8.00 - 9.50	P	0.020 max.
Mo	0.85 - 1.05	S	0.010 max.
V	0.18 - 0.25	N	0.030 - 0.070
Nb	0.06 - 0.10	Al	0.04 max.
Si	0.20 - 0.50	Fe	balance
Mn	0.30 - 0.60		

(wt. %)

All specimens in the present study were made from material which had undergone the same initial heat treatment: normalization for 1 h at 1038°C, air cooled to room temperature; tempering for 1 h at 760°C, air cooled to room temperature. The resultant structure is fully tempered martensite. Two series of specimens were investigated. In the first the steel had been aged 5000h under no load at one of several aging temperatures. The specimens in the second set had either been crept for an extended period at 649°C or fatigued at that temperature. Table 2 gives details of the test conditions.

TABLE 2. Sample Test Conditions.

Sample No.	Test Conditions
H 394	Normalized 1h, 1038°C, AC; Tempered 1h, 760°C, AC.
NOTE: All samples listed below were given this initial heat treatment.	
A 482	Aged at 482°C, 5000h
A 538	Aged at 538°C, 5000h
A 593	Aged at 593°C, 5000h
A 649	Aged at 649°C, 5000h
A 704	Aged at 704°C, 5000h
C 9	Crept at 649°C, 21028 h σ = 62 MPa
F 12	Fatigued at 649°C, $\Delta\varepsilon_T$=0.5% 10,000 cycles, no hold times
F 14	Fatigued at 649°C, $\Delta\varepsilon_T$=0.5% 7930 cycles, 30 s tension hold

The neutron scattering experiments were performed at Oak Ridge National Laboratory and at the National Bureau of Standards. The somewhat different characteristics of the two scattering instruments plus the use of a number of specimen-to-detector distances made it possible to measure the scattered intensity over a wide range of scattering vector. (The magnitude q of the scattering vector is defined as $q = 4\pi \sin \theta/\lambda$, where θ is the Bragg angle and λ is the neutron wave length.) Neutrons interact with matter through their magnetic moments, and therefore there are additional contributions to the scattering in the case of ferromagnetic specimens. Because of their comparatively large size, ferromagnetic domains are a source of scattering by multiple refraction. This unwanted complication was eliminated by magnetizing the specimens during the SANS measurements in a direction perpendicular to the neutron beam. That magnetic saturation of the specimens did indeed eliminate refraction effects was checked by comparing a curve of scattering cross section vs scattering vector obtained with neutrons of one wave length with the curve taken after changing the wave length to another value. If scattering occurs solely by diffraction, the scattering cross section $d\Sigma/d\Omega$ depends on the neutron wave length λ only through the scattering vector q and the two curves are identical, whereas if refractive effects are appreciable $d\Sigma/d\Omega$ depends on both q and λ [6]. A comparison of $d\Sigma/d\Omega$ vs q curves taken at two wave lengths showed that scattering from the magnetized Fe9Cr1Mo specimens was in fact controlled by diffraction. It is necessary to know $d\Sigma/d\Omega$ over all values of q, from zero to infinity, before the maximum amount of information can be extracted from the scattering curves. In the present experiments measurements of $d\Sigma/d\Omega$ typically extended over the q range $\sim 0.04 < q < 2.0\ nm^{-1}$. The missing data can be obtained by extrapolation of the measured curves at the low q end and at the high q end using known asymptotic forms of the scattering cross sections [1-3]. The volume fraction of the scattering centers then can be calculated from the integration of $d\Sigma/d\Omega$ over all of reciprocal space and their size distribution can be determined from an inversion of the scattering data.

RESULTS AND DISCUSSION

SOURCES OF SCATTERING - The ferritic stainless steel of the present study contains a number of features which may give rise to neutron scattering, e.g., carbides, voids, grain boundaries, surface imperfections, dislocations, magnetic domains. It has been shown that grain boundaries and surface imperfections make little contribution to the scattering intensity, especially in oxidation resistant material [7,8]. To verify that this conclusion is valid for the Fe9Cr1Mo steel, the scattering was measured from a control sample made from a steel similar to that used in the rest of the tests except that its carbon content was extremely low. The scattering from the carbide-free sample was subtracted from that of the other samples before the data were analyzed for carbide volume fractions, size distributions, etc. However the scattering from the control sample was so low that this subtraction procedure had little effect on the final results.

The modified Fe9Cr1Mo steel is highly resistant to void formation. Only under extreme conditions of fatigue cycling with a tension hold have grain boundary cavities been observed [9]. Extensive examination by HVEM of the specimens used in the present study failed to reveal the presence of any voids. Dislocations generally contribute only weakly to SANS (e.g., [10]).

The scattering from dislocations will be enhanced somewhat in the present case because of magnetoelastic coupling with the strain field of the dislocations.

As has been mentioned, domain scattering was largely eliminated by magnetizing the specimens during SANS measurements. Figure 1 shows the drop in the scattering cross section as the strength of the magnetic field is increased.

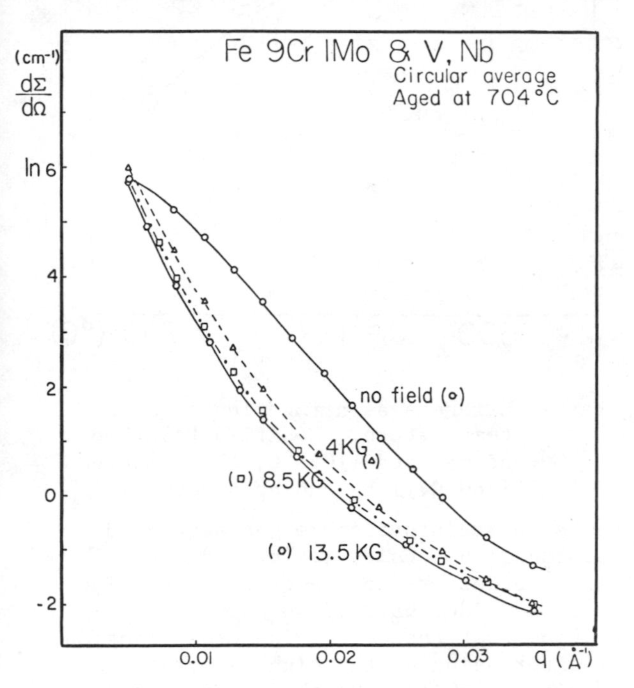

Figure 1. Dependence of $d\Sigma/d\Omega$ on strength of the magnetizing field. Specimen was modified Fe9Cr1Mo aged 5000 h at 704° C. Neutron wave length λ = 0.85 nm.

Inhomogeneities, which normally give rise to neutron scattering through the interaction of the neutron magnetic moment with the nuclear magnetic moment of the atoms in the samples, cause additional scattering in the case of ferromagnetic materials. This additional scattering varies as $\sin^2\alpha$, where α is the angle between the direction of magnetization of the specimen and the direction of the scattering vector $\vec{q}$ [1]. (The vector $\vec{q}$ is parallel to the change in direction of a neutron as the result of scattering.) This dependence of the scattering on α is illustrated by the isointensity curves in Figure 2. The direction of the magnetic field was horizontal, and it can be seen that the scattering is a maximum in the vertical direction. Figure 3 shows the dependence of $d\Sigma/d\Omega$ on q for scattering parallel and perpendicular to the direction of magnetization in two thermally aged specimens. In the analyses which follow the scattering intensities were radially averaged in order to improve the statistical reliability of the data.

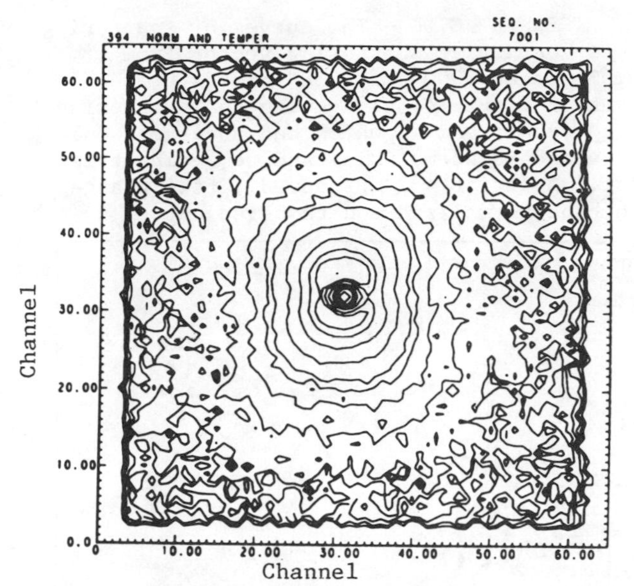

Figure 2. Isointensity curves for scattering from a magnetized specimen of normalized and tempered modified Fe9Cr1Mo. The direction of the ~ 28 KG magnetic field was horizontal. λ = 0.48 nm.

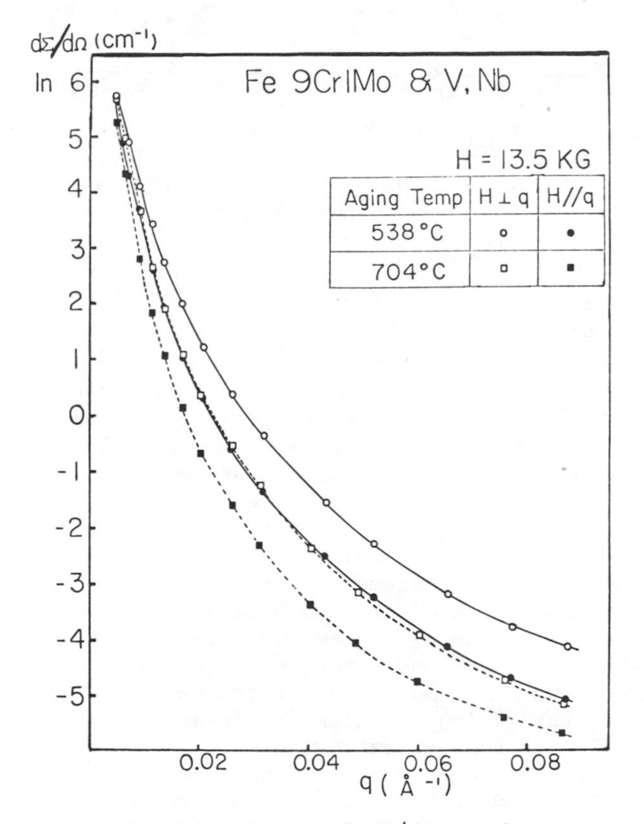

Figure 3. Dependence of $d\Sigma/d\Omega$ on the orientation of the scattering vector with respect to the magnetic field H for specimens of modified Fe9Cr1Mo aged 5000 h at 538° or 704°C. H = 13.5 KG; λ = 0.85 nm.

THERMAL AGING - The curves of scattering cross section vs scattering vector from samples aged 5000 h at 538° or 704°C are given in Figure 4, along with the scattering curve from a sample which had undergone only the standard normalizing and tempering treatment. Aging at 482° produced little change from the normalized and tempered curve.

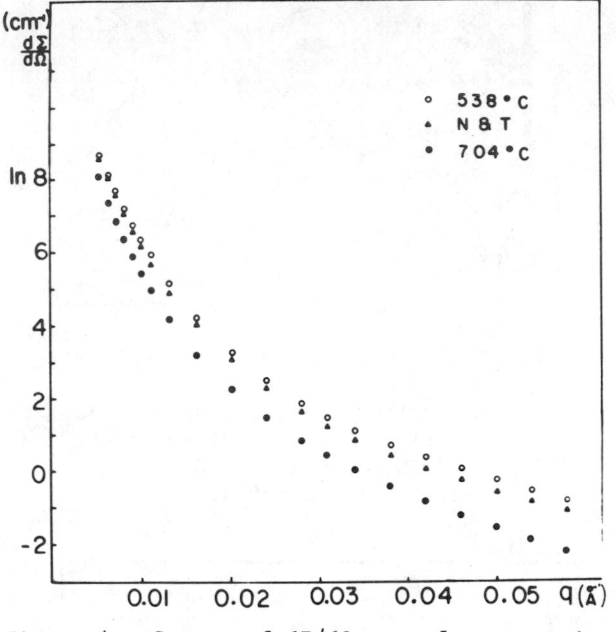

Figure 4. Curves of $d\Sigma/d\Omega$ vs q for magnetized specimens of modified Fe9Cr1Mo aged 5000 h at 538° or 704° C or given only initial normalization and tempering treatment. H = 4KG; λ = 0.48 nm. The scattering from the low carbon specimen has already been subtracted.

Scattering from samples aged at temperatures between 538° and 704° fell between the curves shown for these two temperatures. Scattering in this temperature range dropped monotonically with increasing aging temperature. In Fig. 4, the scattering from the low carbon control sample already had been subtracted. It is interesting to compare the SANS results with hardness measurements. As shown in Fig. 5, hardness after 5000 h of aging is a maximum when the aging temperature is 538°C. Sikka et al. [4] and Vitek and Klueh [11] have found that ∼ 538°C is the tempering temperature which produces the greatest hardness in one hour of tempering.

From the discussion of the preceding section it may be concluded that carbides are the major source of scattering in the modified Fe9Cr1Mo steel. Figure 6a shows an HVEM micrograph of the starting microstructure. The lath structure is clearly visible. Carbides lie along prior austenite and lath boundaries while smaller carbides are contained in the matrix. The changes produced by 5000 h of aging at 704°C are evident in Fig. 6b. A more nearly

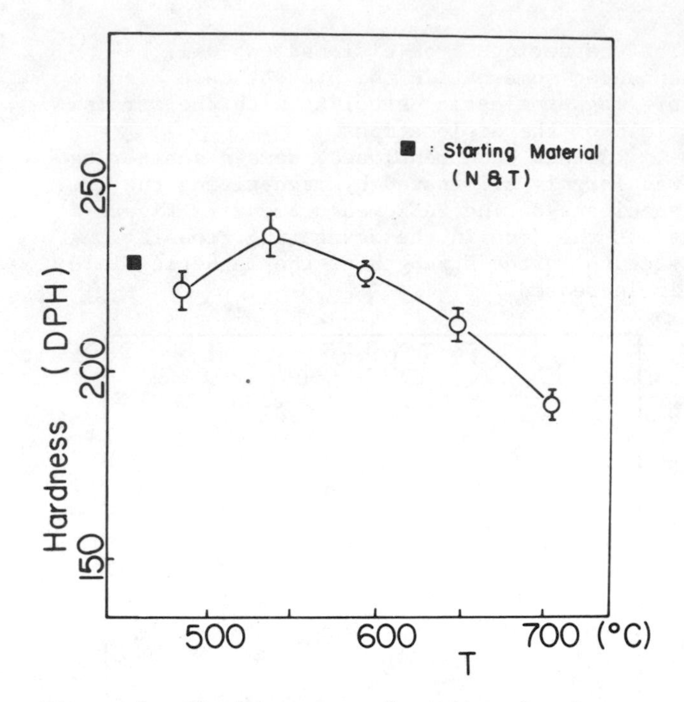

Figure 5. Hardness as a function of aging temperature. Modified Fe9Cr1Mo normalized (1038°C, 1h) tempered (760°C, 1 h) and aged 5000 h.

equiaxed subgrain structure has developed, the dislocation density is low and the matrix carbides are no longer evident. The boundary carbides have grown. It is not surprising that the scattering cross section drops after aging at the high temperatures. In previous work [11,12] the matrix carbides have been identified as MC type and the grain boundary carbides as $M_{23}C_6$. Using extraction techniques Jones [12] was able to study the evolution in the composition of the carbides with aging and with deformation at elevated temperatures. He found that aging ∼ 1000 h at 538°C produces little compositional change in the carbides. Most of the MC carbides are niobium rich while the $M_{23}C_6$ carbides are rich in chromium and iron. Vitek and Klueh [11] reported finding $V_xNb_{1-x}C$ carbides with x varying from 0 to 1. They found the average composition of the $M_{23}C_6$ carbides to be (in wt pct) 63 pct Cr, 29 pct Fe, 7 pct Mo. The initial tempering at 760°C produces a fine dispersion of MC carbides, but it is possible that prolonged additional aging at 538°C causes further precipitation of the small carbides. If so, the peaks in the hardness and in scattering cross sections after aging at 538°C have the same origin. To test this idea an analysis of the scattering data was carried out. As a first step, values for $d\Sigma/d\Omega$ at low q were found by a Guinier extrapolation to q = 0 and at high q by a Porod extrapolation to $q = \infty$ [1-3]. (See Fig. 7.) The total surface area of the scatterers was calculated from the Porod extrapolation. Their volume fraction was obtained by integrating $d\Sigma/d\Omega$

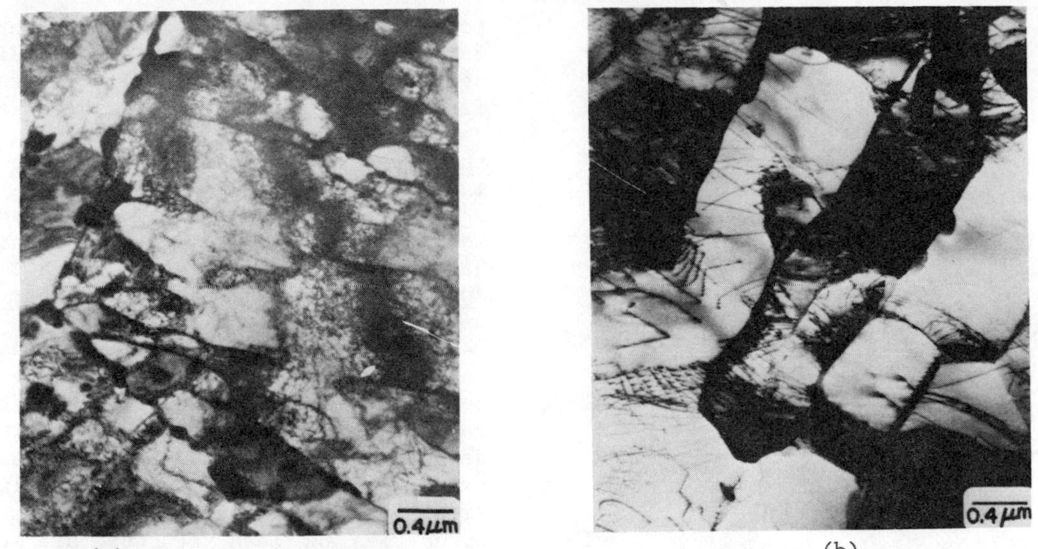

(a) (b)

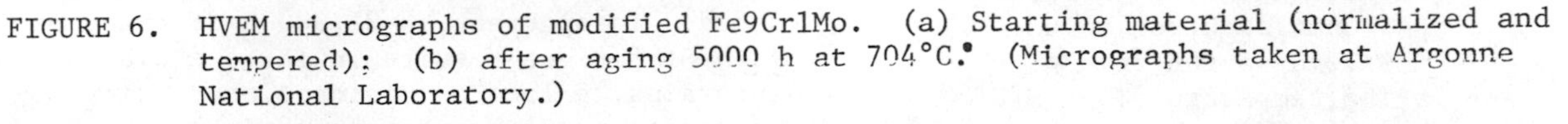

FIGURE 6. HVEM micrographs of modified Fe9Cr1Mo. (a) Starting material (normalized and tempered); (b) after aging 5000 h at 704°C. (Micrographs taken at Argonne National Laboratory.)

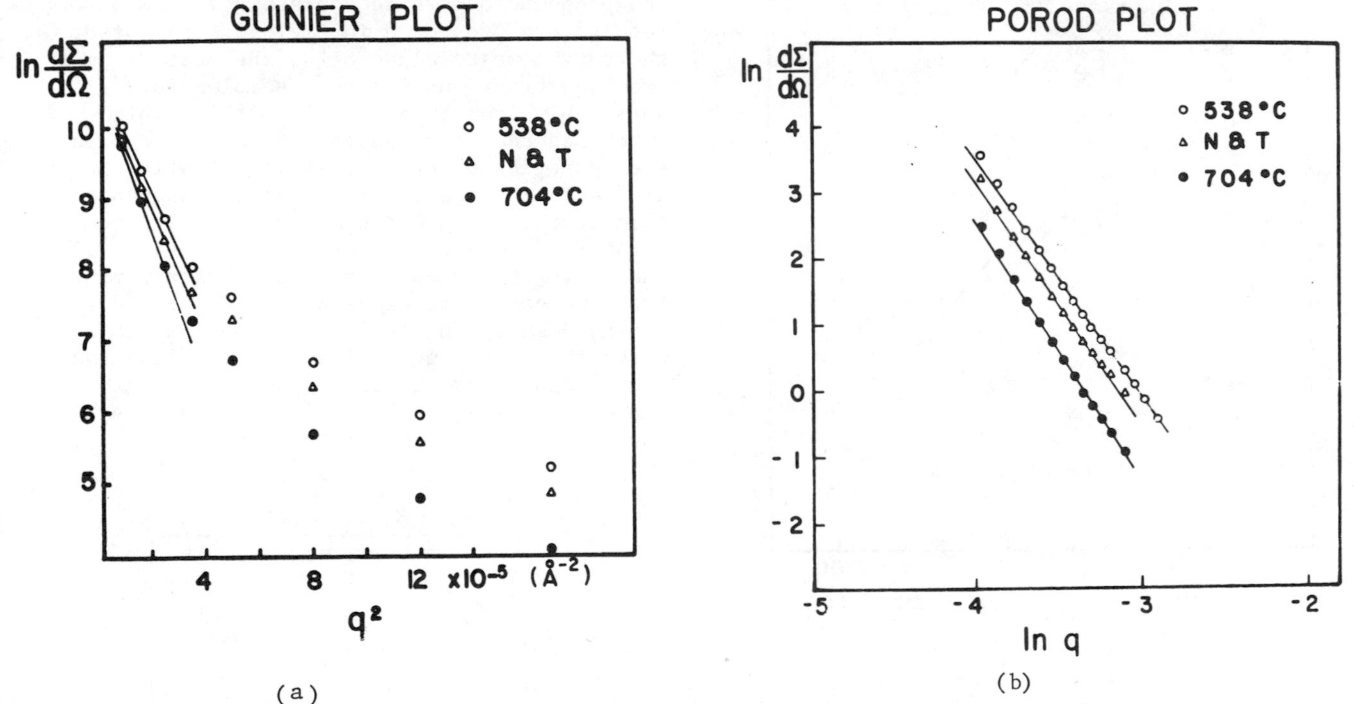

(a) (b)

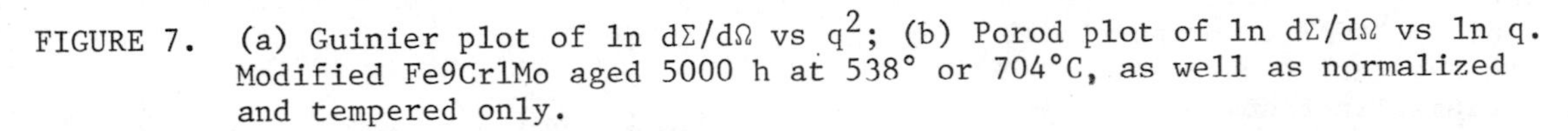

FIGURE 7. (a) Guinier plot of ln $d\Sigma/d\Omega$ vs q^2; (b) Porod plot of ln $d\Sigma/d\Omega$ vs ln q. Modified Fe9Cr1Mo aged 5000 h at 538° or 704°C, as well as normalized and tempered only.

over all of reciprocal space. The results are shown in Figs. 8 and 9. It can be seen that aging at 538°C produces a maximum in both surface area and volume fraction of the scatterers. Absolute values for these quantities are not given because of uncertainty in the scattering contrast [1-3] between carbide and matrix. During aging the MC carbides, which have their particular value for scattering contrast with the matrix, are disappearing while $M_{23}C_6$ carbides, with another scattering contrast, are growing. Fortunately the nuclear scattering contrast between (NbV)C and matrix and between the $M_{23}C_6$ carbides and matrix are reasonably similar, so a constantly changing carbide population does not destroy the value of the SANS analysis.

The method of Fedorova and Schmidt [13] was used to invert the scattering data to obtain a size distribution of the carbides. The results are given in Fig. 10. The SANS measurements show that prolonged aging at

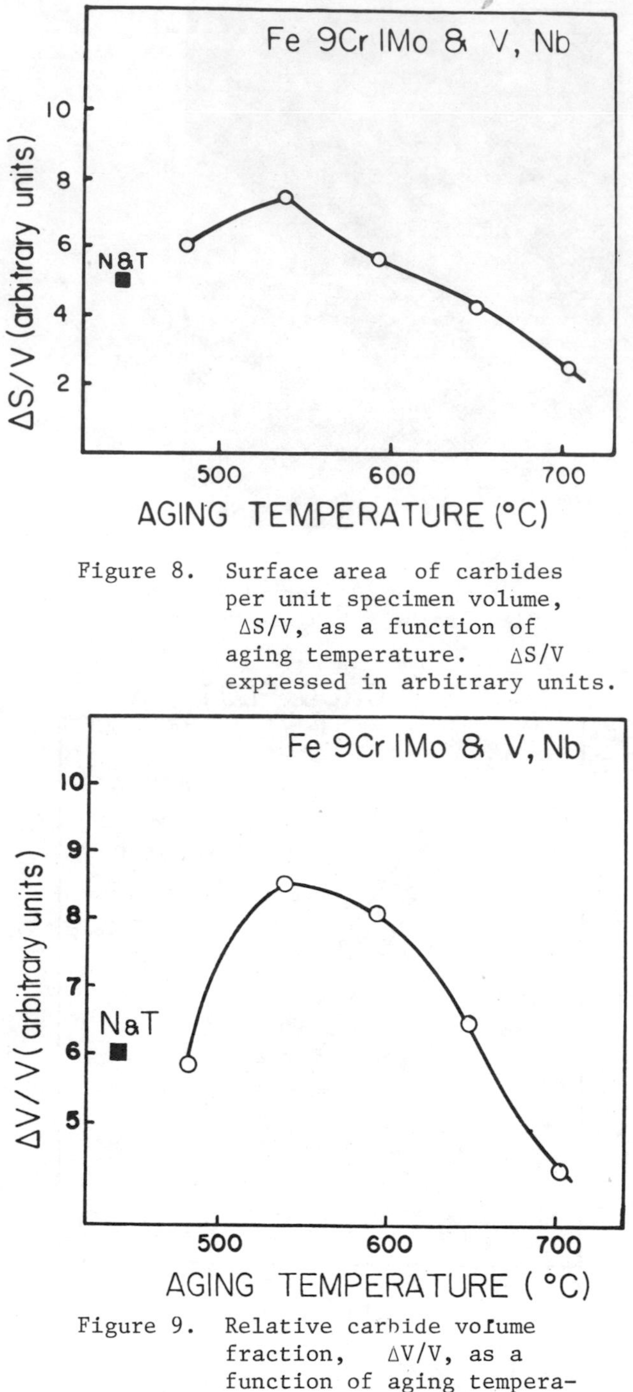

Figure 8. Surface area of carbides per unit specimen volume, ΔS/V, as a function of aging temperature. ΔS/V expressed in arbitrary units.

Figure 9. Relative carbide volume fraction, ΔV/V, as a function of aging temperature. ΔV/V is proportional to the actual carbide volume fraction.

538°C does indeed increase the population of small carbides whereas aging at 704°C displaces the distribution curve down and somewhat to the right. While the curves of Fig. 10 are subject to some uncertainty because of the constantly changing carbide composition, their general trend is expected to be correct.

FATIGUE AND CREEP - The ability of SANS to detect the microstructural changes associated with high temperature deformation was explored by looking for alterations in scattering cross sections produced by creep and fatigue. Two of the specimens studied had been cycled at 649°C in fully reversed fatigue through a total strain range of 0.5%. In one case the cycling was continuous; in the other a 30 s hold was introduced at the maximum tensile strain. The stress amplitude which developed as fatiguing proceeded is shown in Fig. 11. A very slight initial hardening is followed by softening. A third specimen had been crept at 649° for ∿ 21,000h under a stress of 62 MPa. The fatigue tests were terminated before failure. No damage in the form of microcracking or cavitation could be detected in any of the creep or fatigue specimens. The microstructures which develop at 649°C as the result of aging, creep, and fatiguing are compared in Fig. 12. After 5000 h of aging the prior austenite boundaries and lath structure are still evident and little subgrain development is noted. In contrast, ∿ 21,000 h of creep or 72 h of fatigue results in large equiaxed subgrains, a lowered dislocation density, and pronounced carbide coarsening. The SANS results are presented in Fig. 13. Except at the very lowest values of q, the scattering from the crept and fatigued samples is noticeably less than that from the sample which has had only the initial normalization and tempering treatment. This situation undoubtedly is a reflection of the disappearance of the MC carbides during the high temperature deformation. At the low q end of the curve the order is reversed. Scattering from the coarse carbides of the crept and fatigued specimens is largely confined to very low scattering angles. It is clear that significant microstructural changes occur, and can be detected by SANS, after only a few hours of high temperature fatigue.

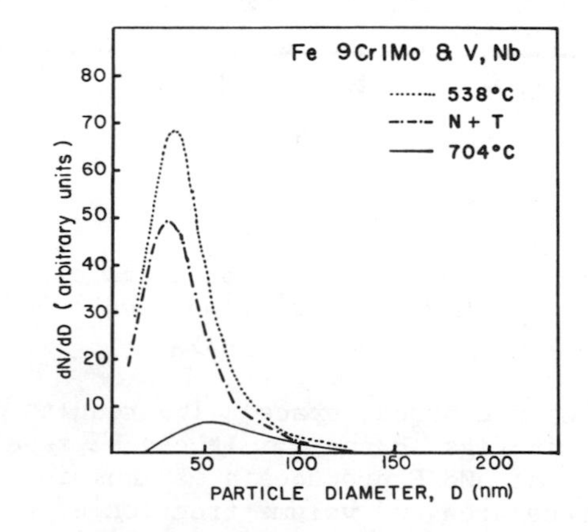

Figure 10. Size distribution dN/dD of the carbides as a function of their diameter D. Modified Fe9Cr1Mo aged 5000h at 538° or 704°C, as well as normalized and tempered only.dN/dD expressed in arbitrary units.

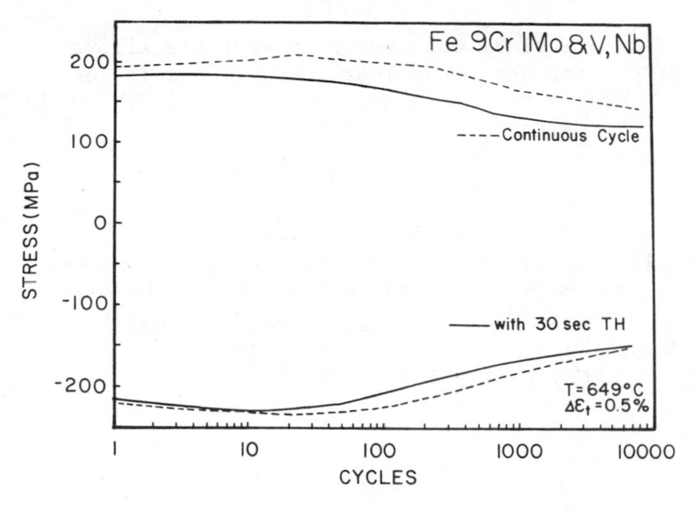

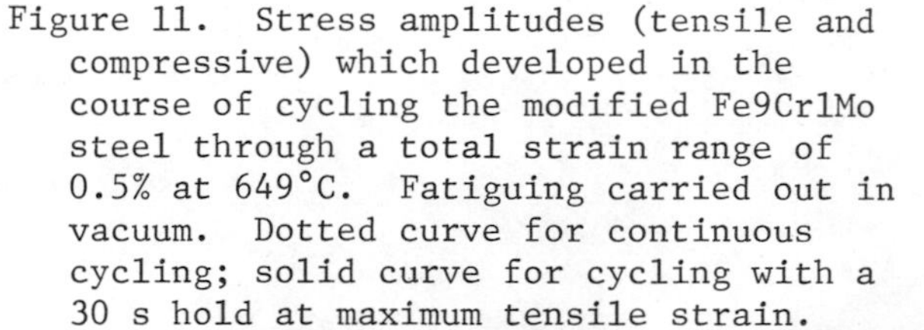
Figure 11. Stress amplitudes (tensile and compressive) which developed in the course of cycling the modified Fe9Cr1Mo steel through a total strain range of 0.5% at 649°C. Fatiguing carried out in vacuum. Dotted curve for continuous cycling; solid curve for cycling with a 30 s hold at maximum tensile strain.

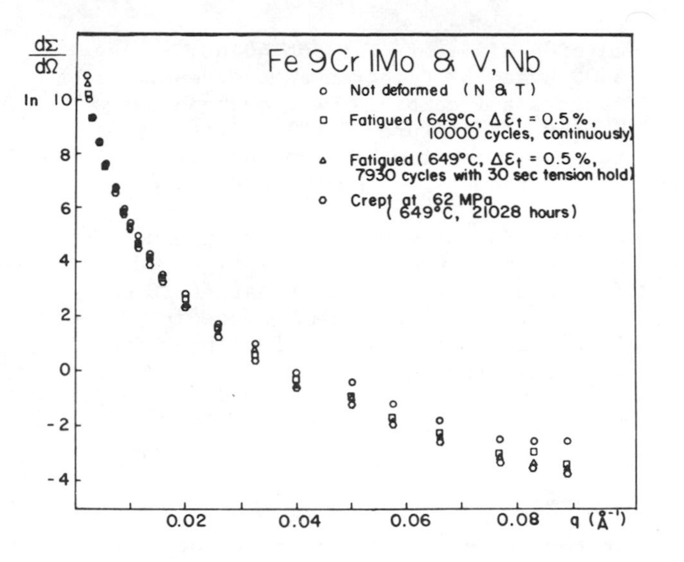

Figure 13. Curves of $d\Sigma/d\Omega$ vs q for specimens of modified Fe9Cr1Mo which have undergone various types of deformation. A magnetic field of ∿28 KG was applied to the specimens during the SANS measurements. λ = 0.48 nm.

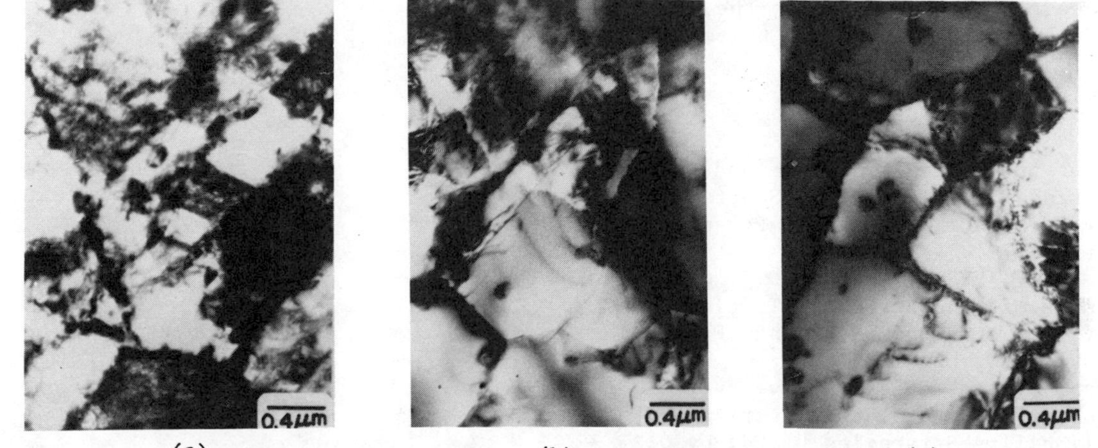

FIGURE 12. HVEM micrographs of modified Fe9Cr1Mo after exposure at 649°C under various conditions. (a) Aged 5000 h; (b) fatigued 7930 cycles with 30 s tension hold at maximum tensile strain (72 h of fatiguing) through a total strain range of 0.5%; (c) crept 21,028 h under a stress of 62 MPa. (HVEM micrographs taken at Argonne National Laboratory.)

SUMMARY AND CONCLUSIONS

SANS measurements have been made on specimens of a ferritic stainless steel, modified Fe9Cr1Mo, which had undergone prolonged aging at one of a number of temperatures or been subjected to high temperature creep or fatigue. Complementary studies of the various microstructures have been carried out by high voltage electron microscopy. Analysis of the data has led to several conclusions.

1. The neutron scattering results generally are in accordance with the HVEM observations. Statistical data on the microstructural features gathered over a large volume of material (∿500 mm^3) can be obtained quickly by SANS. Heterogeneities in the size range from a few nanometers up to about a micrometer can be detected and quantified by this technique.
2. In the case of specimens exposed to an elevated temperature for a long time, the scattering cross section is a maximum when the aging temperature is ∿538°C. This also is the aging temperature which produces the greatest hardness. The SANS data were analysed to obtain carbide size distribution curves. It was found that the distribution produced by the initial normalizing and tempering treatment peaks at a

carbide diameter of about 40 nm. Aging 5000 h at 538°C increases the density of these small carbides whereas aging at a higher temperature (704°) shifts the peak to a somewhat larger size and significantly reduces the number density of the carbides.

3. Changes in the microstructure produced by high temperature creep or fatigue can be detected by SANS. Cycling at 649°C causes a larger microstructural change in 72 h or less than does aging for 5000 h at the same temperature.

ACKNOWLEDGMENTS

The HVEM portion of this work was carried out at the Argonne National Laboratory. We are greatly indebted to Dr. A. Taylor, Mr. A. Philippides, and Mr. E. Ryan for their expert assistance.

Extensive use was made of the facilities of Northwestern University's Materials Research Center, supported by the NSF-MRL program, grant number DMR 8216972.

This research was sponsored by the United States Department of Energy, grant number DE-AC02-81ER10960.

REFERENCES

1.) "Treatise on Materials Science and Engineering, Vol. 15 Neutron Scattering," (Ed. by G. Kostorz), Academic Press, New York, 1979.

2.) Glinka, C.J., H.J. Prask and C.S. Choi, "Mechanics of Nondestructive Testing," (Ed. by W.W. Stinchcomb), Plenum Press, New York, 1980, pp. 143-164.

3.) Weertman, J.R., "Nondestructive Evaluation: Microstructural Characterization and Reliability Strategies," (Ed. by O. Buck and S.M. Wolf), The Metallurgical Society of AIME, Warrendale, PA, 1981, pp. 147-168.

4.) Sikka, V., C.T. Ward and K.C. Thomas, "Ferritic Steels for High-Temperature Applications," (Ed. by A.K. Khare), ASM, Metals Park, OH, 1983, pp. 65-84.

5.) Booker, M.K, V.K. Sikka and B.L.P. Booker, "Ferritic Steels for High-Temperature Applications," (Ed. by A.K. Khare), ASM, Metals Park, OH, 1983, pp. 257-273.

6.) Pizzi, P., "Symposium on Fracture Mechanics of Ceramics," Vol. 3 (Ed. by R.C. Bradt, D.P.H. Hasselman and F.F. Lange), Plenum Press, New York, 1978, pp. 85-98.

7.) Roth, M., J. Appl. Cryst. 10, 172-76 (1977).

8.) Yoo, M.H., J.C. Ogle, B.S. Borie, E.H. Lee and R.W. Hendricks, Acta metall. 30, 1733-42 (1982).

9.) Matsuoka, S., S. Kim and J.R. Weertman, "Topical Conference on Ferritic Alloys for use in Nuclear Energy Technologies," The Metallurgical Society of AIME, Warrendale, PA, in press

10.) Page, R., J.R. Weertman and M. Roth, Acta metall. 30, 1357-66 (1982).

11.) Vitek, J.M. and R.L. Klueh, Met. Trans. A 14A, 1047-55 (1983).

12.) Jones, W.B., "Ferritic Steels for High-Temperature Applications," (Ed. by A.K. Khare), ASM, Metals Park, OH, 1983.

13.) Fedorova, I.S. and P.W. Schmidt, J. Appl. Cryst. 11, 405-11 (1978).

EFFECTS OF CARBON CONTENT ON STRESS AND TEMPERATURE DEPENDENCES OF ULTRASONIC VELOCITY IN STEELS

J. S. Heyman and S. G. Allison
NASA Langley Research Center
Hampton, VA 23665

K. Salama and S. L. Chu
University of Houston
Houston, Texas 77004

ABSTRACT

The effects of carbon content on stress and temperature dependences of ultrasonic velocity have been studied in the four steel alloys AISI 1020, AISI 1045, AISI 1095 and ASTM A533B, through a cooperative effort between NASA/LaRC and the University of Houston. The stress dependence measurements are made at room temperature using the pulsed phase locked loop interferometer operating at 10 MHz. The results of these measurements indicate that the acoustoelastic constant in these alloys increases significantly with decreasing carbon content, but remains unchanged when other heavy alloying elements are added to the alloy. The temperature dependence measurements are performed using a pulsed-echo-overlap system operating at 10 MHz. The results indicate that the temperature dependence of ultrasonic velocity as well as its variation with applied stress do not change with changing carbon content. The changes in the temperature dependence with stress, however, are found to vary when other heavy alloying elements are present.

INTRODUCTION

The nondestructive measurement of residual stresses in solids is one of the more challenging quests for materials characterization. Although surface stress measurements can be performed using x-ray diffraction methods, there are no methods currently available to nondestructively measure residual stress in the bulk. Ultrasonic methods appear to hold the best promise in determining bulk stresses in both crystalline and non-crystalline materials[1]. These methods are based on the anharmonic nature of solids, where changes in ultrasonic velocity are linear functions of applied stress and unknown stresses are determined when the velocity in the absence of stress and appropriate values of third-order elastic constants are known independently[2]. The measured velocity, however, is found to strongly depend on microstructural variables which makes it necessary to develop a calibration between velocity and stress using a specimen of the same material where unknown stresses are to be determined.

The temperature dependences of elastic constants of a solid are also due to the anharmonic nature of the crystal lattice, and a measure of the temperature dependence of ultrasonic velocity can therefore be used for the evaluation of bulk stresses. Experiments performed on aluminum alloys[3-4] show that, in the vicinity of room temperature, ultrasonic velocity decreases linearly with increasing temperature, and the slope of this linear relationship varies considerably when the specimen is subjected to stress. The results also show that the relative changes in the temperature dependence of the velocity as a function of stress in several aluminum alloys are the same indicating the insensitivity of the measurement to alloy composition and perhaps other metallurgical variables[5].

In this paper, the effects of carbon content on the stress as well as the temperature dependences of ultrasonic longitudinal velocity have been investigated in four steels (AISI 1020, AISI 1045, AISI 1095 and ASTM A533B) and the effect of applied stress on the temperature dependences of the velocity in two alloys (AISI 1020 and AISI 1045). These studies are made in order to examine the degree of sensitivity of the acoustoelastic constant (AEC) and the temperature dependence of ultrasonic velocity to variations in carbon content and other alloying elements in steel alloys. The studies are also aimed at examining the possibility of a relationship between these ultrasonic quantities and the amount of ferrite phase present in these alloys.

Table I - Chemical Composition of the Steel Alloys AISI 1020, AISI 1045, AISI 1095 and ASTM A533B

Steel	C%	Mn%	Ni%	Mo%
AISI 1020	0.18 - 0.23	0.43 - 0.50	——	——
AISI 1045	0.43 - 0.50	0.60 - 0.90	——	——
AISI 1095	0.90 - 1.03	0.30 - 0.50	——	——
ASTM 533B	0.25	1.15 - 1.50	0.40 - 0.70	0.45 - 0.60

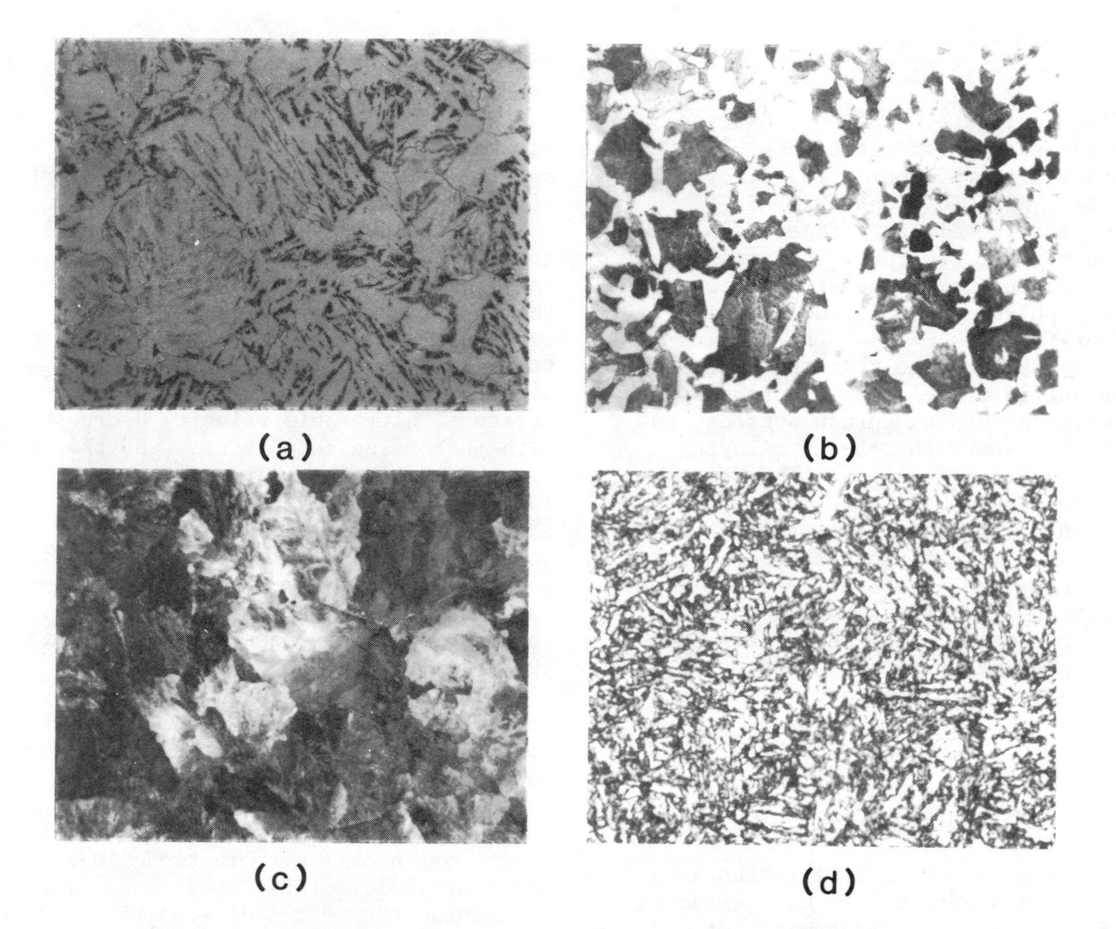

Fig. 1 - Microstructure of Steel Alloys. (a) AISI 1020, (b) AISI 1045, (c) AISI 1095, (d) ASTM A533B

EXPERIMENTAL

Test Specimens - Four steel alloys namely AISI 1020, AISI 1045, AISI 1095 and ASTM A533B are chosen for this investigation. The major alloying elements in these steels[6] are included in Table I. From the Table, it is seen that the three carbon steel alloys contain similar amounts of Mn and the only element which varies in these alloys is the amount of carbon. Also from the table, it is seen that the amount of carbon in ASTM A533B steel is about equal to that of AISI 1020, and the difference in composition between these two alloys is the extra amount of Mn, Ni and Mo added to ASTM A533B steel. Figure 1 displays representative metallographs for these four alloys. The white areas in the metallographs represent the pure ferrite phase, while the dark areas represent pearlite which consists of 88% ferrite and 12% carbide.

The specimens used in this investigation are manufactured using these steels in the form of threaded rods 2.54 cm in diameter and 20 cm long. The central section of each rod was machined to a square cross-section of 2.0 cm x 2.0 cm rounding all corners using ASTM standard practices. These flat surfaces are ground to be smooth and parallel within ±0.005 mm.

Pulsed Phase Locked Loop - The AEC's for the steel samples were measured using a pulsed phase locked loop (P^2L^2) system described in detail elsewhere[7]. A block diagram of the P^2L^2 is shown in figure 2. The basis of the measurement system is a phase feedback scheme using a voltage controlled oscillator (VCO). The VCO output is gated to produce a tone burst of several cycles to drive a broadband transducer. The returning echo is amplified and phase detected using the VCO as a reference. A logic system samples the phase signal at a preselected point and causes the frequency of the VCO to change until quadrature is achieved. Once locked, the P^2L^2 maintains the quadrature condition with the change in frequency related to the change in sample properties given by[8]:

$$\left(\frac{\Delta F}{F}\right) = \left(\frac{\Delta V}{V}\right) - \left(\frac{\Delta L}{L}\right) \quad (1)$$

where L is the sample length. The normalized change in frequency, $\Delta F/F$, is called the natural velocity in the sample. With the natural velocity, one does not have to measure the change in sample length during the measurement.

Figure 3 shows a block diagram of the experimental arrangement for measuring AEC's. The samples were placed in an insulated chamber (oven) and stressed within the elastic range by a MTS-810 fatigue machine using adaptor rods. The load and frequency data were taken with a lab computer on the IEEE-488 bus. The AEC is determined by dividing the change in stress by the change in normalized frequency. Since the AEC tests are run from an electronic 60-second ramp driving the load piston of the MTS machine, the data represent nearly adiabatic conditions.

Pulsed-Echo-Overlap System - The temperature dependence of ultrasonic velocity was measured using the pulse-echo-overlap method which has been fully described elsewhere[9]. Figure 4 displays the experimental system used in this investigation. Pulses of, approximately, 1μsec duration of variable pulse-repetition rate are generated by the ultrasonic generator and impressed on a transducer of a fundamental frequency of 10 MHz which is acoustically bonded to the specimen. The reflected rf echoes are received by the same transducer, amplified and displayed on the screen of an oscilloscope. Two of the displayed echoes are then chosen and exactly overlapped by critically adjusting the frequency of the C.W. oscillator, and the division factor on the decade divider. This frequency, f, accurately determined by the electronic

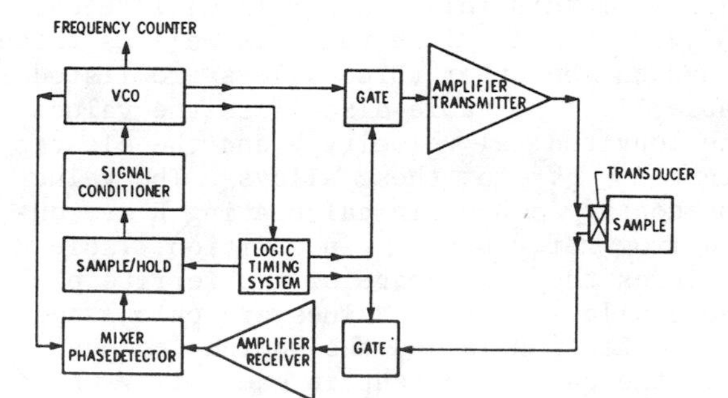

Fig. 2 - Block diagram of the pulsed phase locked loop ultrasonic system.

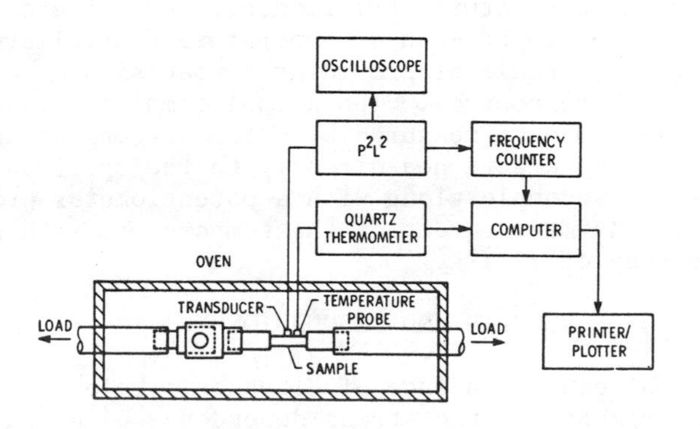

Fig. 3 - Diagram of the system for measuring acoustoelastic constant.

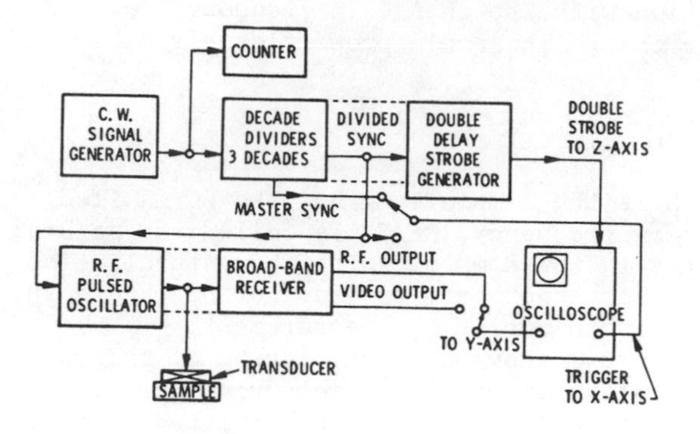

Fig. 4 - Pulse-echo-overlap system.

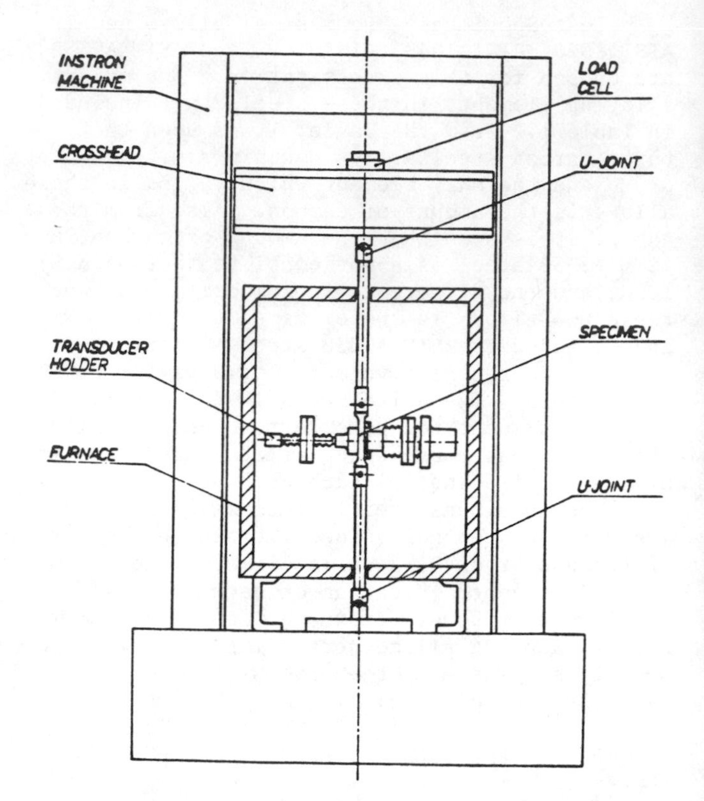

Fig. 5 - Loading system used for the application of tensile stress in the measurements of the temperature dependence of ultrasonic velocity.

counter, is employed to compute the ultrasonic velocity using the relation $V = 2\ell f$, where ℓ is the length of the specimen. An X-cut quartz undamped transducer is used for the generation of the longitudinal waves. This system is capable of measuring changes in the ultrasonic velocity to an accuracy of better than 1 part in 10^5.

The velocity measurements are made while the specimen was subjected to various amounts of tensile stresses using the arrangement shown in Fig. 5. In this arrangement, the specimen is gripped in an Instron machine where a predetermined load is applied and its value is kept constant during the entire velocity-temperature run. The temperature control system is designed to enclose the specimen and gripping assemblies in order to ensure stabilized temperature for the whole specimen during the time required for the velocity measurements. A test furnace along with recirculating blower and plenum system is used to provide a uniform temperature performance. The furnace is also equipped with a temperature controller which is capable of providing a precision temperature control. The actual temperature of the specimen is measured by a copper-constantan thermocouple attached directly to the specimen. The thermocouple along with a potentiometer provides the measurement of the temperature with an accuracy of ±0.1°K.

RESULTS AND DISCUSSIONS

Stress Dependence of Ultrasonic Velocity - Measurements of the stress dependence of ultrasonic longitudinal velocity in the steel alloys AISI 1020, AISI 1045, AISI 1095 and ASTM A533B were performed at room temperature using the pulsed-phase-locked-loop system. The velocity measurements were made at 10 MHz while the specimen is subjected to tensile stress applied in a direction perpendicular to that of wave propagation. A typical example of the relative change in the ultrasonic velocity expressed as $\Delta F/F$ as a function of applied stress obtained on the alloy AISI 1045, is shown in Fig. 6. The results indicate that the ultrasonic velocity in this alloy decreases linearly with stress and the slope of this relationship (AEC) is equal to -37.47×10^4MPa. This value as well as those obtained on the other three alloys are listed in Table II. The table also lists the values of the longitudinal velocity V and the elastic constant $E = \rho V^2$ for these alloys. The values of the density ρ used in calculating E are obtained from reference 6. In addition, Table II includes the percentage of the ferrite phase in these alloys. These values are calculated using the Lever Rule and the nominal composition of the carbon content in each alloy which are listed in Table I.

From the results in Table II, one can see that the variations in both the ultrasonic longitudinal velocity and the elastic constant are very small in the four alloys investigated in this paper. There is a tendency, however,

Table II - Values of longitudinal ultrasonic velocity, elastic modulus, acoustoelastic constant and percentage of ferrite phase in the steel alloys AISI 1020, AISI 1045, AISI 1095 and ASTM A533B.

Property	AISI 1020	AISI 1045	AISI 1095	ASTM A533B
Velocity (m/s)	5889	5883	5910	5800
Elastic Modulus (10^{11}Pa)	2.721	2.715	2.736	2.641
Acoustoelastic Constant (10^{10}Pa)	41.8	37.5	32.5	41.0
Ferrite Phase %	97	93.3	85.8	96.3

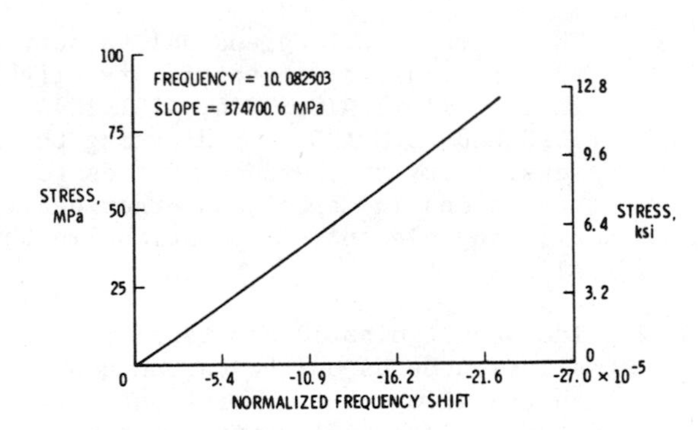

Fig. 6 - Normalized frequency shift as a function of applied tensile stress for a typical AEC measurement with transverse ultrasonic propagation. The above figure is for 1045 carbon steel.

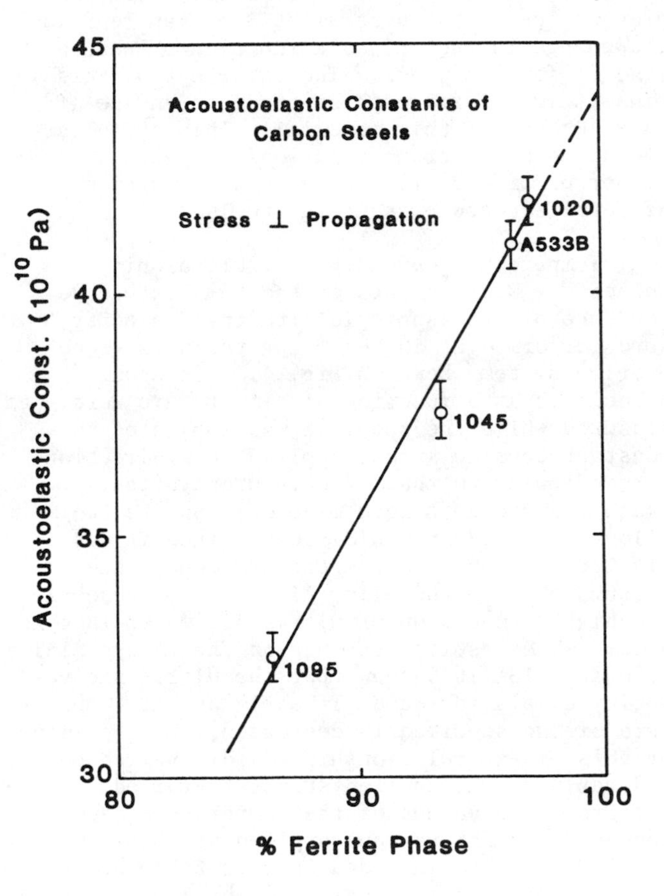

Fig. 7 - The acoustoelastic constants as a function of % ferrite phase in the steel alloys.

for a small decrease in both quantities as the amount of carbon in the alloy is increased. No systematic changes in the ultrasonic longitudinal velocity in plain carbon steels could be observed by Shyne et al[10]. The decrease is more emphasized in the case of the alloy ASTM A533B which contains the heavy alloying elements Ni and Mo in addition to C. Nevertheless it is reasonable to assume that both ultrasonic velocity and elastic constant in these alloys remain unchanged with the additions of carbon (up to 0.95%) or small amounts of heavy alloying elements.

On the other hand, the results listed in Table II indicate that the AEC is sensitive to the addition of carbon but not to the addition of heavy alloying elements. The values of the AEC in the two alloys AISI 1020 and ASTM A533B steel are almost equal, while the values of the AEC decrease as the amount of carbon is gradually increased in AISI 1045 and AISI 1095. The decrease in the AEC is about 20% when the carbon content is increased from 0.20% to 0.95%. Figure 7 displays a plot of the AEC's vs percentage of ferrite phase for the four alloys listed in Table II. From this plot it seems that a straight line can represent the change in the AEC as a function of percentage of ferrite phase. The intercept of this line at 100% ferrite phase predicts an AEC of 45×10^4MPa for this material. This AEC clearly indicates a strong relationship between higher-order elastic constants and the amount of ferrite phase present in steels.

Temperature Dependence of Ultrasonic Velocity - Measurements of the temperature dependence of ultrasonic longitudinal velocity were performed at 10 MHz using the pulse-echo-overlap system shown in Fig. 4. Ultrasonic velocities as a function of temperature were measured while the specimen was subjected to constant tensile stress applied in a direction perpendicular to that of wave propagation. These measurements were made only on the two alloys AISI 1020 and AISI 1045, since it was difficult to perform temperature dependence measurements on the alloy AISI 1095 because of the high attenuation of ultrasonic waves in this alloy. The results obtained on the alloys AISI 1020 and AISI 1045 show that the ultrasonic velocity always increases linearly as the temperature of the specimen is decreased, and the slope of this linear relationship (dV/dT) varies considerably as the applied stress is varied. Table III lists the values of the temperature dependence (dV/dT) at various applied stresses up to 117.3 MPa. Also included in this table are the values[11] of (dV/dT) previously obtained on the steel alloy ASTM A533B. These values are also measured while stress is applied in a direction perpendicular to that of wave propagation.

From the results listed in Table III, it can be seen that the values of (dV/dT) at zero applied stress in the three alloys AISI 1020, AISI 1045 and ASTM A533B are almost the same, with the value obtained on ASTM A533B is the smallest and that obtained on AISI 1045 is the largest. The difference between the two quantities, however, is about 3% which lies within the experimental error ±2% estimated for these measurements. The results in the table also show that in the three alloys, (dV/dT) increases with the increase of the applied tensile stress, and the increase in each of the two alloys AISI 1020 and AISI 1045 is almost the same, but larger than that obtained on ASTM A533B.

The results of Table III are plotted in Fig. 8 to demonstrate the effects of applied tensile stress on the temperature dependences of the ultrasonic longitudinal velocity for AISI 1020, AISI 1045 and ASTM A533B. From the figure, it is apparent that the data points of each alloy can be represented by a straight line which indicates a linear relationship between (dV/dT) and stress. This confirms earlier findings on aluminum and copper alloys[12]. The plots in the figure also indicate that the straight lines representing the data points of the alloys AISI 1020 and AISI 1045 have similar slopes (-5.85×10^{-4} m/s.k/MPa), which is larger than that of the line representing the behavior of ASTM A533B steel (-3.44×10^{-4}m/s.k/MPa). The chemical composition of the later alloy include the heavy alloying elements 0.4-0.7% Ni and 0.45-0.6% Mo in addition to carbon and magnesium.

CONCLUSIONS

From the above discussions, the following conclusions can be drawn:

1. The ultrasonic longitudinal velocities and the elastic constants vary slightly in the alloys AISI 1020, AISI 1045, AISI 1095 and ASTM A533B, indicating the insensitivity of these quantities to variations in carbon and other heavy alloying elements compositions in these alloys.

2. The acoustoelastic constant in these alloys depends on the carbon content in the alloy, but remains unchanged when other alloying elements are added. The acoustoelastic constant varies linearly as a function of percentage of ferrite phase, which demonstrates strong relationship between acoustoelastic constants and second phase presence.

3. Similar to ultrasonic velocity and elastic constant, the temperature dependence of the velocity in the alloys AISI 1020, AISI 1045 and ASTM A533B is insensitive to carbon and heavy alloying elements compositions.

4. The temperature dependence of ultrasonic velocity varies linearly with

Table III - Effect of applied tensile stress on the temperature dependence of ultrasonic longitudinal velocity in the steel alloys 1020, 1045, and A533B. Stress is applied perpendicular to wave propagation.

Stress (MPa)	-dV/dT (m/s.k)		
	1020	1045	A533B
0	0.633	0.641	0.624
23.5	0.655		
30.4			0.636
35.2		0.658	
60.8			0.638
70.4	0.673	0.682	
93.8		0.693	
109.4			0.662
117.3	0.700	0.710	
133.8			0.670

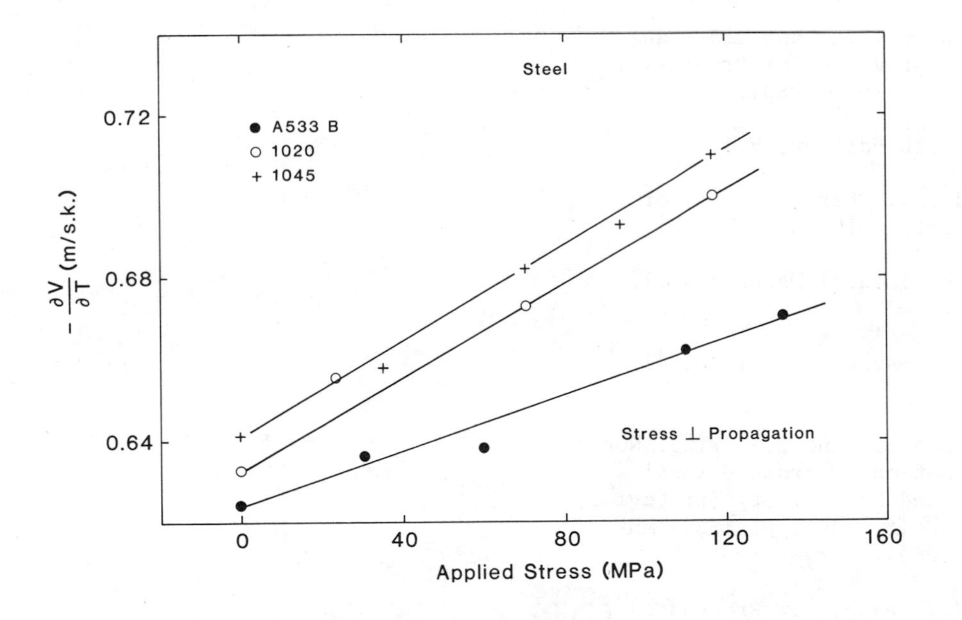

Fig. 8 - The temperature dependence of ultrasonic longitudinal velocity as a function of applied stress in the steel alloys.

stress in the alloys AISI 1020 and AISI 1045, and the slope of this linear relationship is equal in these two alloys. The value of the slope in these alloys is larger than that found in the alloy ASTM A533B.

ACKNOWLEDGEMENT

This research is part of a cooperative program between NASA Langley Research Center, Federal Railroad Association, Department of Transportation and the University of Houston.

REFERENCES

1. Proceedings of Workshop on Nondestructive Evaluation of Residual Stress, NTIAC-72-2 (1976).

2. James, M.R. and O. Buck, Critical Reviews in Solid State and Materials Science, P. 61 (1980).

3. Chern, E.J., J.S. Heyman and J.H. Cantrell, Jr., Proceedings IEEE Ultrasonics Symposium (1981).

4. Salama, K., C.K. Ling, and J.J. Wang, Experimental Techniques 5, 14 (1981).

5. Salama, K., A.L.W. Collins and J.J. Wang, Proceedings DARPA/AF Review of Progress in Quantitative NDE, p. 265 (1980).

6. Metals Handbook, 9th Edition, Vol. 1.

7. Heyman, J.S. and E.J. Chern, Journal of Testing and Evaluation 10, 202, (1982).

8. Heyman, J.S., Experimental Mechanics 17, 183 (1977).

9. Papadakis, E.P., J. Acoust. Soc. Am. 42, 1045 (1967).

10. Shyne, J.C., N. Grayeli and G.S. Kino, Nondestructive Evaluation: Microstructural Characterization and Reliability Strategies, O. Buck and S.M. Wolf, eds., p. 133, TMS-AIME, New York (1980).

11. Salama, K. and J.J. Wang, New Procedures in Nondestructive Testing, P. Höller, ed., p. 539, Springer-Verlag, Berlin and New York (1983).

12. Salama, K. and C.K. Ling, J. Appl. Phys. 51, 1505 (1980).

DETERMINATION OF STRAIN DISTRIBUTIONS AND FAILURE PREDICTION BY NOVEL X-RAY METHODS

Sigmund Weissmann
Dept. of Mechanics and Materials Science
College of Engineering
Rutgers University
Piscataway, New Jersey 08854

William E. Mayo
Dept. of Mechanics and Materials Science
College of Engineering
Rutgers University
Piscataway, New Jersey 08854

ABSTRACT

A number of X-ray diffraction methods are described which make it possible to determine the distribution of elastic and plastic strains emanating from stress raisers. The elastic strain distribution is determined by measurements of reflected intensities and the distribution of plastic strains by X-ray double crystal diffractometry using a computer-aided rocking curve analyzer (CARCA). Applying X-ray Pendellösung topography to a silicon crystal which is used as a model material, it is shown that residual elastic strains are due to the constraint imposed by the strain hardened microplastic zones and that the degree of strain hardening in the plastic zones governs the magnitude of residual strains. Extending the results derived from single crystal studies to polycrystalline materials and applying the CARCA method to commercial aluminum alloys, cycled in air, and in corrosive environment, the accrued damage is determined and a satisfactory estimate of the time to failure is made nondestructively.

INTRODUCTION

When a crystalline material is deformed elastically in such a manner that the strain is uniform over relatively large distances, the spacings of the lattice planes change from their strain-free value to a new value governed by the magnitude of the applied strain. The uniform macrostrain causes a shift of the X-ray diffraction lines to new positions and it is this geometrical shift of line positions which forms the basis of conventional X-ray measurements of residual strains. In technological materials, however, stress raisers are often present, such as cracks, notches, inclusions, voids and microholes. The strains emanating from such stress raisers are nonuniform and the strain gradients associated with these stress raisers are frequently the determining factors of the fracture of the material. The determination of strain gradients emanating from stress raisers is, therefore, one of the principal concerns of fracture mechanics. By applying calculations based on continuum mechanics many attempts have been made to evaluate analytically stress concentrations but such calculations were carried out successfully only for a limited number of stress concentration raisers in isotropic materials (1,2,3,4). The problem becomes even more complicated when plastic zones are generated by the stress raisers and when the constraints of work hardened plastic zones on the residual elastic strains have to be taken into account. This is the case frequently encountered in technological applications of metals and alloys.

In view of these considerations there appears to be a great need to carry out nondestructive measurements of strain gradients on a microscopic scale, to extend the measurements over relatively large distances and to correlate the experimental results with theoretical calculations. This paper addresses itself to this important problem of bridging the gap between micro and macromechanics and it attempts to establish a link between the viewpoints of the materials scientist and the mechanicist. To achieve this objective a series of novel X-ray methods were developed which allow for a close correlation between experimental measurements and calculations based on continuum mechanics. It will be shown that these methods: a) characterize and determine quantitatively strain gradients and strain distribution emanating from stress raisers, viz. notches, holes and inclusions; b) characterize and measure experimentally notch tip plasticity and the effects of the interaction of plastic zones on the fracture process of materials; c) disclose residual elastic strains caused by the constraint of the strain hardened plastic zones; d) are useful in elucidating strain relaxation processes resulting from annealing; e) permit failure prediction in commercial alloys cycled in air and corrosive environments.

For the elucidation of strain distributions induced by mechanical deformation dislocation-free silicon crystals were used as a model material. It will be shown that fundamental principles emerged from these single crystal studies which, it is hoped, will contribute to a better understanding of the interrelationship between plastic zones and residual elastic strains. It will be further shown that the logical extension of these results applied to polycrystalline materials led to the development of an X-ray method which is capable of determining the accrued damage in cycled, commercial aluminum alloys and of carrying out successfully nondestructive failure predictions. By contrast to the conventional X-ray stress measurements which attempt to evaluate the microstructural damage in the material by measuring the average, maximum residual elastic strains, the success of this novel X-ray method rests on identifying and determining the maximum plastic strains associated with work hardened plastic zones. It will be shown that the latter govern the residual elastic strains by constraining them and can, therefore, be regarded as a reliable criterion for failure prediction.

DETERMINATION OF ELASTIC STRAIN CONCENTRATIONS AND STRAIN GRADIENTS EMANATING FROM STRESS RAISERS

PRINCIPLE - Unlike conventional strain measurements by X-ray diffraction, the method described here does not rely on the geometrical line shift resulting from the variations of the interplanar spacings (variation of d-spacings) which yield only strain averages, but is based on local intensity measurements. The method utilizes the high sensitivity of the reflected intensity to variations of strain gradients, or curvatures, in perfect crystals. It has been shown theoretically (5,6) and experimentally (5, 6,7,8) that the intensities reflected from an elastically bent perfect crystal, viz. silicon crystal, are directly proportional to the curvatures of the reflecting planes. Due to the action of stress raisers, viz. notches, holes or inclusions, the lattice curvature, induced externally by an applied moment, becomes locally enhanced. The enhancement is a manifestation of the strain gradient developed by the stress raiser. To measure the strain gradient quantitatively intensity measurements are carried out from point-to-point. By selecting appropriate crystal orientation and reflection geometry and carrying out intensity measurements from various reflections of crystal planes information is also gained on anisotropic effects of strain concentration (6).

ELASTIC STRAIN MEASUREMENTS AND RESULTS

Measurements were performed on dislocation-free silicon crystals at room temperature. These crystals are perfectly elastic up to fracture and hence serve as excellent models for the elucidation of elastic properties of materials in general and brittle materials in particular. The specimens were elastically bent to a curvature of 0.1 m^{-1} about an axis parallel to the symmetry plane of the stress raiser. The bending device and crystal were mounted on a modified Lang camera. In the Lang method (9) a collimated X-ray beam impinges on a crystal in transmission. The crystal is placed at an appropriate angle to the incoming beam so that a set of transverse planes can satisfy the condition of Bragg reflection. The transmitted reflected beam is recorded on the film while the transmitted primary beam is prevented by a stationary screen from striking the film. When the specimen and film are kept stationary, one obtains a section topograph because only the image of the irradiated crystal section is topographically recorded. The distribution of lattice defects can be disclosed, however, over a range that extends beyond that of the crystal section irradiated in a stationary crystal, by coupling the crystal holder and film holder mechanically and moving them to and fro during the exposure. The topographs thus obtained are called projection or traverse topographs.

For measuring the strain gradient emanating from a stress raiser the Lang method was modified in such a way that the crystal was traversed over the region of the stress raiser, accompanied by oscillation so as to register on the film the total range of induced lattice misorientation. Using fine-grained nuclear track plates instead of films to record the intensities, the opacity measurements were carried out from point-to-point, essentially along the line of symmetry passing through the center of the stress raiser configuration (viz. notches, holes, inclusions). Since optical densitometry was considered inaccurate over the range of opacities measured, a method of microfluorescent analysis of the film emulsion was developed (7). After sputtering onto the developed emulsion a thin conductive layer, the topograph was scanned on a scanning electron microscope in the fluorescent mode for the Ag L line. The fluorescent intensity emitted from any point of the topograph is very nearly proportional to the X-ray intensity originally reflected from the corresponding point on the crystal. Sputtering Ti onto the emulsion as the conductive layer enhanced the fluorescence considerably through secondary excitation of the Ag line by Ti K radiation (10).

With the aid of this novel X-ray method combining topographic imaging with analysis of diffracted intensities, it was possible to determine experimentally strain gradients emanating from elastically bent silicon crystals containing a) single (5) and double notches (7), b) single circular hole (8), c) two circular holes at different distances (8), d) single and double inclusion (8). There exist analytical solutions for stress and strain concentration factors for cases a) and b), as calculated by Neuber (1). From these equations the strain concentration factors K_ε were calculated for the respective geometries as a function of distance along the line of symmetry and compared with the results obtained experimentally from measured opacities. Very good agreement was obtained between experiment and theory and typical results are shown in Figs. 1a and 1b. These results confirm the applicability

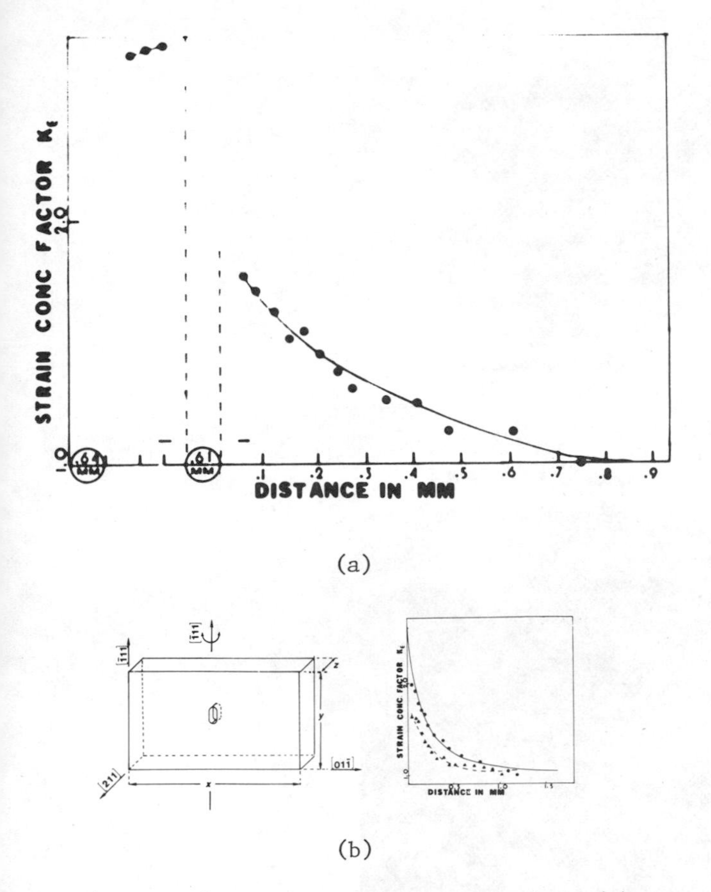

(a)

(b)

Fig. 1 - Strain concentration measured at (a) region of two circular holes, (b) single circular inclusion; upper curve in (b) refers to empty circular hole. (With permission of Int. Union of Crystallography, J. Appl. Cryst., Ref. 8.)

of this method to strain concentration measurements and justify its applicability to cases which have not been evaluated analytically, such as cases c) and d). Thus, it may be seen from Fig. 1a that the strain in the inter-flaw region between two holes (case c) is approximately equal to the superposed additional strains emanating from each hole separately. Inclusions (case d) showed a strain distribution essentially similar to that of empty holes of identical geometry, but the strain concentration factors were lowered by about 20% throughout the region investigated (Fig. 1b).

RELATIONSHIP BETWEEN PLASTIC ZONES, CONSTRAINTS, AND RESIDUAL ELASTIC STRAINS

To elucidate the interrelationship of plastic zones and residual elastic strains in a deformed crystal it was important to characterize both the plastic zones and the elastic strains and to obtain actually a visualization of their structural relationship on a microscopic and macroscopic scale. To obtain this objective advantage was taken of the Pendellösung effect which results from the dynamical interaction between the two X-ray wavefields generated in a perfect crystal, analogous to that which takes place between a pair of coupled pendulum the frequencies of which are nearly equal. The primary beam has all the energy at the surface, but at a certain depth along the ray path this state of affairs is reversed and the secondary beam attains all the energy and so the oscillation continues. This periodic oscillation with thickness in the direction of the energy flow was first developed by P. P. Ewald and is called Pendellösung. Consequently, if a beam is diffracted by a wedge-shaped crystal in transmission, both the direct and the reflected beams, on emergence from the exit surface, consist of two superimposed wave trains of the same frequency but differing slightly in phase velocity. Since in each beam these wave trains are coherent, they give rise to interference fringes, termed Pendellösung Fringes (PF), when recorded on a Lang X-ray topograph such as that shown in the traverse topograph of Fig. 2. Because lattice defects have a profound effect on the dynamical interactions of the wavefields, Pendellösung Fringe topography can be used as a powerful diagnostic research tool to disclose the distribution of plastic zones and residual elastic strains (11). Figure 3 may serve to demonstrate the information which can be obtained from a notched silicon crystal bent at 800°C. Silicon, like other semiconductor crystals, become ductile above 0.6 of the absolute melting temperature and exhibit stress-strain curves analogous to those of metal crystals (12). Accumulations of dislocations such as those shown in the plastic zone at the notch tip, N, and in the plastic zones generated at the surface opposite the notch destroy totally the dynamical interactions of the beam and the images appear as black areas on the topograph (Fig. 3a). A dislocation density of about 10^5 cm^{-2} is usually sufficient to disclose microplastic zones in this manner. Residual elastic strains constrained by the microplastic zones are disclosed by the distortions of the PF patterns with fringe spacing becoming narrower as the strain field increases.

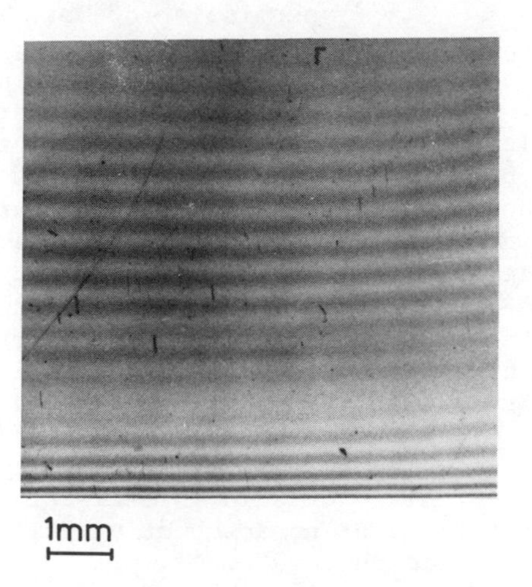

Fig. 2 - X-ray Pendellösung fringe topograph of wedge-shaped silicon crystal. (With permission of Met. Trans., Ref. 11.)

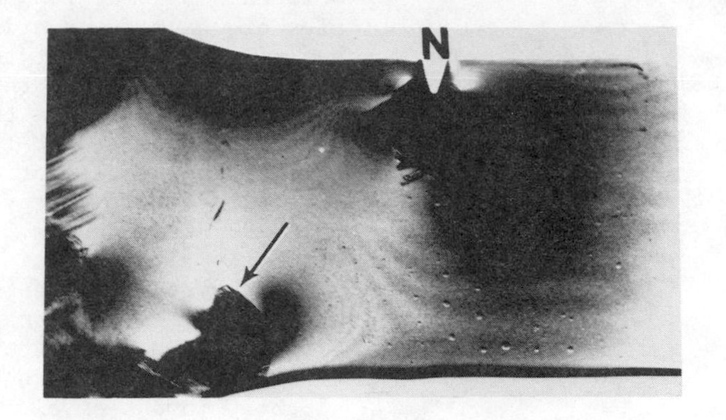

(a)

1 mm

Polygonization detail of 220 reflection

(b)

Fig. 3 - X-ray Pendellösung fringe topograph of wedge-shaped silicon crystal, (a) bent at 800°C; N=Notch (b) polygonization detail of plastic zone after annealing at 1000°C. (With permission of Met. Trans., Ref. 11.)

As may be seen from Fig. 3a the bending and narrowing of the fringe spacings becomes particularly conspicuous when the plastic zone at the notch tip is approached.

Although the changes in the configurations of plastic zones and residual strains as a function of deformation have been studied by PF topography in great detail (11) we shall concentrate here on the phenomenon of strain relaxation induced by high temperature annealing. It will be shown shortly that PF topography applied to such studies is very informative in revealing the mechanism by which strain relaxation occurs and that as a corollary of such investigation a useful criterion for nondescructive failure prediction can be deduced.

The traverse topograph of Fig. 3b shows the effect of annealing at 1000°C for 1 hour on the microstructure of the bent silicon specimen shown in Fig. 3a. It will be seen that upon annealing, dislocation loops emanated from the microplastic zones and penetrated into the areas previously occupied by the elastic residual strains. Because of this dislocation rearrangement in the plastic-elastic boundary zone the elastic residual strains became relaxed and largely disappeared. This was manifested by the restored regularity of the fringe spacings in the PF pattern. It is interesting to note that the white contrast regions, indicated in Fig. 3a by arrows, which are indicative of lattice curvature, exhibited after annealing a three-dimensional polygonization substructure.

The relaxation studies have shown that: (1) relaxation of elastic residual strains by annealing occurs by a reorganization of dislocations at the plastic-elastic boundary zone of the work hardened microplastic regions which constrain the residual elastic strains; (2) the relaxation process involves dislocation climb and hence is dependent on the annealing temperature, on the degree of work hardening of the microplastic zone and on the complexity of the ensuing dislocation arrangement in the plastic zone; (3) the degree of strain hardening in the plastic zones governs the magnitude of the constraint and, therefore, the residual elastic strains; the greater the strain hardening the larger the elastic strains.

The latter perception is particularly important for the nondestructive evaluation of the accrued damage in a material because it suggests at once that it would be more advantageous to analyze the plastic zones which exert the constraining effect on the residual strains rather than the residual strains themselves. Furthermore, it suggests that one should search and localize the plastic zones exhibiting the maximum strain hardening effect. From a microscopic viewpoint such zones can be identified as lattice domains having large accumulation of excess dislocations. Because of these considerations attention will now be focused on the microstructural characterization and analysis of the plastic zones at the notch tips of specimens and the ensuing interaction of these zones on tensile deformation.

ANALYSIS OF NOTCH TIP PLASTICITY IN TENSILE-DEFORMED SILICON

The analysis of notch tip plasticity is based on the concept that the plastic zone generated at the notch tip is caused by dislocation generation and that the interaction of dislocations give rise to local lattice misalignment. The latter can be sensitively determined by X-ray rocking curve measurements using a double crystal diffractometer (DCD) arrangement. In such an arrangement the primary beam is first reflected from a monochromating crystal and then directed toward the test crystal which is to be analyzed. The reflecting power of the test crystal is measured by rotating the crystal through its angular range of reflection. The width, β, at half maximum of the "rocking" curve obtained is taken as a measure of the crystal perfection. The halfwidth values of perfect crystals may extend only over a few seconds of arc while those in imperfect crystals may extend over many minutes or even degrees. If the crystal contains a substructure the corresponding profile of the substructure is multi-peaked (13). By correlating the images obtained from reflection topographs or Berg-Barrett topographs (14) taken in close proximity of the crystal, with the rocking curve analysis of the corresponding lattice misorientation, quantitative information regarding the excess dislocation densities of the crystal can be obtained (15).

In the analysis of the notch tip plasticity each probed region in the plastic zone is considered to be the second crystal in a DCD arrangement. To map the plastic zone it is necessary to carry out rocking curve measurements of small adjacent lattice domains. In this manner the plastic zone becomes characterized quantitatively in terms of a contour map of equal lattice misalignment or equal density of excess dislocations. A computer-aided rocking curve analyzer (CARCA) has been designed and constructed which can perform this task rapidly and efficiently and can give a nearly instantaneous, detailed analysis over a large specimen area (16). Figure 4 will illustrate schematically the salient features in the operation of CARCA. After being reflected from

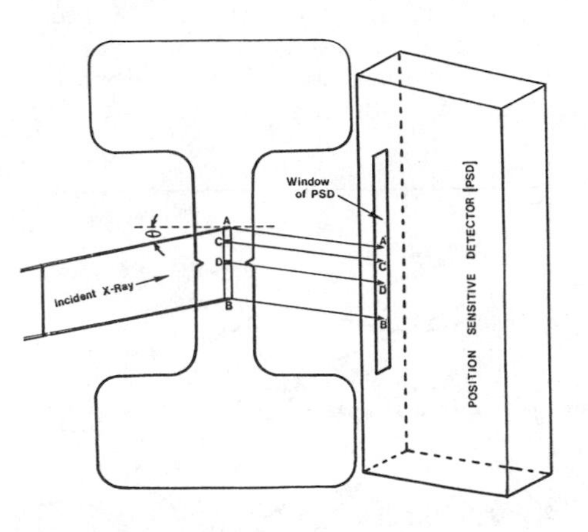

Fig. 4 - Principle of operation of CARCA for single crystals.

the first monochromating crystal the incident beam is made to impinge over a large area of the specimen, coincident with its rotation axis. The specimen is represented by a double-notched test specimen and the extent of the irradiated area is indicated by the letters A and B. The beam reflected from the test specimen is registered by a linear position-sensitive detector (PSD), placed parallel to the rotation axis of the specimen and recorded via a multi-channel analyzer (MCA) and computer on a printout chart. It will be readily seen from Fig. 4 that there exists a point-to-point correlation between the reflecting specimen areas and the positions on the PSD. Thus, the diffracted beam from the area around point C will fall in region C' of the PSD wire and that of specimen area D in region D' of the PSD. Consequently, adjacent specimen areas will diffract the beam onto adjacent sites on the PSD which correspond to adjacent channels of the MCA. Since the resolution of the PSD is about 65 μm, it will be readily seen that an irradiated specimen area of vertical length of 2 cm will yield simultaneously for one specimen setting about 307 diffracted beams from adjacent specimen areas 65 μm in size. Thus, by a single step-rotation of the specimen rocking curves are generated simultaneously at each channel. By translating the specimen on a microscope stage and repeating the rocking curve measurements, a complete map of rocking curves covering the entire specimen area can be obtained in a few minutes time and the data registered are stored on a floppy disk. From the computer printout the halfwidth values (β values) as well as the integrated intensities of the rocking curves are obtained and with the aid of these data and their known location on the specimen area contour maps of equal lattice misalignment (equal β values) are constructed. Figure 5 shows such quantitative mapping of the plastic zone of a single-notched silicon crystal, tensile-deformed at 800°C. Although the β values were determined with a precision of seconds of arc they are given in Fig. 5, for the sake of clarity, rounded up to the nearest value of minutes of arc. If one connects the β maxima of each lobe of the plastic zone one obtains the direction of the maximum plastic strain trajectories. They are indicated by the solid lines.

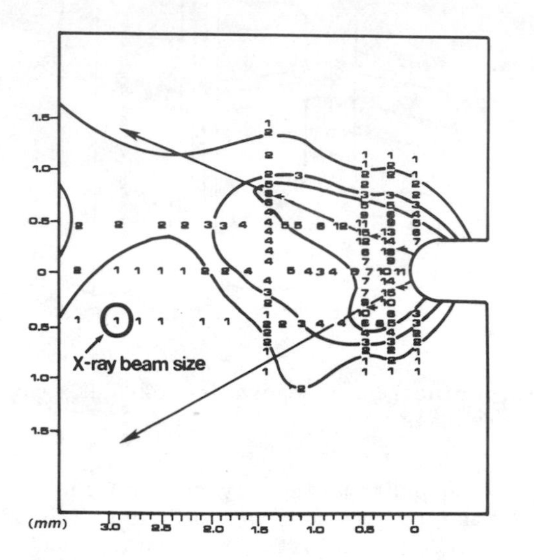

Fig. 5 - Quantitative mapping of plastic zone in regions of equal lattice misalignment, β. (With permission of J. Appl. Cryst., Ref. 17.)

With the aid of these data, complemented by the pictorial visualization of notch plasticity by B-B topography, it was possible to compare the experimental results to theoretical predictions based on continuum mechanics (17). Good agreement between experiment and theory was obtained regarding the shape of the plastic zone, the contribution of the active slip systems to the size of the plastic zone and the direction of the maximum plastic strain trajectory in the zone. Discrepancies between experiment and theory regarding the symmetry relation of the plastic zone lobes and the dislocation density near the notch tip underscored the fundamental differences between micro and macromechanics. They resulted from the interaction of dislocations and concomitant work hardening. These aspects cannot be taken into account in calculations of continuum mechanics.

MAPPING AND ANALYSIS OF PLASTIC ZONES AND THEIR INTERACTIONS IN TENSILE-DEFORMED DOUBLE NOTCHED SILICON

In view of the foregoing results it became apparent that for the nondestructive determination of accrued damage and for failure prediction of deformed materials it is important to obtain information regarding the interaction of plastic zones. It was also recognized that special attention should be given to the work hardening properties of surface layer and bulk since various investigators have shown marked differences in their deformation response (18-21). Furthermore, it was perceived that by mapping quantitatively the regions of largest lattice misalignment a maximum plastic strain trajectory could be found which would outline nondestructively the future fracture path of the material. To obtain these objectives the CARCA method was employed to map and analyze the microplascitity induced in a double-notched silicon crystal, tensile deformed at 800°C, which was again selected as a model material. The CARCA analysis was carried out on the entire specimen area both as a function of tensile deformation and on depth below the surface (16). The semiconductor crystals exhibit the attractive feature for deformation studies that the dislocations are immobile at room temperature so that the dislocation and configuration introduced at elevated temperature remains completely preserved when surface layers are removed by chemical polishing (12). Figures 6a and 6b show the mapping of the plastic zones at the surface in terms of equal lattice misalignment (excess dislocation density). When the

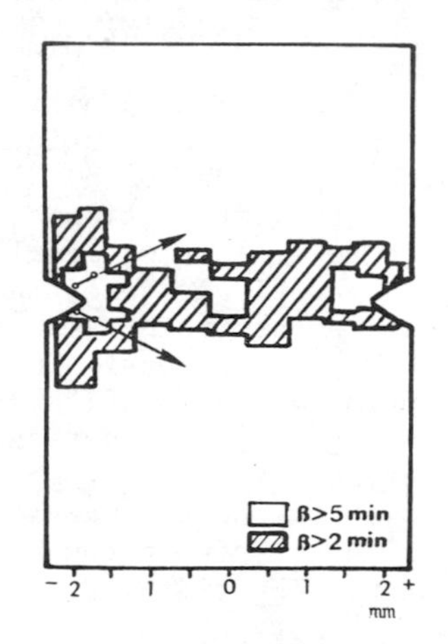

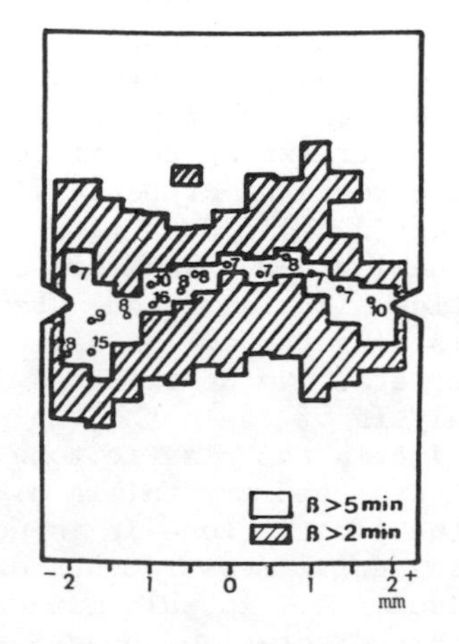

Fig. 6 - Mapping of plastic zones in silicon crystal, tensile-deformed at 800°C (a) 23.4 MPa, ε = 0.83%, (b) 27.5 MPa, ε = 1.04%.

small stress of 23.4 MPa (ε = 0.83) was applied, long-range interaction effects of the strained plastic zones were negligible and the plastic zones formed at the notch tip were similar to those developed in single-notched specimens (Fig. 6a, Fig. 5). They exhibited shape characteristics in agreement with calculations based on continuum mechanics. With increased deformation, viz 27.5 MPa (ε = 1.04%), the lattice misorientation induced in the internotch region of the specimen increased and as a result of long-range interactions of plastic strains the shape characteristics and strain distribution of the plastic zones at the notches became modified (Fig. 6b). While the optical microscope could only reveal at this early deformation stage slip activity around the notches (Fig. 7a) the CARCA method was able to establish clearly the development of a maximum plastic strain trajectory in the internotch region in the direction perpendicular to the stress axis (Fig. 7b). What is remarkable about

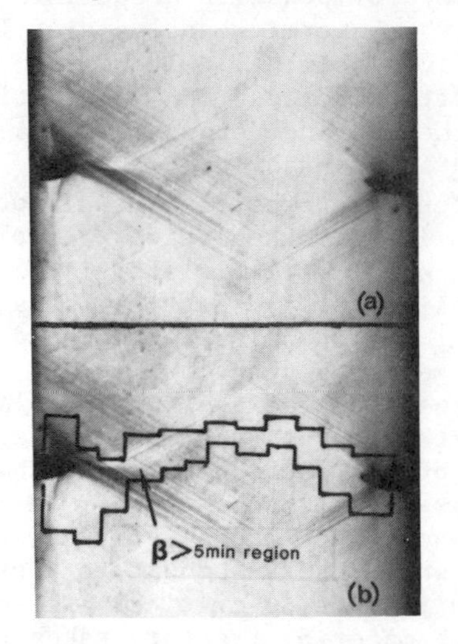

Fig. 7 - Plastic deformation characteristics of notch and internotch zones (a) light microscopy (b) zone mapping by X-ray DCD.

this preferred lattice misalignment in the internotch region is that it outlines the future fracture path as may be seen from the optical micrograph of Fig. 8.

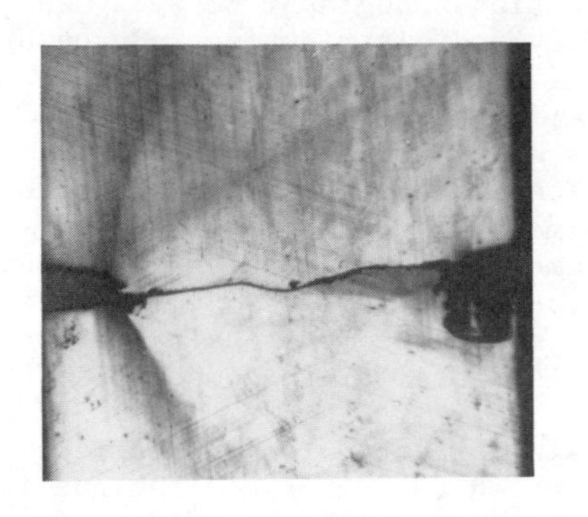

Fig. 8 - Fracture path disclosed by light microscopy.

Figure 9 shows the dependence of the extent and degree of microplasticity on the distance from the surface. It will be seen that after successive removal of the surface layers the extent as well as the severity of lattice misalignment decreased with depth. Thus, at the depth of 450 μm only a single isolated zone in the internotch region exhibited β values larger than 5 minutes of arc.

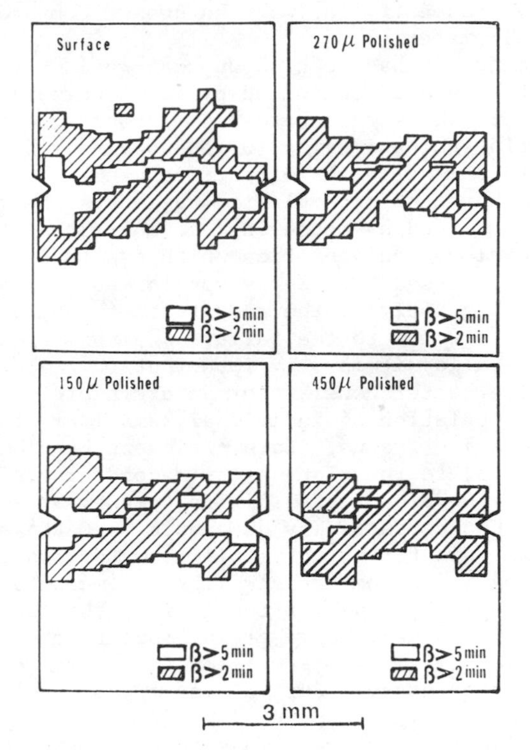

Fig. 9 - Dependence of distribution of microplasticity on distance from surface.

In agreement with X-ray depth profile studies on different materials and findings of a great many other investigators studying surface deformation (11,18-22), the results show clearly that the surface layers in ductile materials show a special propensity for work hardening compared to the bulk material.

SUMMARY OF RESULTS OF SINGLE CRYSTAL STUDIES AND APPLICATION OF DERIVED PRINCIPLES TO FAILURE ANALYSIS OF CYCLED ALLOYS

Using silicon single crystals as a model material, some important results have emerged regarding the distribution of elastic and plastic strains resulting from deformation. It will be shown shortly that these results can function as important guidelines for the determination of the accrued damage in cycled alloys and that on the basis of such determination a nondestructive failure prediction can be carried out successfully. The results of the single crystal studies may be summarized as follows:

a) Strain hardened microplastic zones constrain residual, elastic strains and the degree of strain hardening of the plastic zones governs the magnitude of the residual strains.

b) Accumulation of strain hardened microplasticity is accompanied by an increased dislocation density and the latter is manifested by pronounced lattice misalignment.

c) Large lattice misalignment is closely linked to the fracture of the material and may outline the macroscopic future fracture direction.

d) The surface layers of ductile materials have a propensity for work hardening compared to the bulk material.

Since it has been shown that residual, elastic strains are controlled by the constraining effect of the strain hardened microplastic zones, the principal guideline for the analysis of the accrued damage is the direct measurement of the strain hardened microplastic regions. The latter are determined by measuring accumulations of excess dislocation densities which can be sensitively assessed from X-ray rocking curve measurements. In applying the results of the single crystal studies to the failure analysis of a commercial alloy it is very important to realize that the latter consists of an aggregate of a large population of individual, small crystals: the metallic grains. Each grain can be viewed, therefore, to function as the second crystal in an X-ray double crystal diffractometer arrangement. Consequently, by developing a CARCA system of analysis for polycrystalline materials and applying it to commercial alloys to measure the X-ray rocking curves of the grains, the lattice defects of a very large grain population can be analyzed swiftly and precisely for every desired deformation stage. Such a system was developed and besides other applications (23) was used to determine the accrued damage in cycled Aℓ 2024-T4 alloys.

The schematic drawing of Fig. 10 may serve to illustrate the principle of operation of CARCA

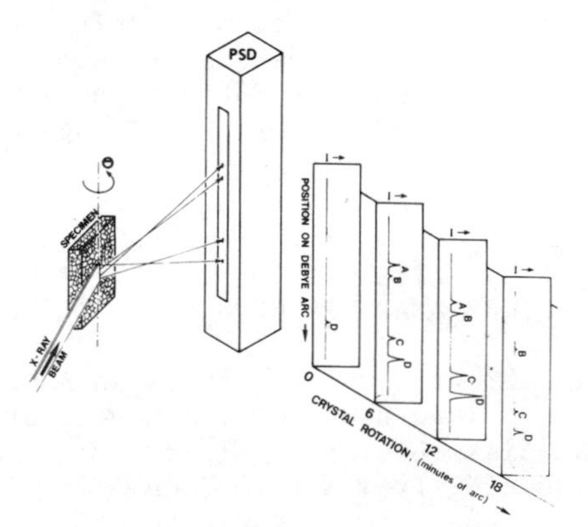

Fig. 10 - Principle of operation of CARCA for polycrystalline materials. (With permission of J. Appl. Cryst., Ref. 8.)

for polycrystalline materials. As in the single crystal arrangement the sample is irradiated by a parallel crystal-monochromated beam. The beams reflected from the individual grains are intercepted by a position-sensitive detector (PSD) which is placed parallel to the rotation axis of the specimen and tangent to the Debye-Scherrer arc of a specific (hkℓ) reflection to be analyzed. Each reflecting grain acts independently as the second crystal of a double crystal diffractometer and the grain reflections are recorded separately by the PSD and its associated multichannel analyzer at each increment of specimen rotation. An on-line minicomputer simultaneously collects these data and applies the necessary corrections. This process is then automatically repeated through the full rocking curve range. The computer carries out the rocking curve analysis of the individual grain reflections as well as that of the entire reflecting grain population. The instrument is provided with a specimen translation device which permits analysis of large sections of solid specimens. Sites of local accumulation of lattice defects can be identified with the aid of the PSD and topographic imaging on a film, placed in front of the PSD.

The intense interactions of dislocations give rise to local accumulations, or plastic zones, which are invariably associated with lattice misalignment and lattice curvature. It is this local lattice misalignment and curvature which CARCA measures quantitatively. Consequently, the accrued damage in a material is assessed in terms of the increases of the average half-width value of the rocking curve of the grains, $\bar{\beta}$, when a large grain population is analyzed. It will be shown that when a critical value, β^*, is reached, which is tantamount to a critical lattice misalignment or critical excess dislocation density, material fracture takes place. The nondestructive failure prediction by CARCA is based, therefore, on the correct measurement of β^*.

DETERMINATION OF ACCRUED DAMAGE AND FATIGUE FAILURE IN CYCLED Aℓ 2024 ALLOY

Pangborn et al. (22) using the DCD method carried out rocking curve measurements coupled with reflection topography to study the dislocation density and distribution induced by tensile deformation in single crystals of silicon, aluminum and gold and by tension-compression cycling in aluminum single crystals and Aℓ 2024 alloys. The measurements of dislocation density were also made at various depths below the surface by removing surface layers incrementally. In all these studies it was demonstrated that there exists a marked propensity for work hardening in the surface layers compared to the bulk material.

Focusing attention on the commercial Aℓ 2024 alloy subjected to high cycle fatigue in air, it was shown that the dislocation density, $\bar{\rho}$, in the surface layer increased rapidly early in the fatigue life but maintained virtually a plateau value from 20 to 90 pct of the life. Beyond 90 pct the dislocation density increased rapidly again to a critical value, ρ^*, at failure. Due to the extended plateau covering most of the fatigue life of the alloy it became quite evident that ρ^* at failure cannot be predicted from the surface analysis of the cycled alloy. Systematic evaluation of the dislocation distribution showed that, with increasing distance from the surface, the excess dislocations density increased more gradually during the life over the range of fractions of life from 0.1-0.9, and a linear relationship between the average rocking curve halfwidth and fraction of life was found. Advantage was taken of this gradual increase and using deeply penetrating molybdenum K_α radiation (capable of analyzing grains representative of the bulk region) the accrued damage and the onset of fatigue failure could be predicted. The critical dislocation density ρ^* was obtained when in the plot of $\overline{\rho}$ vs. N/N_F the ρ value of the bulk approached that of the surface. The physical interpretation of this behavior rests on the concept that due to rapid work hardening in the surface layer, the egression of the dislocations developed in the bulk is impeded by the surface layer and catastrophic crack propagation then sets in when the value of the excess dislocation density in the bulk reaches that of the surface, supporting the concept that the induced excess dislocation structure is the important criterion for failure predictions. Spectral loading studies have shown that the ability to predict failure was not hampered by complex loading histories within the range of stresses investigated.

The same principle of defect structure analysis and failure prediction was applied by Takemoto et al. for Aℓ 2024 cycled in tension-tension (R=0.1) in corrosive 3.5% NaCl solution (24). X-ray rocking curve measurements were carried out as a function of depth distance from the surface and typical results of the dependence of β on depth distance for an alloy cycled with σ=276 MPa, corresponding to the static yield stress, are shown in Fig. 11. It may be seen

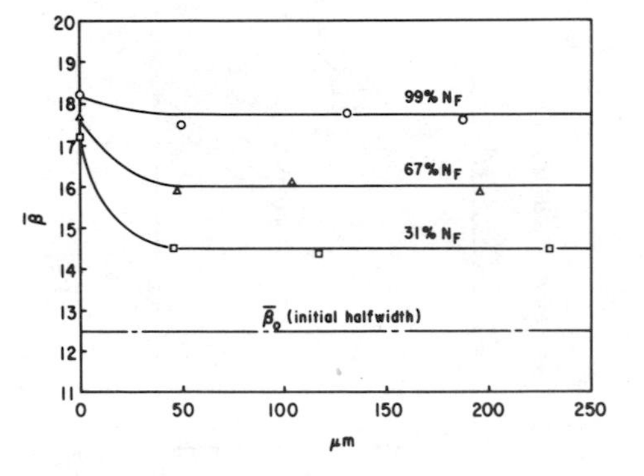

Fig. 11 - Dependence of the average rocking curve halfwidth $\overline{\beta}$ on depth distance from surface for different fractions of corrosion fatigue lives, N_F of Aℓ 2024-T4. Sm=276 MPa, R=0.1, Crkα_1. (With permission of Plenum Press, Ref. 24.)

that the $\overline{\beta}$ values at the surface layer were larger than those in the interior. The $\overline{\beta}$ values declined up to a depth of about 50 μm from the surface and subsequently retained a plateau value throughout the interior of the specimen for each fraction of the life. It should be noted that with cycling the $\overline{\beta}$ value of the surface layer approached rapidly the saturation level, while that in the bulk, represented by the plateau of the curves, increased more gradually with cycling. Furthermore, when fatigue failure set in the $\overline{\beta}$ value in the bulk reached practically the same value as that of the surface layer. It is this characteristic bulk response to cycling, showing a less rapid increase in $\overline{\beta}$ relative to the surface grains, that leads to the determination of the accrued damage by the nondestructive X-ray method.

Figure 12 shows the dependence of the average rocking curve halfwidth, $\overline{\beta}$, on the number of fatigue cycles for various maximum stress levels with R=0.1. As indicated in this figure, the solid lines of the curves pertain to the measurements with CrKα_1 radiation, while the dotted lines refer to those performed with MoKα_1 radiation. The absorption of the specimen is such that for chromium radiation only surface grains can be analyzed. By contrast, the short wavelength of the molybdenum radiation permits analysis of the grains exceeding the 50 μm depth distance from the surface. It will be seen from Fig. 12 that for the maximum stress of 241 MPa with R=0.1 the $\overline{\beta}$ value increased during the first several hundred cycles. This increase was much more pronounced for the surface grains (CrKα_1 radiation), which is in good agreement with the depth profile study of Fig. 11. For subsequent cycling, the $\overline{\beta}$ value of the surface grains maintained virtually a plateau level until about 20,000 cycles when macroscopic crack propagation set in at point A, corresponding to the critical halfwidth value, $\overline{\beta}^*$, of about 18 minutes of arc. By contrast, the $\overline{\beta}$ value of the bulk grains, obtained from MoKα_1 radiation, exhibited a conspicuous ascent during cycling. It should be noted that the two curves pertaining to surface and bulk grains converged at point A.

For the maximum stress of 276 MPa with R=0.1, the two curves pertaining to surface and bulk grains were obtained. They show an identical behavior to that observed for the maximum stress of 241 MPa with R=0.1, namely, a sharp increase in $\overline{\beta}$ value early in the life, different slopes for bulk grains, and the convergence of the two curves at B, where catastrophic failure set in. The critical halfwidth, $\overline{\beta}^*$, obtained was about 18 minutes of arc. The increase of the maximum stress level from 241 to 276 MPa decreased the life by about half, from N_F=20,000 cycles to N_F=11,100 cycles. When with R=0.1 the maximum stress of 310 MPa was applied which was a higher stress than the static yield stress, the $\overline{\beta}$ value at the surface layer no longer exhibited a plateau value but increased with cycling. This behavior is typical for low cycle fatigue when crack propagation takes up most of the life. The $\overline{\beta}$ value obtained from the bulk grain showed a

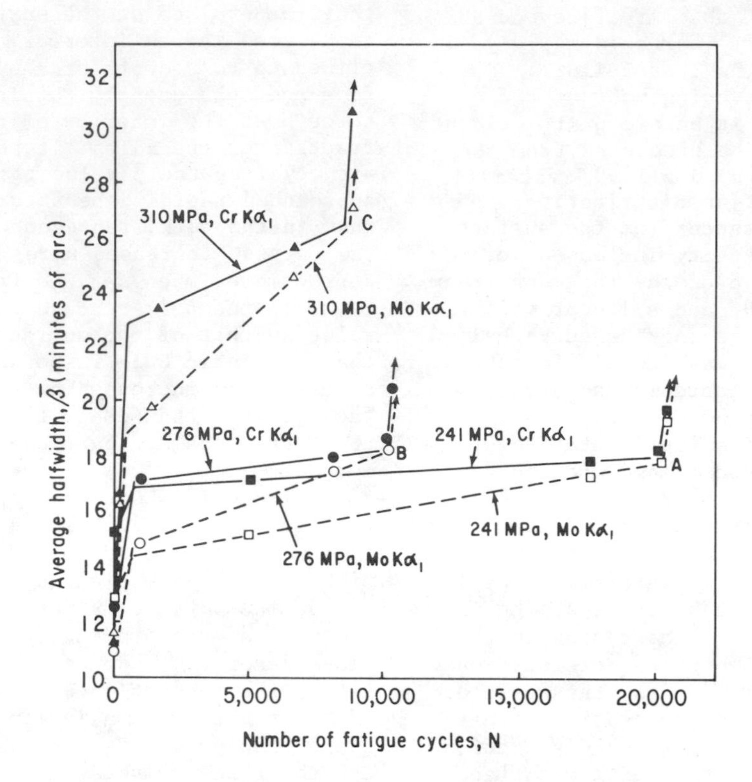

Fig. 12 - Dependence of $\bar{\beta}$ on number of cycles N at various stress levels of Aℓ 2024-T4. (With permission of Plenum Press, Ref. 24.)

similar behavior with a much steeper slope throughout the life. It should be noted, however, that the two curves converged again at C when propagation of a macrocrack set in. For this high stress level, the critical halfwidth value, $\bar{\beta}$*, corresponding to the point C, was about 26 minutes of arc.

The optical micrographs of Figs. 13a and 13b demonstrate the criticality of β* showing the catastrophic crack propagation corresponding to point A and B of Fig. 12, respectively.

It is evident that by using two different wavelengths, viz. CrKα and MoKα, at different stages of cycling and measuring in surface layer and bulk the changes of β, which is tantamount to measuring their excess dislocation densities, the critical β* (critical excess dislocation density) can be predicted. Such predictions were made with an error spread of only ±5 pct on specimens cycled to fracture (24). The rocking curve measurements of high cycle fatigue of Aℓ 2024 in air by Pangborn et al. (22) and those in hostile environment by Takemoto et al. (24) were carried out laboriously by registering the intensity variations of the reflecting grains photographically (13). Using the CARCA method the present work extends the concept of failure prediction to the area of low cycle fatigue and repeats also

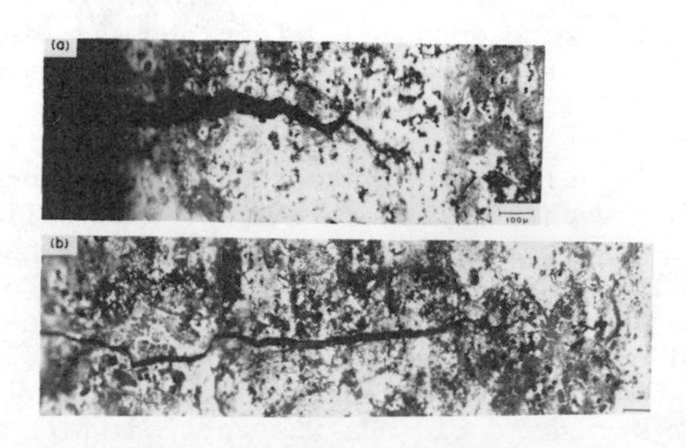

Fig. 13 - Micrographs showing cracks (a) for N=20,000 corresponding to point A in Fig. 12, (b) for N=11,000 corresponding to point B in Fig. 12. (With permission of Plenum Press, Ref. 24.)

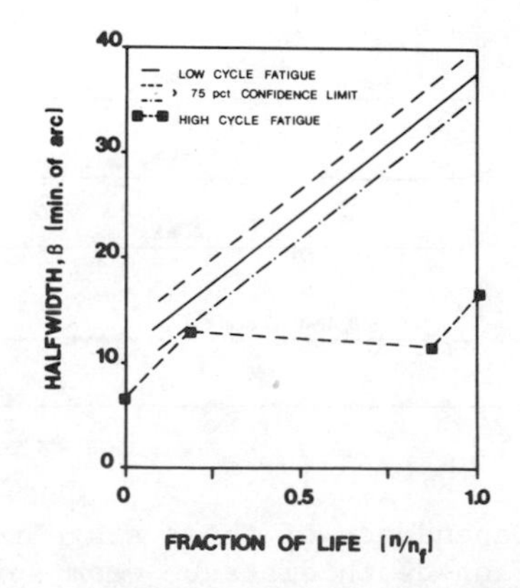

Fig. 14 - Dependence of $\bar{\beta}$ of surface layers on fraction of life, N/N_F for low and high cycle fatigue of Aℓ 2024.

the work of Pangborn et al. to ascertain the common and perhaps also the deviating aspects of fatigue behavior. These fatigue studies were conducted in a tension-tension mode under dynamic strain control. Figure 14 shows the dependence of $\bar{\beta}$ on fraction of life N/N_F for high cycle and low cycle fatigue when the surface layers were investigated by CARCA using CrKα radiation. It will be noted that for low cycle fatigue there exists a straight line dependence starting from very low N/N_F values to fracture. The data of high cycle fatigue, on the other hand, exhibit again the characteristic plateau effect, thus corroborating the results of Pangborn et al. The steady ascent in deformation response for low cycle is not surprising if one considers the fact that crack nucleation starts very early and that most of the life is taken up by crack propagation with its concomitant strong dependence on lattice misalignment and accumulation of excess dislocations. The resistance of the work hardened surface layer to the egression of dislocations originating in the bulk is then easily overcome. This effect was also apparent in corrosion fatigue (Fig. 12) when the alloy was cycled with the high stress amplitude of 310 MPa. It will be noted that both surface layer and bulk exhibit an ascending linear dependence of $\bar{\beta}$ on N/N_F and there is a conspicuous absence of the plateau, characteristically encountered in high cycle fatigue.

If in high cycle fatigue a more penetrating radiation like MoKα was used to extract information well below the surface where saturation of the buildup of excess dislocation has not occurred, a linear relationship between $\bar{\beta}$ and N/N_F was obtained in agreement with the results of Pangborn et al. After correcting for the difference in initial β_o values it can be seen in Fig. 15 that the two fatigue processes, although radically different in strain history, exhibit

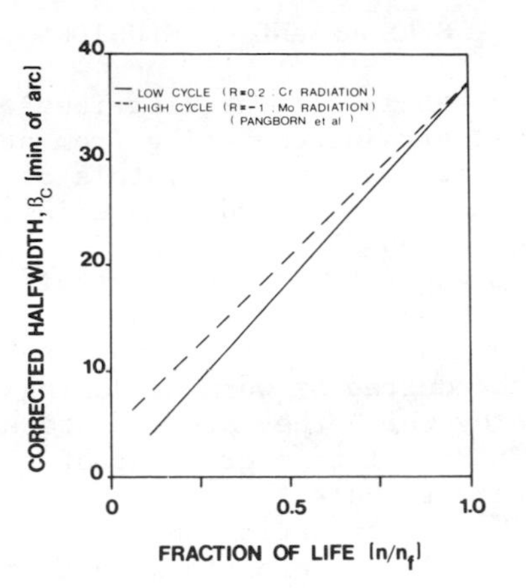

Fig. 15 - Dependence of $\bar{\beta}$ on N/N_F for low cycle fatigue and bulk properties of high cycle fatigue of Aℓ 2024.

similar behavior throughout most of the fatigue life.

The independence of failure prediction on the loading history was demonstrated by a series of spectral loading studies in low cycle fatigue. Keeping the total number of cycles the same for all specimens, N=17,000, the sequence of the different applied strain amplitudes was varied. After cycling, the specimens were examined with CARCA and the remaining fatigue life was estimated on the basis of $\bar{\beta}$ extracted from the analysis of a large grain population. The specimens were then returned to the fatigue machine and cycled to failure at 0.8% maximum strain. The actual remaining life was thus determined and made a comparison of the predicted life possible. The comparisons are summarized in Table I. It will be seen that the agreement between predicted and actual remaining life is quite satisfactory and that CARCA is a useful method for fatigue failure prediction.

Table 1 - Predictability Studies of Aℓ 2024-T4

Type of Loading	History	$\bar{\beta}$ Measured	Predicted Cycles Remaining	Actual Cycles
Monotonic	3,000 @ 1.0% ε_{max}	22.09'	4,550 ± 1,550	6,430
Spectral	2,000 @ 1.0% ε_{max}	19.65'	26,025 ± 5,225	27,285
	+10,000 @ 0.6% ε_{max}			
	+ 5,000 @ 0.8% ε_{max}			
Spectral	10,000 @ 0.6% ε_{max}	22.09'	22,230 ± 3,230	19,842
	+ 2,000 @ 1.0% ε_{max}			
	+ 5,000 @ 0.8% ε_{max}			
Spectral	10,000 @ 0.6% ε_{max}	18.89'	26,200 ± 3,230	22,730
	+ 5,000 @ 0.8% ε_{max}			
	+ 2,000 @ 1.0% ε_{max}			

SUMMARY AND CONCLUSIONS

X-ray studies of the distribution of elastic and plastic strains emanating from stress raisers were performed on single crystals of silicon which functioned as a model material. These studies have shown:

1. Microplastic zones constrain residual elastic strains.
2. The magnitude of the residual strains depend on the degree of work hardening of the plastic zones by which they are constrained.
3. There exists a gradient of plasticity from surface to bulk.
4. The microplastic zones can be mapped out and quantitatively determined in terms of excess dislocation densities by X-ray rocking curve measurements.

The results and ideas which emerged from the single crystal studies were applied to determine by means of computer-aided rocking curve analyzer the accrued damage in commercial Aℓ 2024 alloys, cycled in air, and in corrosive medium. By contrast to conventional X-ray residual stress measurements which depend principally on measurements of variations of interplanar spacings to yield an average maximum strain value from which an average maximum residual stress is deduced, the success of the present method rests on the determination of the maximum plastic strain associated with grains of greatest excess dislocation density. On the basis of this determination successful, nondestructive failure predictions were made for the cycled alloys.

ACKNOWLEDGMENT

The support of the single crystal studies by the Division of Materials Research, Ceramics Program, Metallurgy, Polymers and Ceramics Section of the National Science Foundation is gratefully acknowledged. It is a pleasure also to acknowledge the support of the fatigue studies by the D. W. Taylor Naval Ship R&D Center, Annapolis, Maryland.

REFERENCES

1. Neuber, H., "Kerbspannungslehre", pp. 58-111, Springer, Berlin (1958), (Engl. trans: Edwards Bros., Ann Arbor, Michigan)
2. Neuber, H., Trans. ASME, 28, 544-550 (1961)
3. Muskhelishvili, N. I., "Some Basic Problems of the Mathematical Theory of Elasticity", Noordhoff, Groningen, Holland (1953)
4. Irwin, G. R., J. Appl. Mech. 24, 361-364 (1957)
5. Kalman, Z. H. and S. Weissmann, J. Appl. Cryst. 12, 209-220 (1979)
6. Kalman, Z. H. and S. Weissmann, J. Appl. Cryst. 16, 295-303 (1983)
7. Kalman, Z. H., Chaudhuri, J., Weng, G. J. and S. Weissmann, J. Appl. Cryst. 13, 290-296 (1980)
8. Chaudhuri, J., Kalman, Z. H., Weng, G. J. and S. Weissmann, J. Appl. Cryst. 15, 423-429 (1982)
9. Lang, A. R., J. Appl. Phys. 29, 597-598 (1958)
10. Weissmann, S., Greenhut, V. A., Chaudhuri, J. and Z. H. Kalman, "Quantitative Analysis of Intensities in X-ray Topographs by Enhanced Microfluorescence", J. Appl. Cryst. (1983, in press)
11. Tsunekawa, Y. and S. Weissmann, Met. Trans. 5, 1585-1593 (1974)
12. Alexander, H., Phys. Status Solidi 26, 725-741 (1968); ibid. 27, 391-412 (1968)
13. Weissmann, S. and D. L. Evans, Acta Cryst. 7, 733-737 (1954)
14. Barrett, C. S., Trans AIME 161, 15-64 (1945)
15. Hirsch, P. B., Prog. Met. Phys. 6, 282 (1956)
16. Liu, H. Y., Mayo, W. E. and S. Weissmann, "Mapping and Analysis of Microplasticity in Tensile-Deformed Double-Notched Silicon Crystals by Computer-Aided X-ray Double-Crystal Diffractometry, Mat. Sci. and Eng. (in press)
17. Liu, H. Y., Weng., G. J. and S. Weissmann, J. Appl. Cryst 15, 594-601 (1982)
18. Kramer, I. R. and N. Balasubramanian, Acta Met. 21, 695-699 (1973)
19. Goritskii, V. S., L. S. Ivanova and J. F. Teren'EV, Sov. Phys. Doklady 17, 776-779 (1973)
20. Tabata, T. and H. Fujita, J. Phys. Soc. Jap. 32, 1536-1544 (1972)
21. Kramer, I. R., Trans. Met. AIME 230, 991-1000 (1964)
22. Pangborn, R. N., Weissmann, S. and I. R. Kramer, Met. Trans. 12A, 109-120 (1981)
23. Yazici, R., Mayo, W., Takemoto, T. and S. Weissmann, J. Appl. Cryst. 16, 89-95 (1983)
24. Takemoto, T., S. Weissmann and I. R. Kramer, "Fatigue Environment and Temperature Effects", pp. 71-81, Plenum Press, New York (1983)

THE USE OF ULTRASONIC SIGNAL ANALYSIS EVALUATION OF CARBON STEEL FOR DEFORMATION INDUCED MICROSTRUCTURAL DAMAGE

G. H. Thomas
Sandia National Laboratories
Livermore, California 94550

S. H. Goods
Sandia National Laboratories
Livermore, California 94550

A. F. Emery
University of Washington
Seattle, Washington 98195

ABSTRACT

Ductile fracture of metals and alloys generally occurs by the formation and growth of voids or cavities. In this study, ultrasonic attenuation measurements have been shown to be sensitive to the presence of a deformation induced cavity microstrucure. A series of experiments was performed on a deformed carbon steel alloy (AISI 1074) which had been heat treated to produce a spheroidized carbide microstructure. This microstructure insured a uniform distribution of void nucleation sites. The degree of ultrasonic attenuation was determined by spectral analysis of broadband wave forms. The results showed that certain discrete frequency intervals were sensitive to the density of cavities and possibly to their size distribution. The attenuation encountered was primarily caused by Rayleigh scattering from the cavities. This was confirmed by a post-deformation annealing which reduced the number of cavities and caused the ultrasonic signal to regain its strength.

INTRODUCTION

There is significant interest in a nondestructive technique for detecting and measuring cavitation in metals. Ultrasonic attenuation has been shown to be a viable method for detecting cavitation in an aluminum alloy, 6061-T6 [1]. In this work, the procedure was extended to study the relationship between ultrasonic signal loss and cavity formation in steel. The steel was heat treated to form a spheroidized carbide microstructure. These carbides act as void nucleation sites when the steel is subjected to deformation.

The mechanism causing cavitation induced attenuation is that of scattering. The ultrasonic pulse is directed through the material where the voids have formed and is scattered at the void surfaces [2]. Absolute measures of attenuation have been related to the deformation induced microstructure of the specimen and an apparent correlation is presented.

SPECIMEN PREPARATION AND MECHANICAL TESTING

The material used in this study was AISI 1074 steel bar stock. The stock was subjected to a duplex heat treatment in order to produce a spheroidized microstructure. Specimen blanks were first austenitized at 805°C for 2 hours and then oil quenched to room temperature. A spheroidized microstructure was then produced by tempering the as-quenched material at 720°C for 2 hours, followed by air cooling. The blanks were then machined to the final specimen dimensions. The specimens had a 5 centimeter gage length with a rectangular cross-section measuring 6.3 x 9.5 millimeters.

Tensile tests were performed on an electromechanical test frame to strain levels near fracture. After testing, specimen cross-sections were carefully measured at various locations along the neck. From these measurements, the reduction in area as a function of specimen location was computed.

ULTRASONIC DATA ACQUISITION AND REDUCTION

Once the tensile samples were deformed to strains near failure (50% reduction of cross sectional area at the center), they were machined flat and parallel and then scanned ultrasonically. A computerized ultrasonic data acquisition system is shown schematically in Figure 1. The system consisted of: a focused 25 MHz broad band transducer; a Panametric 5052 PRX pulser/receiver; a Biomation 8100 analog to digital converter; and a PDP 11/34 minicomputer. The x-y scanner in this system moved the transducer along the necked region and acquired R-F waveforms at 1.27 millimeter intervals. The stored R-F waveforms consisted of the front surface reflection and at least six back wall

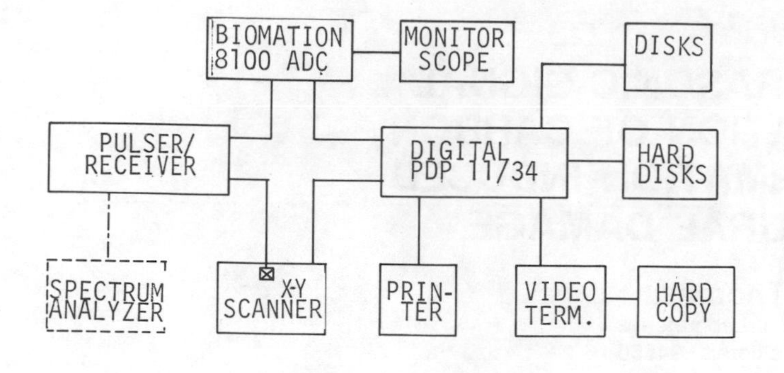

Fig. 1 - Block Diagram of the Computerized Ultrasonic Data Acquisition and Signal Processing System

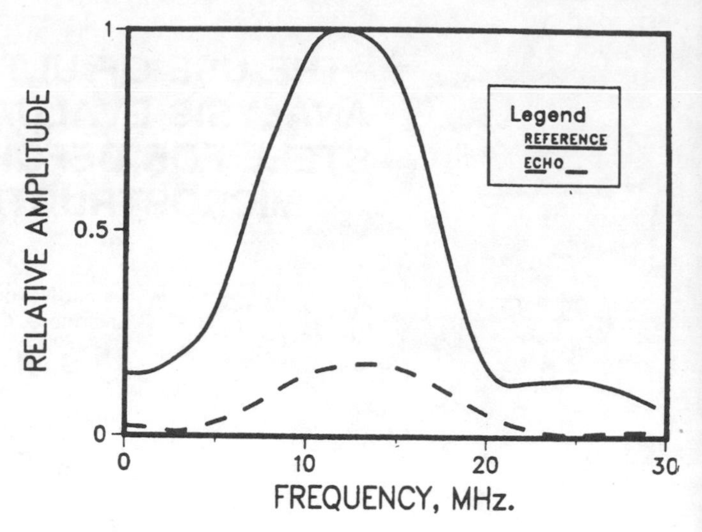

Fig. 3 - Examples of Fast Fourier Transforms for Reference and Back Wall Echo

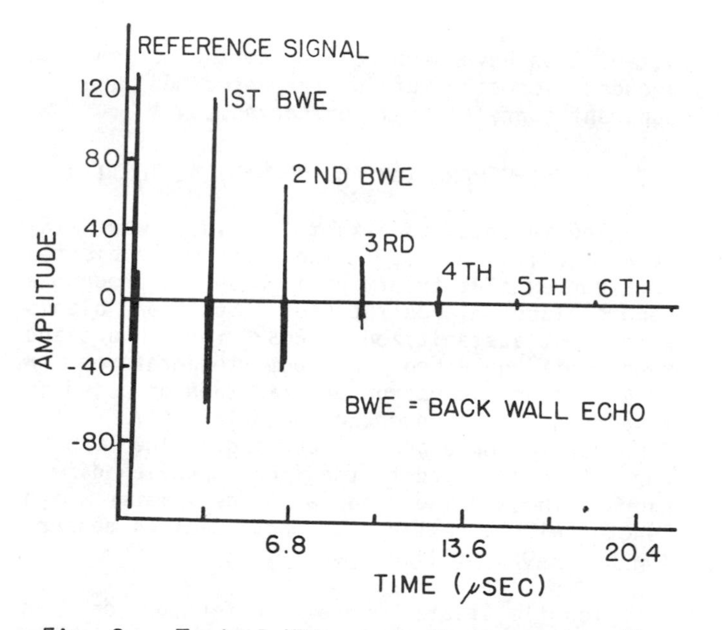

Fig. 2 - Typical Ultrasonic Response From The Steel Tensile Specimen. Attenuation Calculated by Dividing Peak-to-Peak Amplitude of Each Echo by the Peak-to-Peak Amplitude of the Reference

reflections, Figure 2. A Fourier Transform was then calculated for each of these backwall reflections. The frequency spectrum for the first surface reflection and each back wall echo was divided into five megahertz intervals and the area under the curve for each interval was calculated. A normalized value for ultrasonic attenuation for each of the five megahertz intervals was determined by dividing the area under the curve for the particular interval for each echo by the correspoinding area for the front reflection, see Figure 3. In this way the attenuation was calculated at various locations along the neck where the absolute value of the strain was known.

MICROSTRUCURAL ANALYSIS

After ultrasonic examination, the deformed gage sections of the tensile specimens were sectioned and mounted for metallographic analysis. These sections were ground, polished to a 0.05 μm finish and lightly etched using a 3.0% nital solution.

The prepared specimens were then examined by scanning electron microscopy. A computer controlled automated image analysis routine was employed to count and size the number of cavities at different locations (corresponding to various strain levels) within the gage length. In this way, the variation of void fraction with specimen strain could be quantitatively measured.

The technique employed a digital scan generator to locate and size the cavities. Cavity detection was based upon comparing a digitized video signal level (from the backscattered electron signal) to preset upper and lower threshold values. Upon locating the center of the cavity, it was sized through the use of eight rotating diameters. Values of the maximum, minimum and average diameter of each cavity were stored in memory along with the calculated area fraction. The data could then be plotted using a histogram, examples of which accompany the following text.

RESULTS

A correlation between ultrasonic attenuation and void nucleation and growth in spheroidized 1074 steel has been determined. In Figure 4a the ultrasonic signal strength of an as-deformed specimen is plotted as a function of location along the gage length of a tensile specimen. Also plotted in Figure 4a is the reduction in cross-sectional area (RA) as a function of location (the increase in RA resulted from necking or inhomogeneous plastic strain). The figure shows that the ultrasonic signal strength decreased with increasing strain in the specimen.

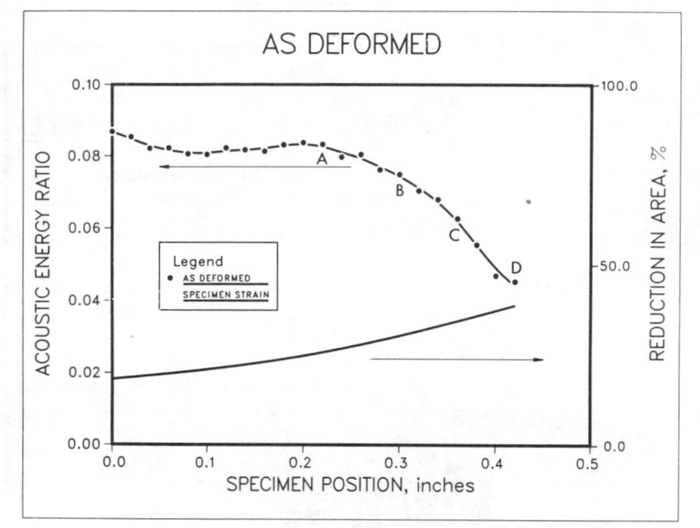

a) As Deformed

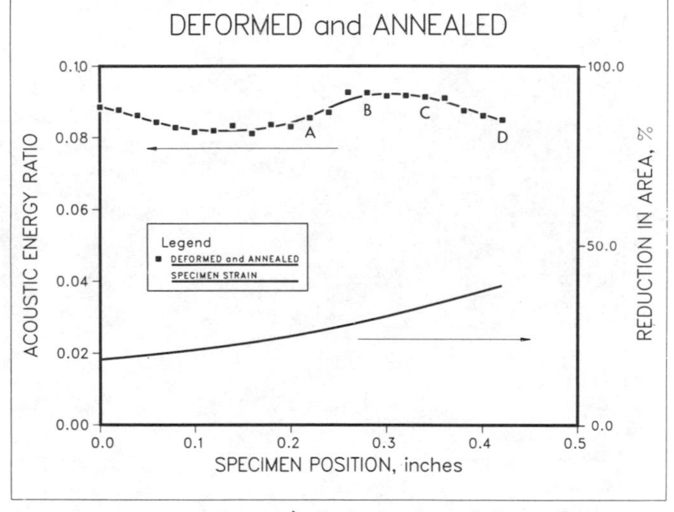

b) Deformed and Annealed

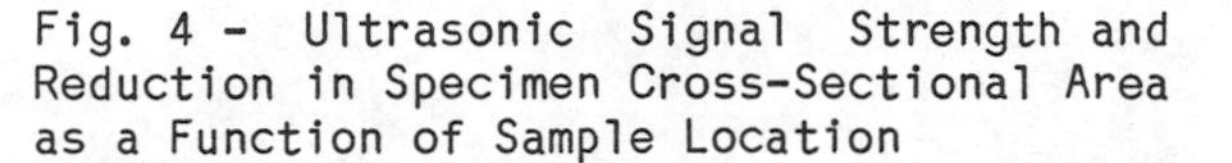

Fig. 4 - Ultrasonic Signal Strength and Reduction in Specimen Cross-Sectional Area as a Function of Sample Location

Metallographic analysis revealed that the decrease in signal strength corresponded to an increase in void area fraction. Figures 5 and 6 are scanning electron micrographs of the microstructure taken from locations A and D (as indicated in Figure 4) of the as-deformed specimen. Along with the accompanying histograms, these figures clearly reveal that the number and area fraction of cavities increased with increasing strain. Thus the increase in void area fraction can account for the loss in acoustic energy shown in Figure 4a by increasing the ultrasonic signal scattering.

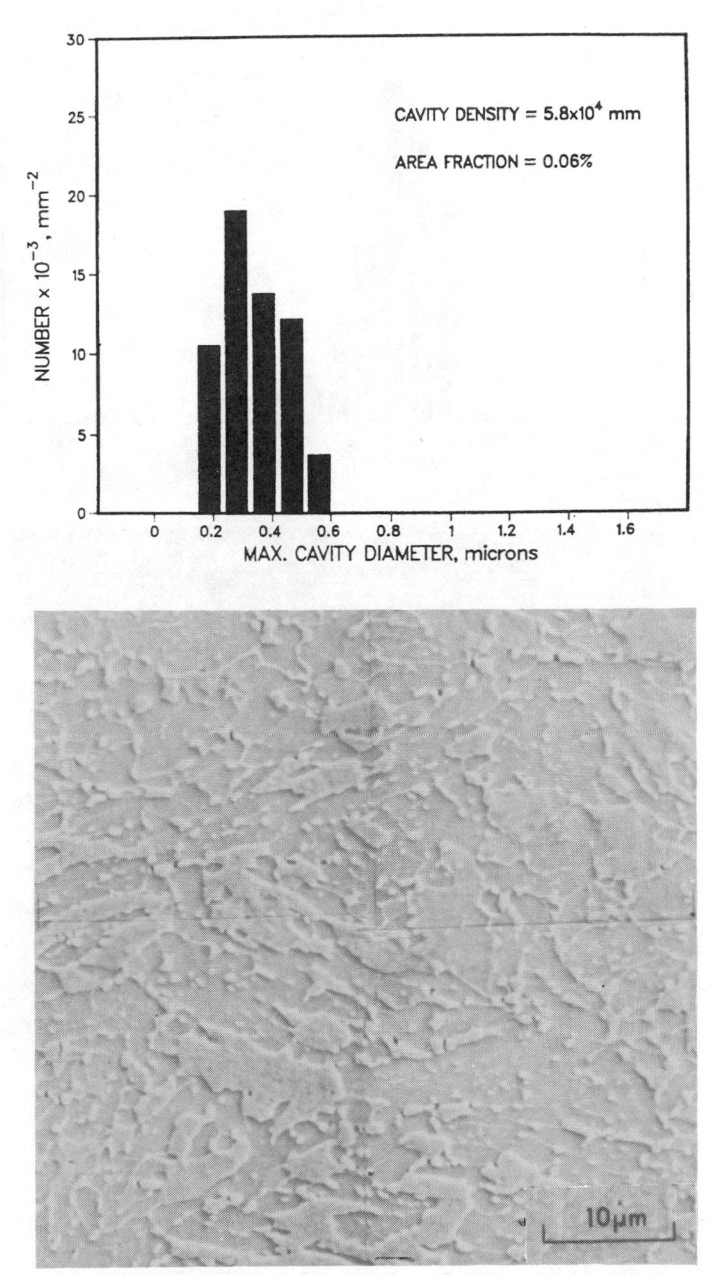

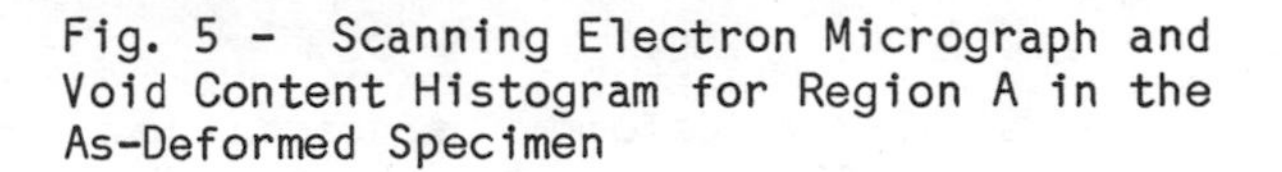

Fig. 5 - Scanning Electron Micrograph and Void Content Histogram for Region A in the As-Deformed Specimen

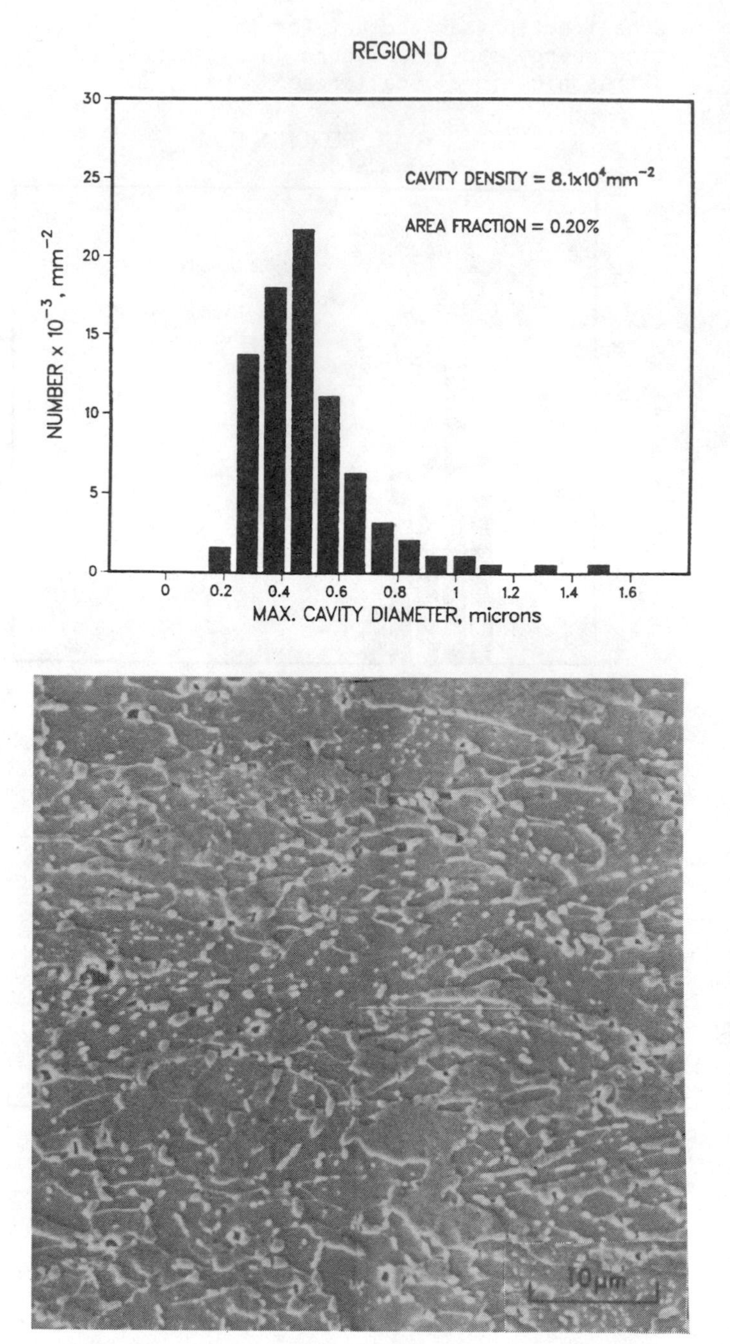

Fig. 6 - Scanning Electron Micrograph and Void Content Histogram for Region D in the As-Deformed Specimen

A similarly deformed specimen was then annealed at 675°C for two hours. This heat treatment resulted in a significant decrease in void area fraction in the region of highest strain (region D) while only marginally affecting the microstructure in regions of low strain (A). The effect of the annealing procedure on the cavity morphology is apparent in Figure 7 which shows the microstructure and distribution for region D in the heat treated specimen.

Analysis of the cavity size and distribution revealed that the 675°C annealing resulted in a five fold decrease in the void area fraction in region D when compared to the as-deformed specimen.

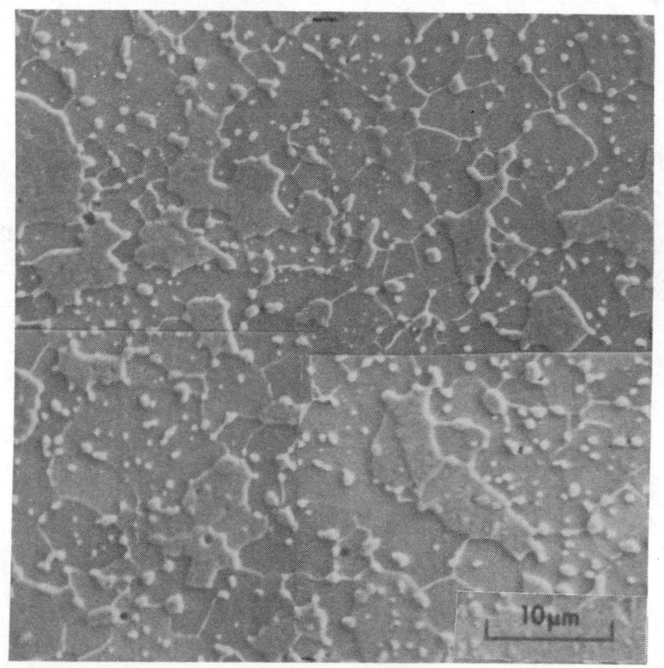

Fig. 7 - Scanning Electron Micrograph and Void Content Histogram for Region D in the Annealed Specimen

The sample was then ultrasonically scanned and the attenuation computed. These results are shown on Figure 4b. It is clear that the post-deformation annealing has resulted in an increase in signal strength in the region of greatest reduction in area. This further suggests that the observed attenuation was caused by scattering from the cavities.

CONCLUSIONS AND FUTURE EFFORTS

The results indicate that ultrasonic attenuation is sensitive to the population of deformation induced cavities in the 0.1 μm to 1.5 μm size range. Increasing the extent of cavitation resulted in a decrease in signal strength.

Future efforts will focus on separating the influence of grain size variation, specimen thickness, strain level and strain inhomogeneities on the level of attenuation.

ACKNOWLEDGEMENTS

This work is supported by the U. S. Department of Energy under contract DE-AC04-76DP00789.

REFERENCES

1. Thomas, G. H., S. H. Goods, and A. F. Emery, "Detection of Strain Induced Microstructural Changes in Aluminum 6061-T6 Using Ultrasonic Signal Analysis", Review of Progress in Quantitative Nondestructive Evaluation, Vol. 2, D. O. Thompson and D. E. Chimenti, Eds., Plenum Press, New York, 1983
2. Krautkramer, J., H. Krautkramer, Ultrasonic Testing of Materials, Springer-Verlag, New York, pp. 107-113, 1977

EVALUATION OF SURFACE MACHINING DAMAGE IN STRUCTURAL CERAMICS

B. T. Khuri-Yakub
Edward L. Ginzton Laboratory
Stanford University
Stanford, California 94305

Y. Shui
Edward L. Ginzton Laboratory
Stanford University
Stanford, California 94305

D. B. Marshall
Rockwell International Science Center
1049 Camino dos Rios
Thousand Oaks, California 91360

ABSTRACT

The grinding and polishing of structural ceramics has been found to leave subsurface cracks in the range of 10-20 μm in depth. These subsurface cracks, which are closed at the surface due to residual stress, are responsible for the fracture of ceramic parts in at least 70% of the samples tested. We developed an acoustic nondestructive evaluation technique to locate and size subsurface cracks, and a fracture mechanics model to predict failure stress based on the knowledge of the size of these subsurface cracks. Our experimental results indicate that our error in crack sizing is less than 20%, and our error in fracture stress prediction is less than 10%.

INTRODUCTION

Cutting, grinding, and polishing operations leave subsurface damage in brittle materials. In high-quality ceramics, such as hot pressed silicon nitride, the subsurface machining damage can be the major strength reducing factor. In some earlier work[1,2] we found that machining operations leave subsurface long slot-like cracks. We simulate the machining damage by pushing a Knoop indentor into the surface of a sample and then dragging the indentor to generate a subsurface crack of the required lateral extent.

From the results of this work, we find that the fracture mechanism developed in reference 1 still holds. Namely, the long slot-like cracks are closed at the crack mouth due to residual stress, and when the samples are stressed to fracture, a single half-penny shaped crack grows from the precrack and leads to failure. The cracks observed on a fracture surface are shown schematically in Fig. 1(a), and an actual photograph of a fracture surface is shown in Fig. 1(b). Notice that there is evidence of stable crack growth from the precrack to a half-penny shaped crack. For a precrack depth C_0, the sample will break when the half-penny shaped crack reaches a depth of C_d. The crack at fracture is elliptical in shape and has a semi-major axis parallel to the surface of length C_ℓ. Typically, C_ℓ/C_d is equal to 3. We find that the fracture stress is given by:

$$K_{IC} = 2.4 \sqrt{\frac{a}{\pi}}\, \sigma_F \qquad (1)$$

where K_{IC} is the fracture toughness of the material ($K_{IC} = 4.7\ MPa\sqrt{m}$ for hot pressed Si_3N_4) and

$$a = \sqrt[3]{C_d^2 C_\ell} \qquad (2)$$

Experimentally we find the simple relationship:

$$a \simeq 4.6\ C_0 \qquad (3)$$

We see then that our measurement technique has to address the problem of sizing a long, slot-like subsurface crack.

The theoretical reflection coefficient of long, slot-like, subsurface cracks to a surface acoustic wave has already been calculated by Achenbach and Brind.[3] Figure 2 shows the result of their calculation for three cracks with different closure depths at the mouth of the crack versus crack depth. For cracks under consideration in this work, the different closure depths would correspond to different amounts of surface residual stress. Notice that at low kb (Fig. 2), the reflection coefficient of the cracks is quite different which makes it possible to measure the depth of the contact area at the crack mouth, and possibly the residual stress.

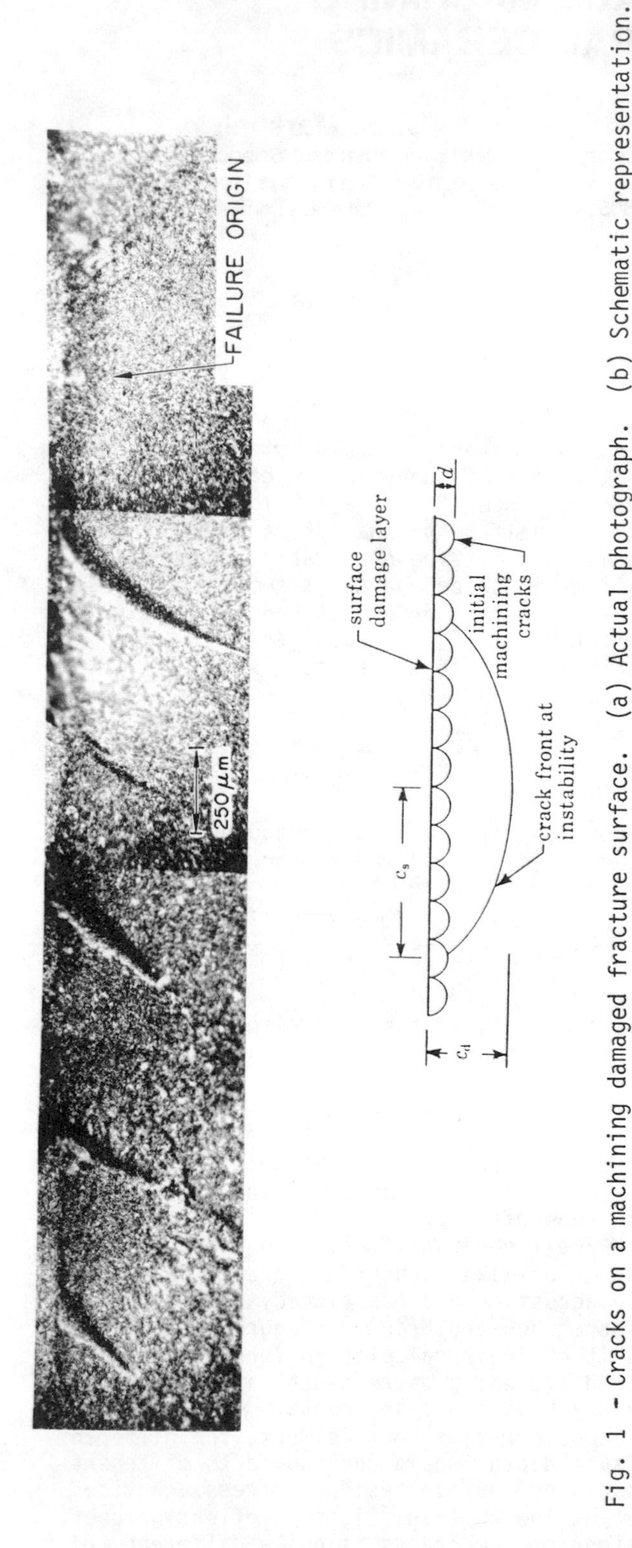

Fig. 1 - Cracks on a machining damaged fracture surface. (a) Actual photograph. (b) Schematic representation.

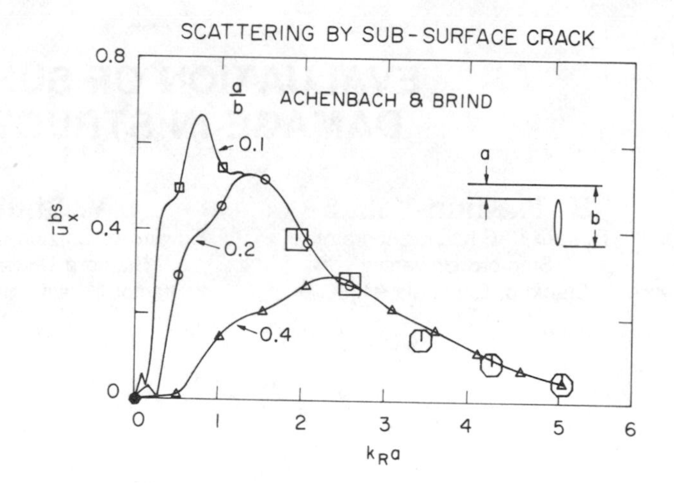

Fig. 2 - Surface wave reflection coefficient of subsurface cracks[2]

We use an immersion pulse-echo technique to measure the reflection coefficient of the cracks, as described in reference 2. Basically, a longitudinal transducer operating at a center frequency of 50 MHz is incident on the sample at the critical angle for Rayleigh wave excitation. If no crack is present in the sample, the Rayleigh wave, which is leaky due to the presence of the water, reradiates its energy back into the water as a longitudinal wave. If a crack is present on the sample, a reflected leaky Rayleigh wave sends its energy back to the transducer which detects the presence of the crack.

The transducer is excited with a sharp pulse in order to obtain a broadband excitation spectrum. The reflection coefficient of the crack is measured versus frequency, along with the reflection coefficient of a 90° corner. The corner reflection is used to divide out the frequency response of the transducer, electronics and propagation path by standard Wiener filtering techniques.[4] Figure 3 shows the result of one such measurement. Our measurements are made in the region of low kb, and we notice the presence of a null in the measured reflection coefficient that corresponds to a null in the theory for the case of $(a/b) = 0.2$, and occurs at $kb = 1.25$. The null observed is not due to a resonance phenomena; it is due to the destructive addition of the reflected shear and longitudinal components of the surface wave.

Scratches made with six different loads on the Knoop indentor, as shown schematically in Fig. 4, were used in our experiments. All the measurements showed nulls in the reflection coefficient of the cracks corresponding to $kb = 1.25$ and $a/b = 0.2$. Thus, knowing the frequency of the null, and the surface wave velocity on the sample, it is possible to calculate the precrack depth b. Figure 5 shows a graph of the acoustically-predicted precrack sizes as compared to the actual crack sizes measured on the fracture surface. We find that we consistently underestimate the crack size.

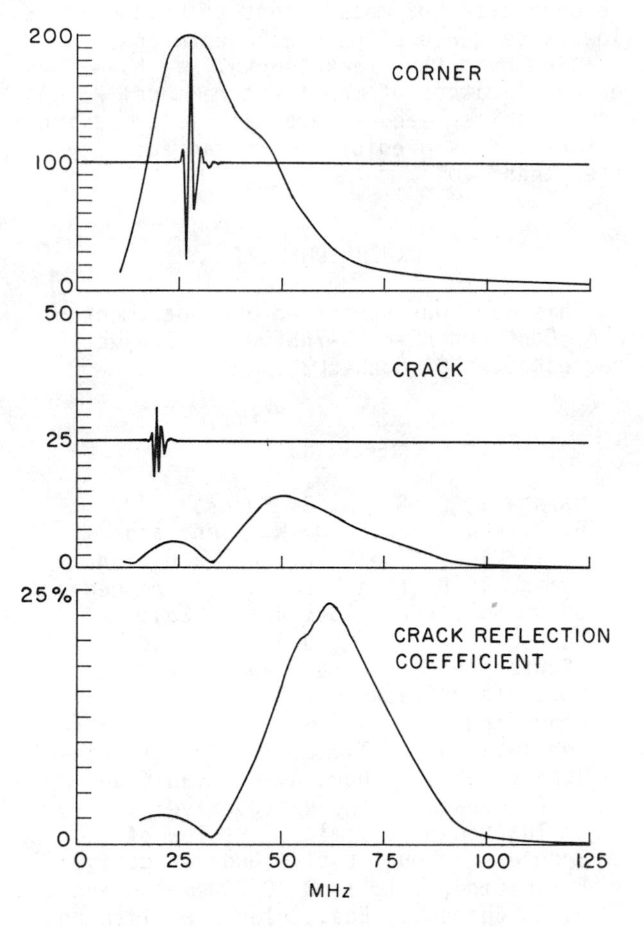

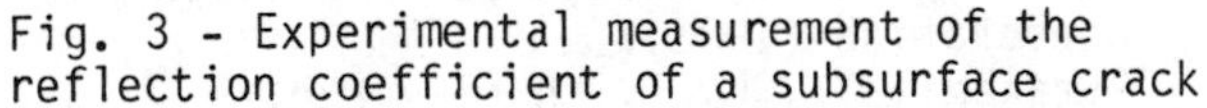
Fig. 3 - Experimental measurement of the reflection coefficient of a subsurface crack

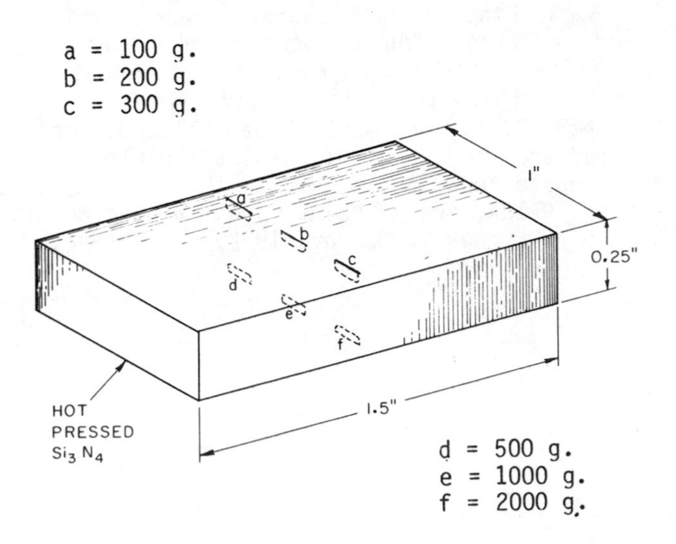

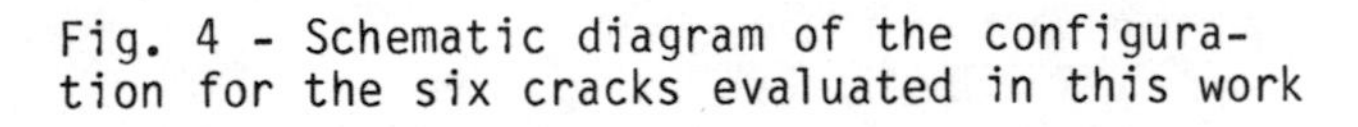
Fig. 4 - Schematic diagram of the configuration for the six cracks evaluated in this work

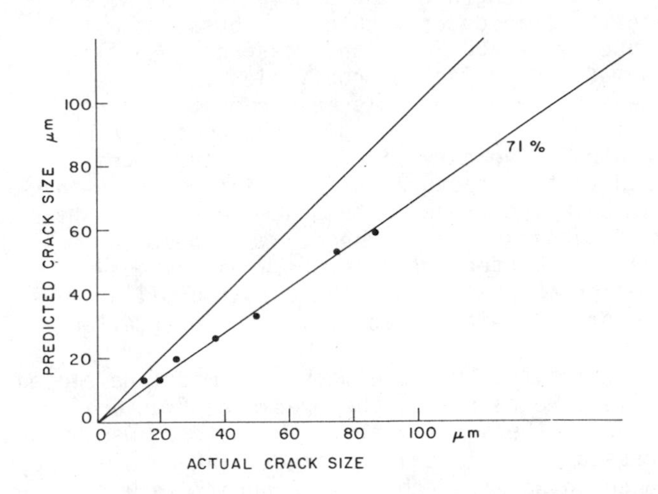

Fig. 5 - Comparison of acoustically-predicted crack size and measured crack size

A close examination of the fracture surface in Fig. 1b shows that the precrack is not a uniform slot of constant depth; instead, it shows a series of cracks of various depths that make the overall precrack. Consequently, when carrying out a measurement with an acoustic beam width of 1 mm, we will average the crack size over the insonified region. It is important to restate that the longitudinal and shear components of the reflected surface wave will cancel at some frequency for a subsurface crack of varying depth, and thus yield a null in the frequency spectrum, as shown in Fig. 3. This averaging effect is responsible for the underestimation of the precrack size. We are presently developing a set of focused transducers that will allow us to probe surface cracks with an acoustic beam 0.1 mm wide.

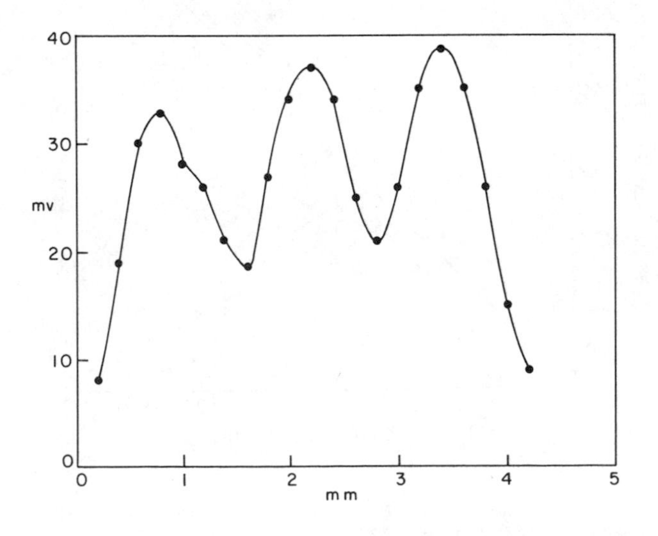

Fig. 6 - Reflection coefficient of 1000 g. crack at 50 MHz vs. distance along crack length

We carried out a separate measurement to determine the variation of the crack depth across the length of the sample. Here, we excite the transducer with a tone burst at a frequency of 50 MHz , and measure the amplitude of the reflection coefficient. Figure 6 shows the result of this measurement for the crack made with a 1000 gm load on the Knoop indentor. The curve in Fig. 6 is to be compared to the fracture surface picture shown in Fig. 1b. The close correspondence between the two figures is clearly seen. We expect to have much better agreement between the measured crack profile and the actual profile when probing with an acoustic beam of smaller width.

Using the fracture mechanics model developed earlier, we calculate the predicted fracture stress of the six samples. The samples were then stressed to failure in a three-point bending jig, placed in an MTS machine . Figure 7 shows a comparison between the actual and predicted fracture stresses for the six samples under study, with the solid line representing zero percent error in predicting the fracture stress. For small cracks the error in predicting the exact fracture stress is due to an underestimation of crack size as discussed earlier, and to some uncertainty in the relationship between a and C_0 of Eq. (3). For the larger cracks, the crack depth is about 5% of the total thickness of the sample which affects the stress intensity factor at the crack tip . The change in stress intensity factor has not been taken into account in the present fracture model. Notice that we still obtain excellent agreement between the predicted and actual fracture stresses. The reduction in error in fracture stress prediction over crack sizing is due to the square root dependence of the fracture stress on crack size.

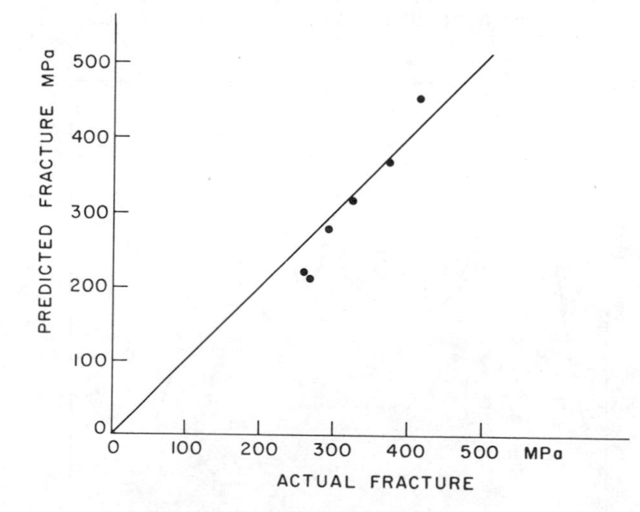

Fig. 7 - Comparison of acoustically-predicted and actual fracture stresses

We conclude that machining damage leaves surface cracks which are closed at the mouth due to the presence of residual stress. Cracks, when stressed to fracture, exhibit slow crack growth into a single, half-penny shaped crack. We attribute this behavior to the presence of surface residual stress that is induced during crack formation. We have developed a set of measurement techniques that allow us to measure the crack depth and its variation over the crack length. We have further confirmation of the fracture model we developed earlier, and we are capable of making fracture stress prediction with an accuracy of better than 10% .

ACKNOWLEDGEMENT

This work was supported by Department of Energy Contract DE-AM03-76SF00326, Project Agreement DE-AT03-83ER45020.

REFERENCES

1. Marshall, D. B., A. G. Evans, B. T. Khuri-Yakub, J. W. Tien, and G. S. Kino, "The Nature of Machining Damage in Brittle Materials," Proceedings of R. Soc. Lond. A385:461 (1983).
2. Achenbach, J. S., and R. J. Brind, "Scattering of Surface Waves by a Subsurface Crack," J. Sound and Vibration, 76:43 (1981).
3. Khuri-Yakub, B. T., G. S. Kino, K. Liang, J. Tien, C. H. Chou, A. G. Evans, and D. B. Marshall, "Nondestructive Evaluation of Ceramics," Review of Progress in Qualitative Nondestructive Evaluation, Vol. 1, D. O. Thompson and D. E. Chimenti, Eds., Plenum Publishing Corporation (1982).
4. Murakami, Y., B. T. Khuri-Yakub, G. S. Kino, J. M. Richardson, and A. G. Evans, "An Application of Wiener Filtering to Nondestructive Evaluation," Appl. Phys. Lett. 33:685 (1978).
5. Shah, R. C., and A. S. Kobayashi, "On the Surface Flaw Problem; Physical Problems and Computational Solutions," J. L. Swedlow, Ed., p. 79-124, presented at Annual Winter Meeting of ASME, New York, Nov. 1972.

A NONDESTRUCTIVE NEAR SURFACE X-RAY DIFFRACTION PROBE

R. A. Neiser
Virginia Polytechnic Institute and State University
Blacksburg, VA 24061

K. S. Grabowski
Naval Research Laboratory
Washington, DC 20375

C. R. Houska
Virginia Polytechnic Institute and State University
Blacksburg, VA 24061

ABSTRACT

Synchrotron x-ray sources provide a capability to probe much smaller near surface zones, mostly due to their longer, tunable wavelengths and high intensities. The use of wavelengths up to about 5 angstroms (Å) and asymmetrical x-ray optics will allow near surface zones thicker than 80 Å to be examined more effectively, while the source's high intensity will allow a vast amount of data to be collected within a reasonable time period. This technique will be able to identify phases, examine interplanar spacing gradients, and measure texture and grain size as well as other near surface features. Variable beam size with computer controlled slits will allow collection of statistically representative data, or detailed analyses of diffraction spots from individual grains if a microbeam is used.

INTRODUCTION

In the last two decades considerable effort has been devoted to developing techniques capable of characterizing the composition and structure of near surface zones by x-ray diffraction.[1,2] Other important examples of near surface probes include ion back scattering, low energy electron diffraction (LEED), x-ray photoelectron spectroscopy (XPS), and Auger spectroscopy combined with ion milling. The advent of synchrotron radiation (SR) facilities allows for a major development in non-destructive x-ray diffraction probes for near surface zones. This diffraction technique is not to be confused with the total reflectance work of Marra, Eisenberger and Cho[3] which is ideally suited for the examination of smooth interfaces extending from a monolayer to about 80 Å. With total reflection, the effective penetration distance, δ, is given by

$$\delta \approx \frac{\lambda}{2\pi(\theta_c^2-\theta_i^2)^{1/2}} \qquad (1)$$

where λ is the x-ray wavelength, θ_c the critical angle (around 0.25°), and θ_i is the experimental angle of incidence. It is apparent that an uncertainty of $\pm\Delta\theta_i$ will introduce an uncertainty of $\pm\Delta\delta$ which can be large as θ_i approaches θ_c. Also, the distorted wave approach[4] used to obtain Eq. (1) assumes an ideally smooth interface, making it important to reduce surface roughness and waviness to a level well below 80 Å if δ is to be well defined. Finally, restricting the angle of incidence to a fixed θ_i, restricts the angular degrees of rotational freedom which limits exploration in reciprocal space.

For many studies, it is desirable to examine regions greater than 80 Å with a well defined δ and without limiting the overall angular range in reciprocal space. This may be accomplished with more conventional angles of incidence that are not constrained to fixed values corresponding to total reflection. We propose that δ be considerably reduced by selecting the longer x-ray wavelengths extending up to about 5 Å which give substantially higher absorption. A sizeable absorption data bank[5], which determines δ, is typically accurate to within a few percent. Finally, optimum diffraction geometry may involve asymmetrical arrangements that further decrease δ without overly sacrificing high resolution.

Several excellent treatises describing x-ray SR facilities already exist[6,7] so that only a brief discussion of the pertinent features will be given here. X-ray SR is generated by circulating packets of electrons at high energies (typically a few GeV) in a storage ring. The x-rays produced are highly directional, consist of a continuous spectrum, and are very intense. After leaving the ring the x-ray beam is focussed and monochromated by a series of mirrors and single crystals before reaching the sample (see Figure 1). The angle at which the white radiation impinges on the single crystals of the monochromator is variable and allows the radiation

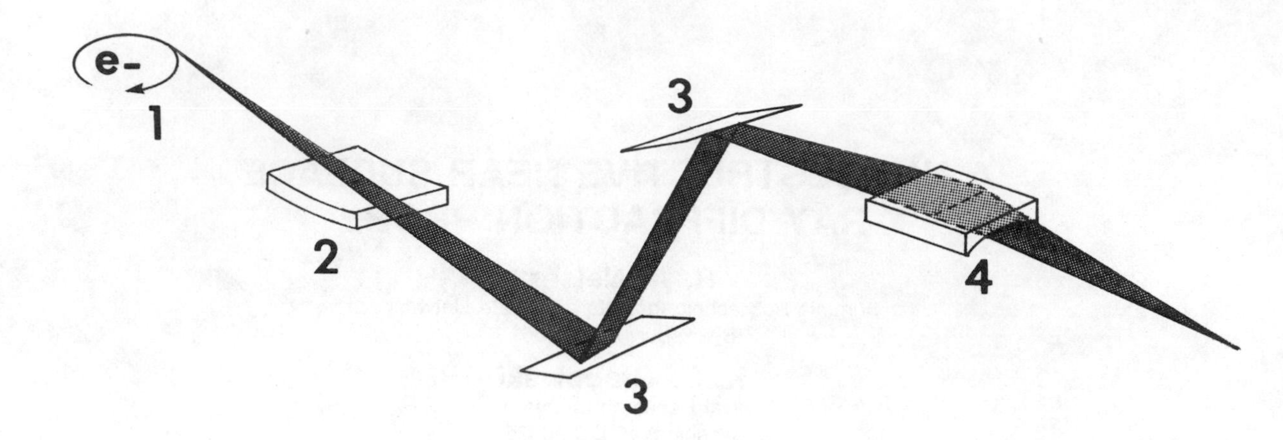

Fig. 1. Typical optics for an X-ray SR beam line. The elements include: 1) the storage ring, 2) a total reflectance, parabolic collimating mirror, 3) a fixed exit position, double crystal monochromator and 4) a total reflectance, ellipsoidal focussing mirror.

to be tuned. Slits which further define the beam are not shown in Figure 1.

SR offers improvements over conventional sources when designing a near surface probe. With a double crystal monochromator the wavelength is tunable over the range of 1 Å to 5 Å while the spectral width of the monochromated radiation is approximately 1/3 the natural width of a typical $K\alpha_1$ line (see Figure 2). Since no $K\alpha_2$ component exists with SR the spectral broadening at a synchrotron facility is reduced even further below that of conventional x-ray tubes. Several authors[8,9] have compared the greatly increased intensity of SR to conventional sources, however, experimental conditions strongly affect the results they obtain. The high brightness and intrinsic collimation of SR allows plane crystal monochromators to be used without the substantial loss of intensity associated with monochromated conventional sources. Additionally, the near surface probe requires tunable radiation unavailable to conventional sources except as the weak bremsstrahlung continuum.

A new generation of diffractometers and related hardware accompanies the development of the synchrotron source. For instance, new diffractometers feature an extended 2θ range of at least 170° as well as improved angular precision. Some beam lines are developing line and area detectors to greatly increase the data collection rate while others will use computer driven slits to control the area illuminated on the sample and limit the axial divergences of the x-ray beam.

A wide range of samples can be studied using the near surface probe. Potential applications include the determination of the structure and distribution of implanted atoms in a substrate, residual strain gradients due to wear or grinding, the structure of as deposited and diffused thin films and near surface oxidation states and composition gradients.

WAVELENGTH CONTROL

The most obvious advantage of employing the longer wavelengths available with SR is the decrease in penetration depth caused by the increase in the linear absorption coefficient, μ. It is convenient to define a penetration depth, δ_x, as the depth beyond which the fraction, x, of the total diffracted intensity originates. For an asymmetrical diffraction geometry

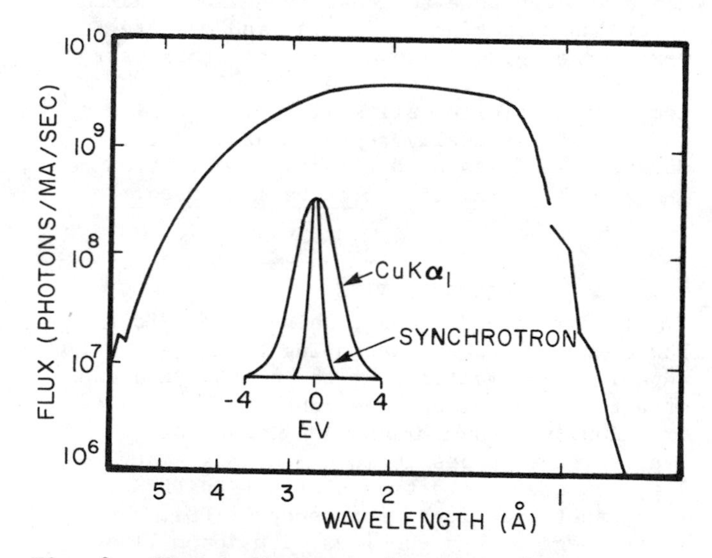

Fig. 2. Photon flux as a function of wavelength for the Naval Research Laboratory's beam line at the National Synchrotron Light Source.* The inset compares the spectral width of SR to Cu $K\alpha_1$ radiation.

*Reprinted from: Kirkland, J. P., D. J. Nagel and P. L. Cowan, Nucl. Instrum. Methods 208, 49 (1983) with the permission of the authors.

$$\delta_x = \frac{-\ln x \cos\chi}{\mu}\left[\frac{1}{\sin(\theta+\omega)} + \frac{1}{\sin(\theta-\omega)}\right]^{-1} \quad (2)$$

where ω is the rocking angle, χ is the angle out of the plane of the paper and θ is the Bragg angle (see Figure 3). A useful diffracted signal is possible with x = 0.9. In this case 10% of the diffracted intensity comes from the material located between the surface and $\delta_{0.9}$. As an example, consider $\delta_{0.5}$ and $\delta_{0.9}$ for a Cu sample with 2θ = 150° and symmetrical geometry, i.e., ω=χ=0° (see Figure 4). At long wavelengths, measurable diffracted intensity can come from a depth as small as 600 Å. Shallower depths can be examined by tilting the sample to lengthen the absorption path. For instance, at 2θ = 150° changing from the symmetrical geometry to a 3° exit angle (θ - ω = 3°) decreases the penetration depth by a factor of 10.

As longer wavelengths are used to reduce the penetration depth, reflections will be pushed to higher angles according to Bragg's law

$$\lambda = 2d\sin\theta. \quad (3)$$

The advantages of working with high angle lines in x-ray diffraction are well known. They are potentially useful as a near surface probe as well. In line shape and line shift analyses, high angle lines are preferred over low angle lines because sample effects are enhanced while instrumental broadening is reduced. The shape of an observed Bragg peak can be described by the convolution of two functions:[10] one function is contributed by the sample and the other defined by the instrument.

The instrumental broadening function is composed of a geometrical term and a spectral term. The geometrical contribution, which arises because of beam divergences, is normally larger than the spectral contribution and its magnitude decreases as 2θ increases. The slight spread of wavelengths in the x-ray beam accounts for spectral broadening. The magnitude of this broadening increases as 2θ increases. The net result is that the width of the instrumental broadening function is expected to decrease as 2θ increases. For a given reflection it may be conservatively estimated that there should be a factor of 3 to 6 gain in instrumental resolution by using longer wavelength SR instead of using a shorter, conventional wavelength.

Elaborate procedures have been developed to examine sample broadening, but for present arguments simple equations suffice to demonstrate the advantage of measurements made at high angles. Particle size broadening has been described[11] by the Scherrer equation

$$\Delta 2\theta_{PS} = \left[\frac{0.94\lambda}{\cos\theta}\right]\frac{1}{L} \quad (4)$$

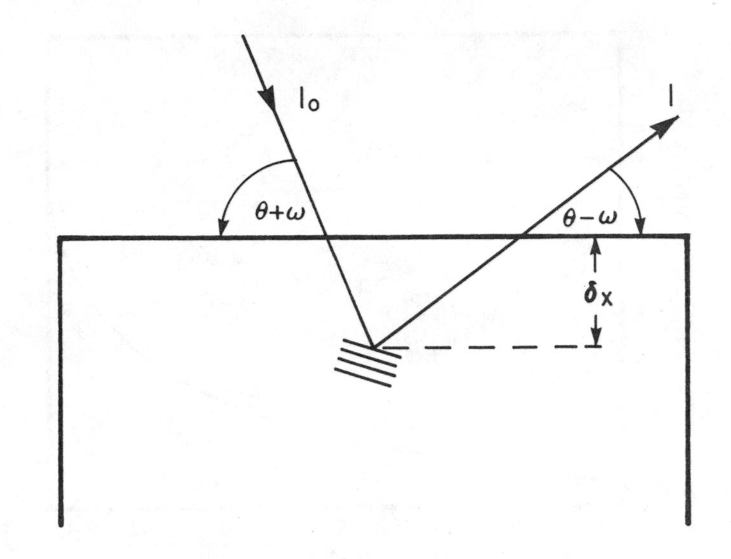

Fig. 3. Asymmetrical diffraction geometry showing the angles, θ and ω, and the penetration depth δ_x.

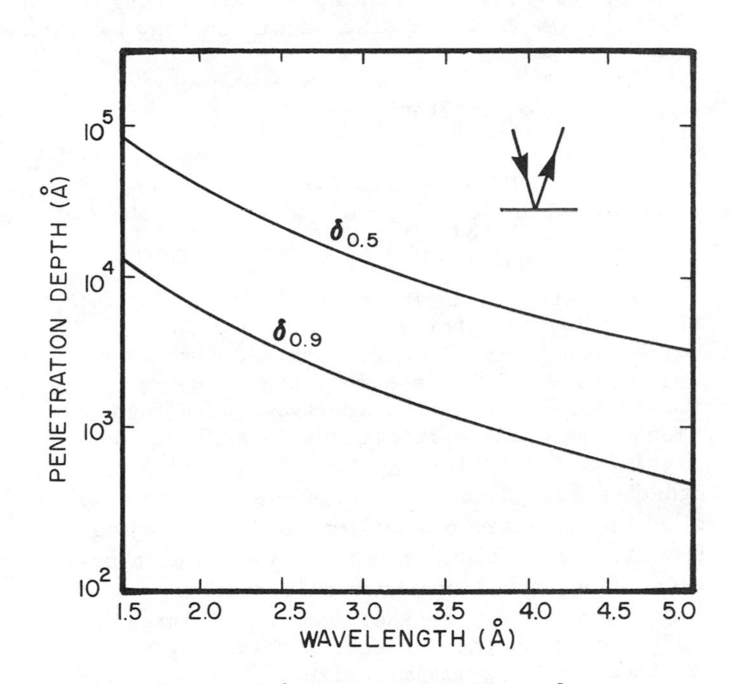

Fig. 4. $\delta_{0.5}$ and $\delta_{0.9}$ as a function of x-ray wavelength for Cu. The inset depicts the diffraction geometry.

where L is the x-ray particle size perpendicular to the diffracting planes. By using Bragg's law Eq. (4) can be expressed in terms of wavelength

$$\Delta 2\theta_{PS} = \left[\frac{0.94\lambda}{(1 - (\lambda/2d)^2)^{\frac{1}{2}}}\right]\frac{1}{L}. \quad (5)$$

Similarly, for uniform and non-uniform strain

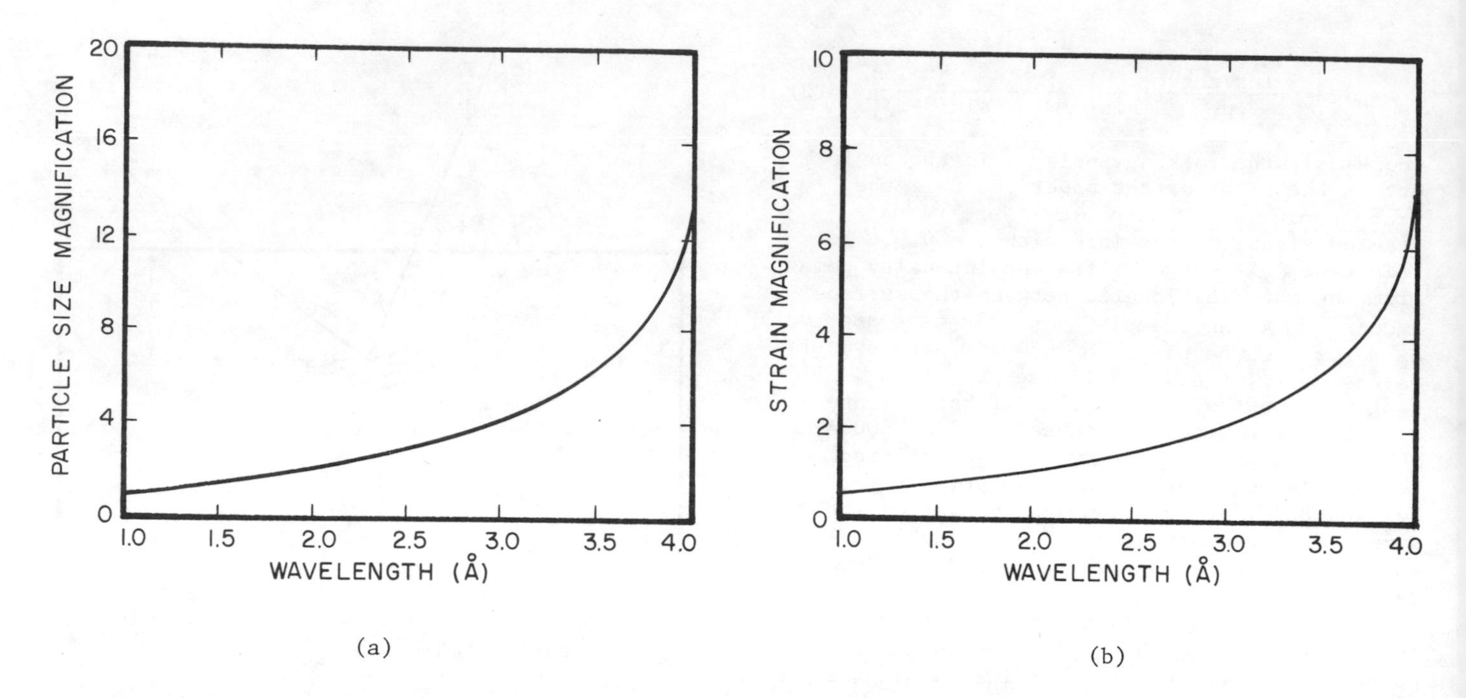

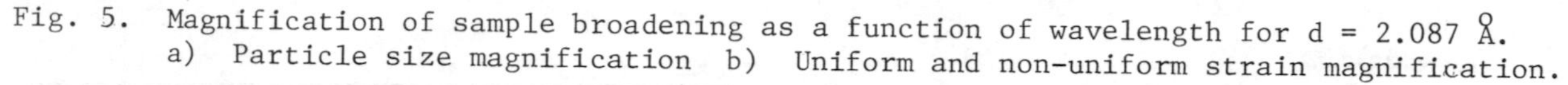

Fig. 5. Magnification of sample broadening as a function of wavelength for d = 2.087 Å.
a) Particle size magnification b) Uniform and non-uniform strain magnification.

$$\Delta 2\theta_d = \left[2\tan\theta\right] \frac{\Delta d}{d}$$

$$= \left[\frac{2}{((2d/\lambda)^2-1)^{\frac{1}{2}}}\right] \frac{\Delta d}{d}. \qquad (6)$$

The bracketed terms in Eq. (5) and Eq. (6) can be viewed as magnification factors that are strong functions of wavelength. Choosing Cu, which has a (111) d-spacing of around 2 Å, the magnification terms are shown as a function of wavelength in Figures 5a and 5b. Both magnification factors go to infinity as λ approaches 2d. On a relative scale the magnification is increased nearly ten fold by changing from 1.5Å radiation, which is available in most x-ray labs, to 4Å synchrotron radiation.

A third advantage of working at large 2θ values is that the correction necessary for surface roughness is minimized. Surface roughness decreases the integrated intensities of Bragg reflections. It has been pointed out[12] that partial correlations between incoming and outgoing absorption paths determine the magnitude of the surface roughness correction. This correction vanishes as θ approaches 90° where the incoming and outgoing x-ray paths travel through almost the same grains and path correlation is nearly complete. Although surface roughness can be an important consideration at intermediate angles, good intensity measurements are possible in the high angle region without going to extremes in surface polish.

The three examples given above illustrate possible applications of a high angle diffraction probe. The tunability of SR allows a number of *hkl* reflections to be examined sequentially around $2\theta = 150^o$, and is depicted schematically in Figure 6. Smaller and smaller d-spacings may be scanned by selecting progressively smaller wavelengths according to Bragg's law. For each *hkl* reflection, sample broadening will be large and instrumental broadening will be diminished, whereas with

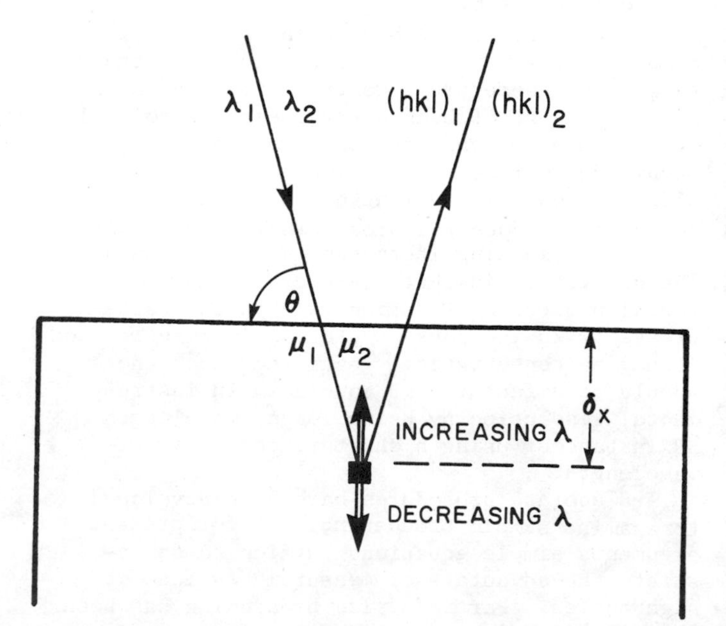

Fig. 6. The high angle diffraction probe using an approximately constant 2θ. δ_x varies for each *hkl* according to Eq. (2).

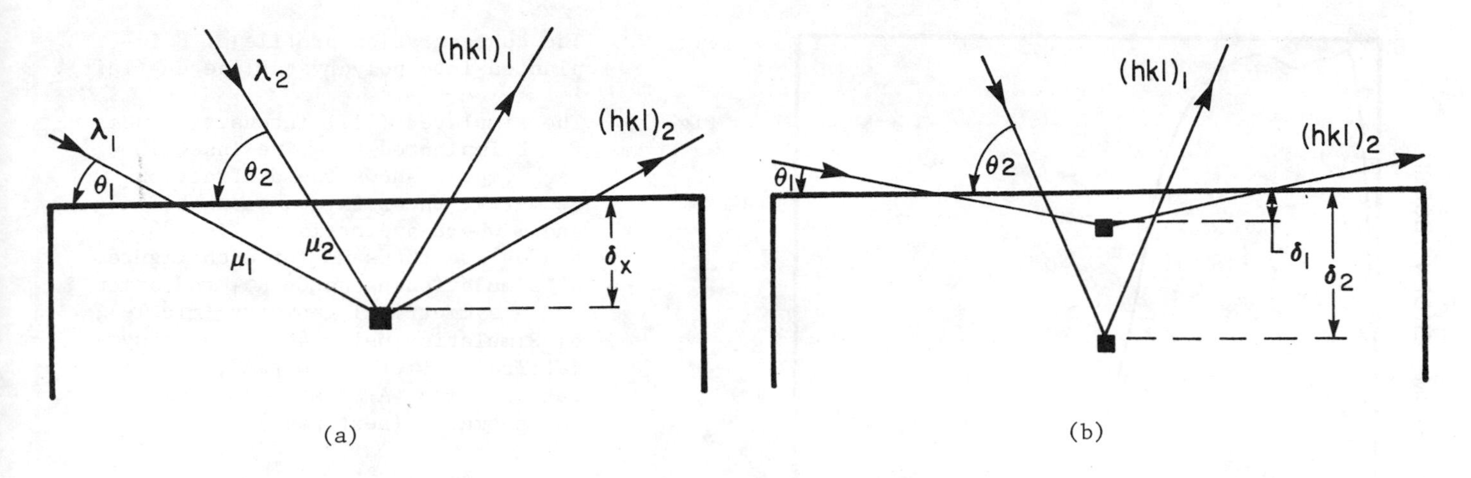

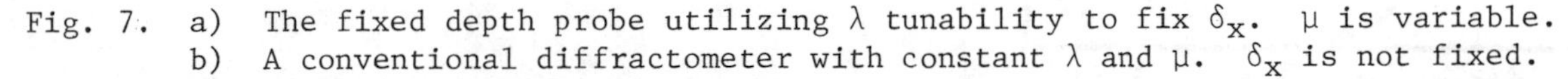

Fig. 7. a) The fixed depth probe utilizing λ tunability to fix δ_x. μ is variable.
b) A conventional diffractometer with constant λ and μ. δ_x is not fixed.

conventional radiation only a few high index reflections will have the most favorable broadening conditions. The use of graduated wavelengths allows various depths to be probed over a well defined zone. The large penetration depths at short wavelengths provide reference diffraction profiles from primarily undisturbed crystalline material while the longer wavelengths yield information about the near surface zone.

The high angle probe allows one to examine a range of depths. Just as important is the capability of studying a series of *hkl* reflections at a fixed penetration depth. Figures 7a and 7b show the advantage of using tunable SR to study near surface zones. δ_1 and δ_2 are the penetration depths for a given value of x

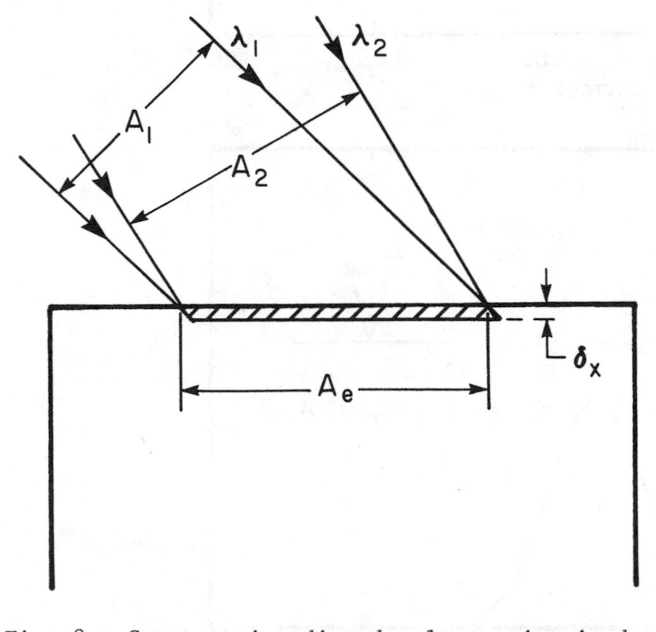

Fig. 8. Constant irradiated volume maintained by keeping A_e and δ_x fixed for each *hkl*.

corresponding to the *hkl* reflections depicted in Figure 7b. With conventional equipment one has little control of the depth to which the beam penetrates the sample. However, employing the tunability of the synchrotron source this depth can be controlled. After choosing a fixed penetration depth, δ_x, the wavelength giving that depth may be calculated with the transcendental equation

$$\left[\frac{-4d\,\delta_x}{\ln x\,\cos\chi}\right]\mu\lambda\cos\omega - \lambda^2 + 4d^2\sin^2\omega = 0\,. \qquad (7)$$

Allowing asymmetrical diffraction geometries dramatically improves the ability to choose shallow penetration depths. In a material with a concentration gradient, a linear absorption coefficient averaged to the depth δ_x, $\langle\mu(\delta_x)\rangle$, should replace μ in Eq. (7).

SLIT CONTROL

Another powerful capability available at some synchrotron facilities is the ability to computer control aperture sizes. Typically the synchrotron source size is 1 mm by 2 mm. By using motor driven slits the dimensions of the beam can be decreased to approximately 25 μm (microns) by 25 μm. In this lower limit the Debye-Scherrer rings of a fine grained, polycrystalline material will break up into spotty microdiffraction patterns since fewer grains are illuminated with the smaller beam size. The defect structure of individual grains can be measured using microdiffraction, rather than the average values measured by conventional diffraction. A classical example of microdiffraction may be found in a paper by Hirsch and Kellar.[13] They studied cold-rolled aluminum with 35 μm and 150 μm microbeams and were able to obtain information about the bending and break-up of grains after cold-rolling. Similarly, d-spacing gradients due to strain or compositional changes can be measured by examining the 2θ broadening of individual grains.

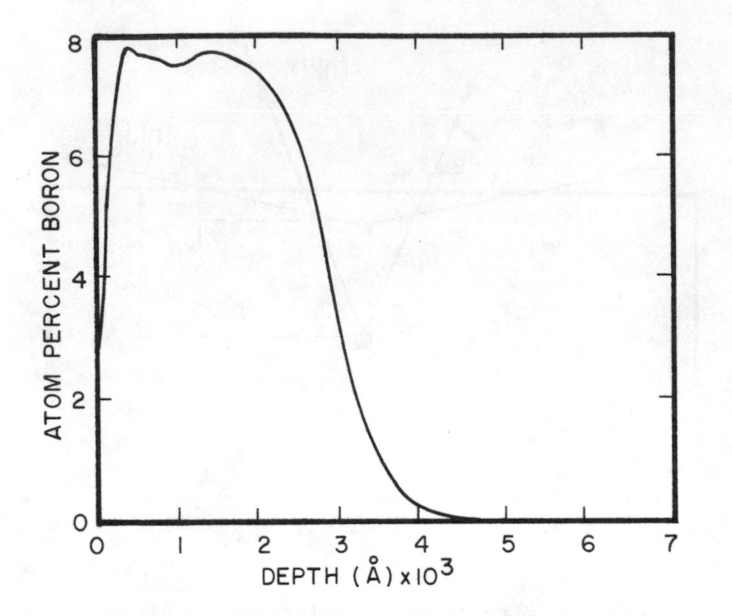

Fig. 9. The concentration profile of B implanted into polycrystalline Cu (left).

Fig. 10. The simulated (111) intensity bands for B implanted Cu. The inset in each figure shows the diffraction geometry, the x-ray wavelength, $\delta_{0.5}$, and ω where appropriate. The 2θ scales are the same for each figure. a) Simulation using Cu Kα_1 radiation and a symmetrical geometry (middle), b) Simulation using 4Å SR and a symmetrical geometry (bottom), c) simulation using 4Å SR and an asymmetrical geometry (next page).

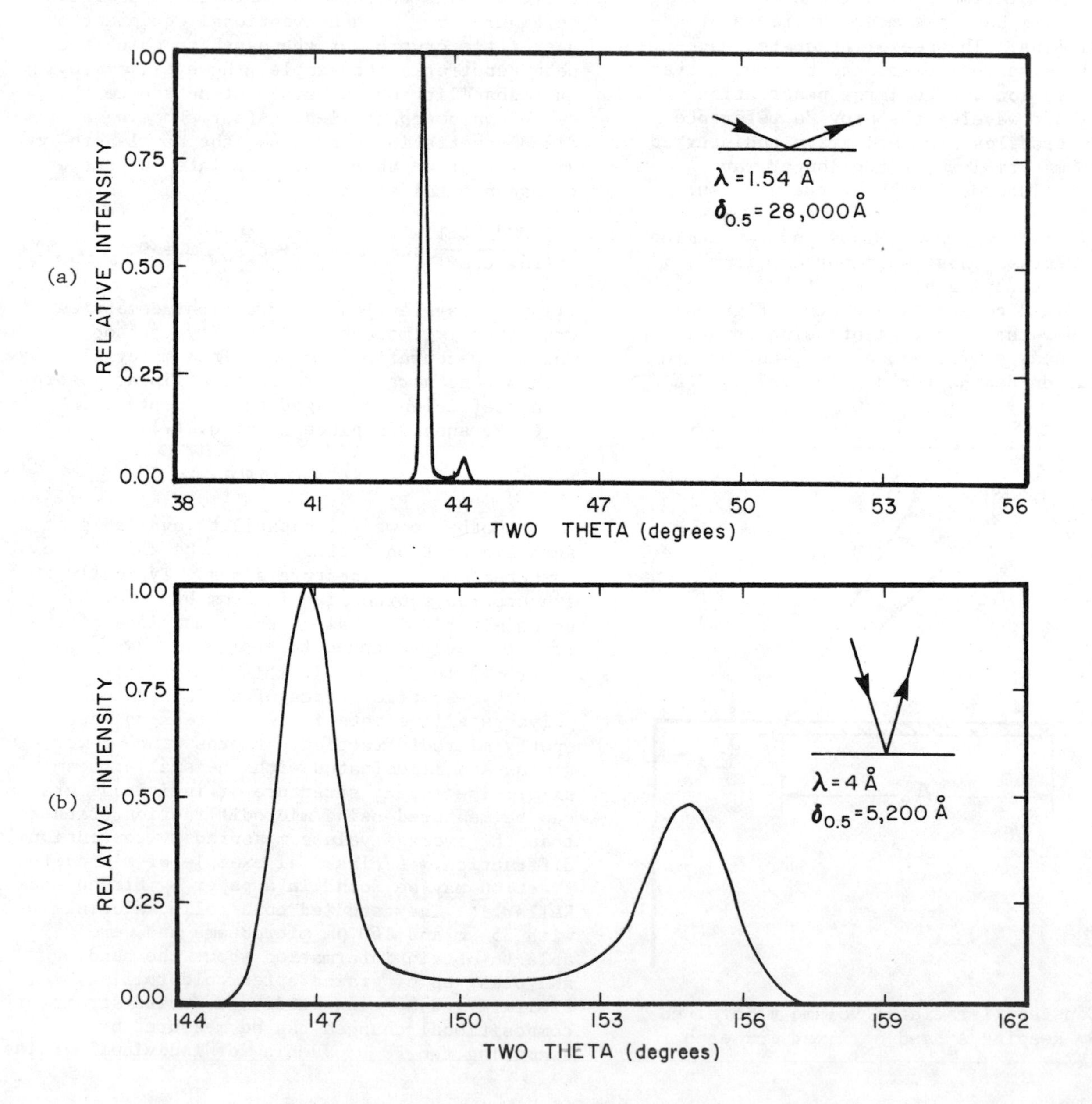

(c)

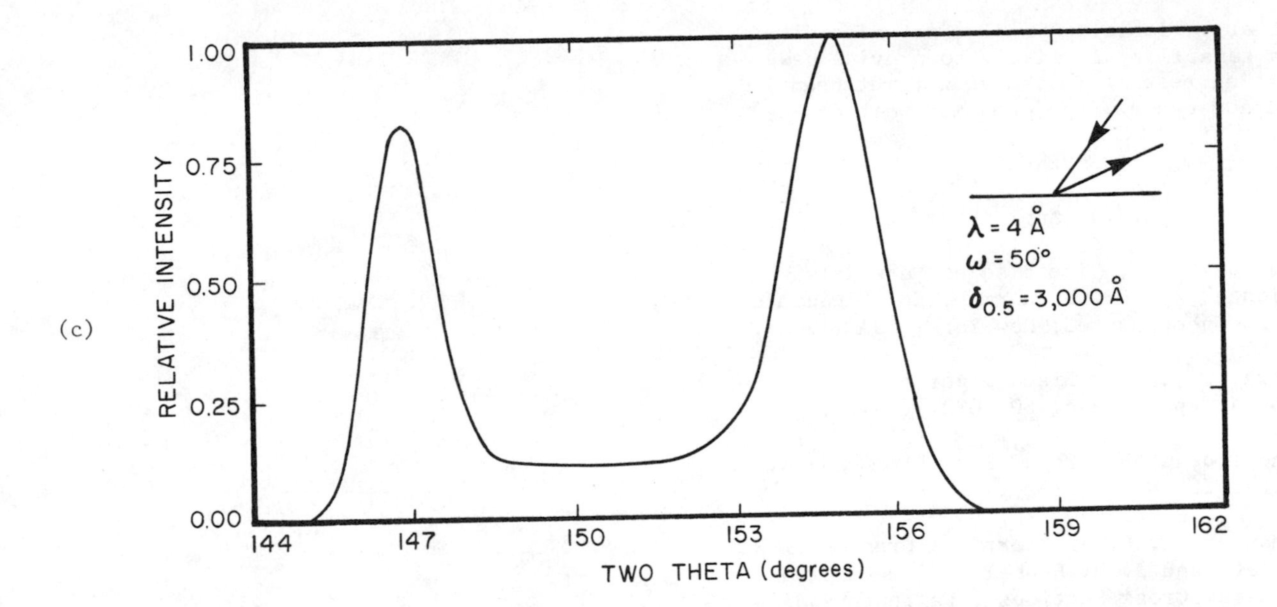

Fig. 10 (Cont'd).

Microdiffraction requires a highly collimated beam which gives very low intensities with conventional sources. The collimation and high brightness of SR are useful features for microbeam x-ray diffraction. Area detectors offer the possibility of fast data collection, with the output displayed on a color graphics terminal in 3 dimensions. For example, intensity variations may be represented as a color change while the abscissa and the ordinate may be any two diffractometer angles.

Computerized aperture selection has applications for standard diffraction as well as for microdiffraction. Typically, the area illuminated on the sample decreases as 2θ increases. Slit control allows the effective area of the beam on the sample, A_e, to be kept constant as 2θ, χ and ω are varied (see Figure 8). If this control is used in conjunction with a constant depth probe the same grains are illuminated for each Bragg reflection. This constant irradiated volume could be useful where gradients exist both with depth and across a sample.

COMPUTER SIMULATIONS

Computer simulations illustrate some of the advantages of using SR as a near surface diffraction probe. The ion implantation of B into a polycrystalline Cu substrate is used as an example. The composition profile is taken to be an error function with a maximum concentration of 8 atom percent over a depth of 2,800 Å (see Figure 9). For the purpose of this simulation the B is assumed to be substitutional. The x-ray intensity diffracted by the implanted Cu was computer calculated using the intensity band technique.[14] The (111) intensity band was calculated for three experimental conditions. The first simulation is the expected diffraction profile using Cu $K\alpha_1$ radiation and a symmetrical diffraction geometry (see Figure 10a). The large peak is from the pure substrate while the small, 5% intensity lobe shifted to the high 2θ side of the substrate peak is diffracted from the implanted layer. The penetration depth, $\delta_{0.5}$, for this example is around 28,000 Å, or 10 times greater than the implanted depth. Figure 10b illustrates the advantage of using a long wavelength. With 4 Å radiation the peak has been shifted to $2\theta = 150^{\circ}$, and the longer wavelength has decreased $\delta_{0.5}$ to 5,200 Å. The intensity from the surface zone is now 50% of the substrate peak and the peak separation has increased from 0.8° to 7.2°. Tilting the sample 50° in ω dramatically increases the near surface diffracted intensity since $\delta_{0.5}$ has been diminished to approximately the implanted depth (see Figure 10c). In addition to the above gains with SR it also offers improved instrumental resolution and higher intensities than conventional x-ray sources. Therefore, microdiffraction has the potential of yielding valuable information about the defect structure of the implanted zone.

CONCLUSION

In conclusion, the near surface diffraction probe outlined here is non-destructive, draws on a large data base that most surface probes do not have, and perhaps most importantly, it is quantitative. The techniques outlined in this paper are not limited to specially prepared, mirror surfaces. They can be used to study real surfaces such as those associated with ion implantation and wear, to name just two. The high angle probe can magnify sample broadening while decreasing instrumental broadening. The depth probe can be made to be constant or variable. Finally, the microdiffraction capability is likely to provide new and exciting results that have not yet been reported and can only be obtained with SR.

REFERENCES

1. Houska, C. R., J. Appl. Phys. 41, 69 (1970).

2. Houska, C. R., "Treatise on Materials Science", p. 63, H. Herman, Ed., Academic Press Incorporated, New York, 19A (1980).

3. Marra, W. C., P. Eisenberger and A. Y. Cho, J. Appl. Phys. 50, 6927 (1979).

4. Vineyard, G. H., Phys. Rev. B: Condens. Matter 26, 4146 (1982).

5. McMaster, W. C., N. Kerr Del Grande, J. H. Mallett and J. H. Hubbell, "Compilation of X-ray Cross Sections", Lawrence Radiation Laboratory (Livermore) Report UCRL-50174, Sec. I (1970); Sec. II, Rev. 1, (1969); Sec. III, (1969); Sec. IV, (1969).

6. Kunz, C., Ed., "Topics in Current Physics - Synchrotron Radiation; Techniques and Applications", Springer-Verlag, Berlin, (1979).

7. Winich, H. and S. Doniach, Eds., "Synchrotron Radiation Research", Plenum Press, New York, (1980).

8. Kunz, C., Ref. 6, p. 19.

9. Gilfrich, J. V., E. F. Skelton, D. J. Nagel, A. W. Webb, S. B. Quadri and J. P. Kirkland, "Advances in X-ray Analysis", p. 313, C. R. Hubbard, C. S. Barrett, P. K. Predecki and D. E. Layden, Eds., Plenum Publishing Incorporated, New York 26 (1983).

10. Warren, B. E., "X-Ray Diffraction", p. 257, Addison-Wesley Publishing Company, Reading, Massachusetts (1969).

11. Warren, B. E., Ref. 10, p. 251.

12. Harrison, R. J. and A. Paskin, Acta Crystallogr. 17, 325 (1964).

13. Hirsch, P. B. and J. N. Kellar, Acta Crystallogr. 5, 162 (1952).

14. Houska, C. R., Metall. Trans. A 14A, 61 (1983).

Subject Index

Acoustic emission
  cooling/heating, caused by, 31, 34, 39, 40
  cracking, 29, 39, 41-47
  dislocations, caused by, 29, 39, 40
  melting/solidification, applications to, 27-40, 41-49
  microstructural correlations, 31-40, 43-47
  solidification, caused by, 43
  statistical analysis of data, 45
Acoustic impedance at solid/liquid interfaces, 23-25, 81-86
Acoustoelastic constants (see Ultrasonics)
Acoustoelastic stress measurements (see Ultrasonics)
Aluminum alloys, 27-40, 54-59, 61-68, 137-144, 147,177, 192-196
Amorphous metals, ultrasonic velocity in, 56-59

Barkhausen noise (see Magnetization transitions)

Carbides, 170, 174
Cast iron (see Iron)
Casting (see Solidification)
Cavity formation (see also Powder Metallurgy-porosity)
  heat treatment, effect of, 200
  plastic deformation, effect of, 170, 197-201
Ceramics (see Magnesium-aluminate spinel; Magnesium oxide; Silicon carbide; Silicon nitride; Yttrium-chromium oxide)
Copper, 208-213
Cracks (see Acoustic emission; Ultrasonics; X-rays)
Creep, 170

Deformation
  elastic strains
    thin films, sources in, 133, 213
    ultrasonic analysis of, 67, 137-144, 147-153, 177-184
    x-ray analysis of, 129-135, 185-192
  magnetoelastic analysis of, 161
  plastic strains, 142, 143, 170-175, 185-196, 197
Dislocations, 29, 39, 40, 72, 134, 135, 144, 170, 171, 187-189, 192-195

Elastic moduli
  aluminum alloys, 59, 65
  amorphous metals, 59
Electrical resistivity, 64-68

Fatigue, 174, 175, 192-195
Ferromagnetic materials, 161-166, 171
Fracture strength, 124, 125, 203, 206

Gold, 192
Graphite, 121-125
Grinding (see Residual stress)

Heat flow (see also Solidification; Ultrasonics)

Heat treatment (see Acoustic emission; Cavity formation; Neutron scattering)
  3-7, 27-30
Indium-gallium-arsenide, 133-135
Indium-phosphide, 134
Interfaces (see Acoustic impedance; Residual stress; Solidification)
Ion implantation (see Residual stress)
Iron, 25, 109, 113

Lattice parameter (see Deformation)

Machining (see Residual stress; Ultrasonics)
Magnetoelastic interactions, 161-166
Magnetization transitions, 161-166
Magnesium-aluminate spinel, 89, 95-98
Magnesium oxide, 93-95
Mechanical properties (see also Creep; Elastic moduli; Fatigue; Ultrasonics)
  of graphite, 122, 125
  of thin films, 65-68
  of silicon nitride, 206
  porosity effects in brittle materials, 122
  powder metallurgically produced steels, 111-117
Melting (see also Solidification; Ultrasonics; Welding)
  acoustic emission, application of, 27-40, 41-48
  electron beam, 27-40
  heat flow, 27-29
  laser, 41-48
  ultrasonics, application of, 13-22, 23-25
Mercury, 14, 24
Microfluorescent analysis, 186
Microstructure
  rapid solidification, 29-31, 34-38, 65-67
  in stainless steel, after welding, 43-49
  in steel, after deformation, 173-175, 200
  in steel powder products, 109, 117, 118
  in titanium weldments, 74-78

Neutron radiography in powder metallurgy, 98-100, 103-108
Neutron small angle scattering
  diffraction, 104
  microstructure characterization, application to, 169-176
  porosity measurements, 105-108
  powder metallurgy, application to, 103-108
  precipitate measurement, 172-176

Neutron small angle scattering (cont.)
  principles, 103-105, 170, 171
  refraction, 104-108
Nickel, 81-86
Nitinol, 156-160
Nuclear magnetic resonance
  powder metallurgy, application to, 98
  tomography, 98

Plastic deformation (see Deformation)
Plexiglass, 95-98
Porosity (see Powder Metallurgy; Ultrasonics)
Porosimetry, 103
Powder metallurgy
  ceramics, 89-100
  green state, 89-100, 105-108
  iron and steel, 111-118
  particle packing, 107
  porosity, 91-100, 105-108, 113, 114
  radiography, application of, 91, 92, 98-100
  sintering, 106, 107, 110-112
  ultrasonics, application of, 93-98, 109-119
Process control, 3, 4, 13-22, 23, 27, 41-49, 51, 52, 68, 103, 106, 108, 109, 161, 165

Quality control, 109, 117, 118

Radiography (see also Neutron radiography; Neutron small angle scattering; X-rays)
  microstructure characterization, application to, 63, 64, 67, 169-176
  microradiography, 91, 92
  powder metallurgy, application to, 91, 92, 98-100, 103-108
  residual stress/elastic strain, 129-135, 156, 213
Recovery/stress relief, 163
Residual stress (see also Deformation; Magnetoelastic interactions; Ultrasonics; X-rays)
  at cracks, 203
  at interfaces, 155-160, 186, 187
  at interference fits, 143, 147, 153, 155-160
  from machining, 158, 163-165, 206
  in microelectronic materials, 129-135
  in railroad ties, 139
  at surfaces, 163, 164, 213
  in weldments, 139, 162-164

Semiconductors (see Indium compounds; Silicon)
Sensors in hostile environments, 4
Silicon, 186-192
Silicon carbide, 89-100
Silicon nitride, 89-100
Small angle neutron scattering (see Neutron small angle scattering)
Solidification (see also Acoustic emission; Casting; Ultrasonics)
  acoustic emission, application of, 27-40, 41-48
  defects due to fissures, hot tearing, etc., 3, 23, 38, 41
  rapid solidification, 27, 41, 48, 51-59, 61, 62, 68

Solidification (cont.)
  solid/liquid interface motion, 14, 23, 24
  at solid/liquid interfaces
    iron, 25
    mercury, 14, 24
    steel, 14, 16, 20, 24, 25
    tin, 14, 23, 24
    water/ice, 14
    Wood's metal, 14
  ultrasonics, application of, 3-11, 13-22, 23-25, 51-59
Steel
  carbon, 4, 15, 24, 110-118, 138, 147-154, 161-166, 177-183
  stainless, 8-11, 15-21, 24, 25, 41-47, 138, 163, 169-175
Stress measurements (see Ultrasonics)
Surface analysis (see also Residual stress; X-rays)
  eddy current, 5
  x-rays, 187-196, 208-213
Synchrotron characteristics, 208

Temperature distributions (see Ultrasonics)
Texture (see Ultrasonics)
Thin Films
  elastic moduli measurements, 54-59, 63-65
  thickness measurement, 132-135
Tin, 23
Titanium alloys, 71-80

Ultrasonics (see also Mechanical properties; Residual stress; Solidification)
  acoustoelastic constants, 177-183
  alloying element/impurity effects on, 67, 68, 74-80, 115
  alloy phase content, application to, 61-68, 181-182
  attenuation, 79, 84-86, 122-125, 197-201
  cracks, interaction with, 203-206
  crystallographic (texture) effects on, 78, 81-86, 139, 143, 150
  elastic moduli determination, 59, 63, 65-68
  guided waves, characteristics of, 52, 53
  machining damage, application to, 206
  mechanical properties, correlation with, 65-68, 111-117, 124, 125, 137, 149-153, 206
  microstructural correlations, 5, 65-67, 78, 90-96, 143, 144, 180-182, 199, 200
  porous materials, coupling into, 93, 94
  powder metallurgy, application to, 93-98, 109-119
  ray tracing, 17-19
  reflection/transmission coefficient
    at cracks, 203-205
    at interfaces, 81-86
    at rough surfaces, 83, 155-159
  residual stress, application to, 78, 137-144, 147-153, 155-160
  scattering
    at interfaces, 13-22, 23-25, 81-86, 95
    Rayleigh, 122-125
  signal processing, 13, 15, 24, 83-86, 197, 198

Ultrasonics (cont.)
solidification/welding, applications to, 13-22, 23-25, 81-86
temperature distribution, application to, 3-11
tomography, 3-11
velocity
anisotropy, 81-86, 140, 141
in composites, 94, 95
dispersion, 96, 143, 144
microstructural effects, 71-80
in powder metallurgical products, 93-95 112-116, 122-125
stress dependence, 137-144, 147-154, 180-182
temperature dependence, 3-5, 9-11, 139, 147, 149, 152, 182
weld contamination, application to, 71-80
Void formation (see Cavity formation)

Welds
acoustic emission, application of, 41-49
arc welding, 15-22, 72
laser welding, 41-49
residual stress at, 162-164
ultrasonics, application of, 13-22, 71-80, 138

X-rays (see also Radiography, Residual stress, Synchrotron characteristics)
crack tip strain fields, application to, 189-196
diffraction, 73, 78, 129-135, 210-213
residual stress/elastic strain, application to, 63, 64, 129-135, 156, 185-192, 213
strain measurement
elastic, 129-135, 185-196
plastic, 185-196
topography, 129-135, 189-196
surface/near surface analysis, 208-213

Yttrium-chromium oxide, 93, 103-108